Concurrent Engineering

To our parents,
Abolfazl Parsaei and Barat Atabaki,
William H. Sullivan and Kathleen A. Glasstone

Concurrent Engineering

Contemporary issues and modern design tools

Edited by

Hamid R. Parsaei

Associate Professor
Center for Computer-aided Engineering
University of Louisville
USA

and

William G. Sullivan

Professor
Department of Industrial and Systems Engineering
Virginia Polytechnic Institute and State University
USA

CHAPMAN & HALL

London · Glasgow · New York · Tokyo · Melbourne · Madras

Published by Chapman & Hall, 2–6 Boundary Row, London SE1 8HN

Chapman & Hall, 2–6 Boundary Row, London SE1 8HN, UK

Blackie Academic & Professional, Wester Cleddens Road, Bishopbriggs, Glasgow G64 2NZ, UK

Chapman & Hall Inc., 29 West 35th Street, New York NY10001, USA

Chapman & Hall Japan, Thomson Publishing Japan, Hirakawacho Nemoto Building, 6F, 1-7-11 Hirakawa-cho, Chiyoda-ku, Tokyo 102, Japan

Chapman & Hall Australia, Thomas Nelson Australia, 102 Dodds Street, South Melbourne, Victoria 3205, Australia

Chapman & Hall India, R. Seshadri, 32 Second Main Road, CIT East, Madras 600035, India

First edition 1993

© 1993 Chapman & Hall

Typeset in 10/12 Times by Interprint Limited, Malta
Printed in Great Britain by the University Press, Cambridge

ISBN 0 412 46510 8

A catalogue record for this book is available from the British Library

Library of Congress Cataloging-in-Publication data available

Contents

Contributors

M. Munir Ahmad, Department of Mechanical and Production Engineering, University of Limerick, Plassey Technology Park, Limerick, Ireland.

Adedeji B. Badiru, Director, Expert Systems Laboratory, Department of Industrial Engineering, The University of Oklahoma, 202 West Boyd, Suite 124, Norman, OK 73019-0631, USA.

Ali Bahrami, Department of Economics and Management, Rhode Island College, Providence, RI 02908, USA.

Bert Bars, Systems Design Laboratory, Department of Mechanical Engineering, College of Engineering, University of Houston, Houston, TX 77204-4792, USA.

David C. Brown, AI Research Group, Computer Science Department, Worcester Polytechnic Institute, Worcester, MA 01609, USA.

J. Browne, Computer Integrated Manufacturing Research Unit, University College, Galway, Ireland.

F. Frank Chen, Department of Engineering Management, College of Engineering, The University of Southwestern Louisiana, Lafayette, LA 70504-2250, USA.

Cihan H. Dagli, Intelligent System Center, Department of Engineering Management, University of Missouri-Rolla, Rolla, MO 65401-0249, USA.

Ray A. Daley, Concurrent Engineering Research Center, Drawer 2000, West Virginia University, Morgantown, WV 26506, USA.

Robert E. Douglas, AI Research Group, Computer Science Department, Worcester Polytechnic Institute, Worcester, MA 01609, USA.

Stephen Evans, The CIM Institute, Cranfield Institute of Technology, Cranfield, Bedford, MK43 0AL, USA.

Michael E. Fotta, Concurrent Engineering Research Center, Drawer 2000, West Virginia University, Morgantown, WV 26506, USA.

Satyandra K. Gupta, Department of Computer Science, University of Maryland, College Park, MD 20742, USA.

Mahendra S. Hundal, Department of Mechanical Engineering, The University of Vermont, Burlington, Vermont 05405, USA.

David G. Jansson, President, Sugar Tree Technology, 1035 Pearl Street, Boulder, CO 80302, USA.

Hyeon H. Jo, Department of Industrial Engineering, University of Louisville, Louisville, KY 40292, USA.

Raghu R. Karinthi, Department of Computer Science, University of Maryland, College Park, MD 20742, USA.

Steven H. Kim, Design Research Institute, 406 Engineering Theory Center, Cornell University, Ithaca, NY 14853, USA.

Andrew Kusiak, Intelligent Systems Laboratory, Department of Industrial Engineering, The University of Iowa, Iowa City, IA 52242, USA.

Angela Locascio, Decision Systems Laboratory, Department of General Engineering, 117 Transportation Building, University of Illinois at Urbana-Champaign, Urbana, IL 61801-2996, USA.

Gary A. Maddux, Director, Center for the Management of Science and Technology, University of Alabama in Huntsville, Huntsville, AL 35899, USA.

Joseph M. Mellichamp, Department of Management Science and Statistics, University of Alabama, Tuscaloosa, AL 35487, USA.

David M. Miller, Department of Management Science and Statistics, University of Alabama, Tuscaloosa, AL 35487, USA.

Farrokh Mistree, School of Mechanical Engineering, College of Engineering, Georgia Institute of Technology, Atlanta, GA 30332-0405, USA.

Eoin Molloy, Computer Integrated Manufacturing Research Unit, University College, Galway, Ireland.

Gary P. Moynihan, Department of Industrial Engineering, The University of Alabama, Box 870288, Tuscaloosa, AL 35487-0288, USA.

Dana S. Nau, Department of Computer Science, University of Maryland, College Park, MD 20742, USA.

James S. Noble, Department of Industrial Engineering, College of Engineering, University of Washington, Seattle, WA 98195, USA.

Michael J. O'Flynn, Department of Mechanical and Production Engineering, University of Limerick, Plassey Technology Park, Limerick, Ireland.

Hamid R. Parsaei, Department of Industrial Engineering, University of Louisville, Louisville, KY 40292, USA.

Pipatpong Poshyanonda, Intelligent Systems Center, Department of Engineering Management, University of Missouri-Rolla, Rolla, MO 64501-0249, USA.

Srinivasa R. Shankar, Institute for Innovation and Design in Engineering, Department of Mechanical Engineering, Texas A&M University, College Station, TX 77843-3123, USA.

Warren Smith, Systems Design Laboratory, Department of Mechanical Engineering, College of Engineering, University of Houston, Houston, TX 77204-4792, USA.

William E. Souder, Center for the Management of Science and Technology, University of Alabama in Huntsville, Huntsville, AL 35899, USA.

Mary S. Spann, Center for the Management of Science and Technology, University of Alabama in Huntsville, Huntsville, AL 36899, USA.

William G. Sullivan, Department of Industrial and Systems Engineering, Virginia Polytechnic Institute, Blacksburg, VA 24061, USA.

Fariborz Tayyari, Department of Industrial Engineering, Bradley University, Peoria, IL 61625, USA.

Deborah L. Thurston, Decision Systems Laboratory, Department of General Engineering, 117 Transportation Building, University of Illinois at Urbana-Champaign, Urbana, IL 61801-2996, USA.

A.R. Venkatachalam, Department of Management Science and Statistics, University of Alabama, Tuscaloosa, AL 35487, USA.

Hsu-Pin Wang, Department of Industrial Engineering, College of Engineering, University of Iowa, 4130 Engineering Building, Iowa City, IA 52242, USA.

Hsu-Hao Yang, Intelligent Systems Laboratory, Department of Industrial Engineering, University of Iowa, Iowa City, IA 52242, USA.

Masataka Yoshimura, Department of Precision Engineering, Kyoto University, Yoshida Honmachi, Sakyo-Ku, Kyoto 606-01, Japan.

Chun Zhang, Department of Industrial Engineering, College of Engineering, University of Iowa, 4130 Engineering Building, Iowa City, IA 52242, USA.

Guangming Zhang, Department of Computer Science, University of Maryland, College Park, MD 20742, USA.

M. Carl Ziemke, Center for the Management of Science and Technology, University of Alabama in Huntsville, Huntsville, AL 36899, USA.

Preface

In the area of computer-integrated manufacturing (CIM), concurrent engineering (CE) has been recognized as the manufacturing philosophy for the 1990s. The mission of CE is to develop high quality products and bring them to the competitive global marketplace at a lower price and in significantly less time. However practicing the CE philosophy requires a large number of design tools during the design phase. This book explores a wide variety of CE topics, including popular tools for making CE a reality.

As technological tasks in the manufacturing environment have become progressively complicated, today's designers are faced not only with increasing complexity of product designs but also with a constantly increasing number of sophisticated design tools. Consequently, the entire design process has become very demanding. A large number of highly specialized design tools have to be developed to meet the technological needs of complex design tasks, such as design analysis for manufacturability, assemblability, cost estimation and engineering analysis. Moreover, these design tools should be integrated into a CAD framework to provide a variety of functions and services to the design community.

Design decisions have to be made early in the product development cycle to have significant impacts on manufacturability, quality, cost, time-to-market, and thus on the ultimate success of the product in the marketplace. This implies that all the information pertaining to a product's life cycle should be used to augment the information for design decisions to achieve the optimized product design for manufacture. This is the underlying philosophy of CE, which entails the concurrent consideration of product design and all its related processes with an organization's manufacturing capabilities and future strategies.

The product development cycle begins with the conception of a need based on the market analysis and research and development (R&D) activities. Conventionally, a series of sequential steps is followed to design the product, identify the process operations, fabricate the parts, assemble the components, and ship the product to the marketplace. Product designers are mainly concerned about their products' performance and functionality and rarely take the process design and/or manufacturing constraints into consideration. This traditional sequential path has not entailed much dialogue between design and downstream value-adding processes except a series of requests for engineering changes. It is true that the cost of incorporating engineering changes increases significantly as the changes are made later in the product's life cycle. This implies that the product designer must be knowledgeable of important manufacturing implications of a design as early as possible along with the structural, functional, and aesthetic requirements.

The philosophy of CE is not entirely novel. Pioneers of the automobile industry, like Henry Ford and Ransom Olds, practiced to a certain extent the philosophy of concurrent engineering. These companies have grown into giant corporations with numerous departments each specialized in a task. This specialized separation contributed to the further development of special functions within the departments. It also caused some detrimental effect to the corporation as a whole, mainly spawned by the lack of communications among those departments. As a result, it now seems to be inevitable for those large corporate organizations to reverse their functional silos by building CE teams.

This book is an attempt to address the central issues associated with concurrent engineering. This volume consists of twenty-five reviewed articles which are grouped into four parts. Part One deals with organizational issues in concurrent engineering and presents six chapters covering various aspects of management's challenge in implementing concurrent engineering. Part Two offers ten chapters regarding proven tools and techniques of concurrent engineering. Part Three contains two chapters on the topic of designing for cost targets set in the marketplace. Current and future research directions of concurrent engineering that encompass artificial intelligence are considered in Part Four.

We are indebted to the following individuals: A. Soundararajan, Hyeon Ho Jo, Lakshmi Narayanan, Jian Dong, Sai S. Kolli, and Patricia Ostaszewski for their assistance and support to make this endeavor possible.

Hamid R. Parsaei
William G. Sullivan

PART ONE

Organization Issues in Concurrent Engineering

CHAPTER 1

Principles of concurrent engineering

Hyeon H. Jo, Hamid R. Parsaei and
William G. Sullivan

1.1 INTRODUCTION

The world-wide competitive economy is forcing us to utilize fully the best equipment and techniques available with efficient control of organizational structure to produce high quality, well-designed products at lower prices and in less time. Increasing awareness is being directed to the product design phase, because the more advanced CIM technologies are of little consequence unless the product design lends itself to the overall system utilizing all available and relevant technologies. For instance, the following statements represent the significance of product design.

- Boothroyd (1988) cites published reports from Ford Motor Company which estimate even though product design accounts for only 5% of total product cost, 70% of the cost is influenced by the design.
- Huthwaite (1987) shows the same percentages of product cost and influence, and analogously calls the influence of the product design as a 'Ripple effect' that can virtually impact every area of an organization. The product design function can therefore be imagined as a rock thrown into the center of a lake, which signifies a corporate organization.
- It is believed that 40% of all quality problems can be traced to poor design (Dixon and Duffey, 1988).
- Suh (1990) believes that as much as 70–80% of manufacturing productivity can be determined at the design stage.
- Gatenby and Foo (1990) estimate that an even higher percentage (from 80 to 90%) of the total life-cycle cost of a product is determined during the design phase.
- An observable statement of a manufacturing executive quoted in a prestigious paper (Whitney, 1990): 'designers make million-dollar decisions every minute without ever knowing it.'

These figures demonstrate that product design should lend itself to all its related manufacturing processes so that corporate objectives can be satisfied through an optimized product design. This also demands that the central dogma of engineering should be design (Tomiyama, 1990). Unfortunately, this appeal has often been overlooked in US industry. Dixon and Duffey (1990) observe that opportunities to regain manufacturing leadership lie in engineering design.

In this context, concurrent engineering (CE) has been recently recognized as a viable approach in which the simultaneous design of a product and all its related processes in a manufacturing system are taken into consideration, ensuring required matching of the product's structural with functional requirements and the associated manufacturing implications. This means that all pertinent information flows should be multi-directional among the design function and all related processes in an organization. Processes influencing product design usually include market analysis, materials procurement, product cost estimation, machining, assembly, inspection as well as the later phases of the product's life cycle such as service and maintenance, and disposal.

A survey of the literature reveals the lack of comprehensive study of the challenging concept of CE. There seems to be a pressing need for an in-depth study of all aspects of CE in order to establish the domain and promises of this exciting specialty area.

1.2 AN OVERVIEW OF CONCURRENT ENGINEERING

1.2.1 Background

The product development cycle begins with the conception of a need based on market analysis and research and development (R&D) activities. Conventionally, a series of sequential steps is followed to design the product, identify the processes, machine the parts, assemble the components, and ship the products to the marketplace. Product designers are mainly concerned about their products' performance and functionality and rarely take process design and/or manufacturing's constraints into consideration. This traditional sequential path has not entailed the dialogue between design and the downstream processes except through a series of standard engineering change orders (Fig. 1.1).

However, it has been recognized that design decisions made early in the product development cycle can have a significant effect on the manufacturability, quality, product cost, product introduction time, and thus on the ultimate marketplace success of the product. Furthermore, it is believed that the corrective cost of engineering change orders increases logarithmically as the orders are placed later in the product's life cycle (Siegal, 1991). This means that the product designer must include manufacturing considerations

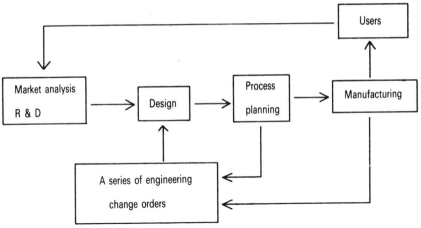

Fig. 1.1   Sequential product development cycle.

as early as possible along with the structural, functional, and aesthetic requirements. In other words, those implications must be *designed in* rather than *inspected in* to avoid the costly iterative process (Burling *et al.*, 1987). Therefore, all organization-wide information should be used to augment design information to arrive at the finalized product design for manufacture.

Concurrent engineering is a philosophy and not a technology. It is the philosophy entailing computer optimized manufacturing (COM) which may be the next generation beyond CIM. That is, the new philosophy of COM may emerge by blending the two existing philosophies of CIM and CE.

1.2.2 History

The philosophy of CE is not entirely novel. Pioneers of the automobile industry, like Henry Ford and Ransom Olds, practiced to a certain extent the philosophy of what we now call concurrent engineering (Evans, 1988; Donovan, 1989). Owing to them and many others, the automobile companies as well as other companies have grown to become giant corporations with numerous departments each specialized in a task. This specialized separation actually contributed to the further development of special functions within the departments. It also caused detrimental effects to the corporation as a whole, mainly spawned by the lack of communications among those departments. For instance, the product design department came up with its own final design and tossed it over the wall to the manufacturing department and expected manufacturing engineers to figure out how to make it. Manufacturing engineers often need to modify the design to manufacture on their shop floor; they often have to adapt the manufacturing conditions to accommodate the design specifications, usually yielding products which are poor in quality, high in price, and introduced to the marketplace late. This practice, however, was good enough to make profits in the era of tranquil prosperity.

As the world became more technologically competitive, people began to realize the importance of efficient utilization of manufacturing resources, i.e., the maximum value obtainable from manufacturing resources. Gladman (1968) emphasized that products should be designed 'right-first-time for production' so that the manufacturing resources were used effectively to enjoy the maximum benefits. He addressed a basic concept of 'Design for Production'; the designer should be provided with all related data from other specialists so that he could modify his design at the design stage to secure manufacturable and economical production. Gladman firmly believed that design for production would indeed have significant meaning in the automated and highly competitive production age. He urged a systematic study of design for production and exerted an influence on formation of Working Group 'O', the optimization group within CIRP (College Internationale de Recherches Pour la Production) to develop a more integral blueprint of production engineering in its modern settings. Encouraged by Gladman and Merchant, a CIRP sub-group on design for economic manufacture was established in 1970 (Chisholm, 1973) for systematic research.

Recognition of the significance of design has come slowly to American industry as compared to European and Japanese industry. During the late 1970s and early 1980s in the US, a few individuals recognized the tremendous benefits that might be provided by more efficient product design for manufacture. Achterberg (1974), Datsko (1978), Boothroyd (1982) stand out as some of the principal pioneers in their efforts to understand and practice the CE philosophy. Now, as US industries compete in the global market, the old ideas of design for manufacture are being rediscovered to restore the manufacturing competitive edge. The US Department of Defense (DoD) has paid special attention to CE, promoted CE studies, and implemented CE in weapon system production (Pennell and Winner, 1989; Pennell *et al.*, 1989]. A government-industry-academia consortium, DARPA Initiative on Concurrent Engineering (DICE), was launched in 1988 by the Defense Advanced Research Projects Agency (DARPA) to encourage the practice of CE in the US military and industrial base. The overall aim of the consortium was to develop an information architecture for CE in which each member working on a project can communicate and coordinate information through a high-speed computer network to support CE practices. Many advanced corporate organizations, such as Hewlett-Packard (Conradson *et al.*, 1988), Motorola (Inglesby, 1989), AT&T (Gatenby and Foo, 1990), Texas Instruments (Hutchison and Hoffman, 1990), and IBM (Pursell, 1988; Correia and Freeman III, 1988; Oliver, 1990) have been precisely honing their product realization processes, fortifying them with the rediscovered philosophy of CE. A goal of the 'Chrysler Technology Center' is known to bring people together under one roof to promote the CE practices (Holt, 1991).

1.2.3 Aliases of concurrent engineering

Many terms have been used to describe similar approaches, including simultaneous engineering, life-cycle engineering, design integrated manufacturing, design fusion, early manufacturing involvement, parallel engineering, concurrent design, design fusion, and design in the large. In Europe, design for production and design for economic manufacture (DEM) have been used instead. Some process-oriented technological terms have been also used to describe the functional concepts of CE. They include design for manufacturabilility (DFM), design for producibility, design for assemblability (DFA), design for testability, design for reliability, design for installability, design for serviceability, and so on. Gatenby and Foo (1990) call these 'Design for X' (DFX), where X stands for the above *ilities* and other related upstream and downstream functions.

1.2.4 Approaches to concurrent engineering implementation

There may be two basic approaches to implementing the concurrent engineering practice: team-based and computer-based approaches. The former approach is human-oriented in that the team consists of designers and individuals from all other related functional areas. Team members are selected for their ability to contribute to the design of product and processes by early identification of potential problems and timely initiation of actions to avoid a series of costly reworks (Pennell *et al.*, 1989). The multifunctional team is crucial for effective implementation. Continuously developed computer technologies, in both hardware and software, have given team members from different departments the ability to work with the same design to evaluate the effects of design attributes. In this context, it is believed that a significant educational program is a must to have each team member fully understand the philosophy of CE. Many corporate organizations have already initiated the educational programs for a philosophical shift to CE. For instance, Motorola recently advertised that many new employees would begin with a 24-week 'Concurrent Engineering Training Program'. Numerous papers have reported case studies in which the team-based approaches were implemented and significant benefits were realized (Dwivedi and Klein, 1986; Schehr, 1987; Conradson *et al.*, 1988; Inglesby, 1989; Goldstein, 1989; Charles, 1989; Hutchison and Hoffman, 1990; Allen, 1990).

While the team-based approach can be readily implemented and is being widely adopted in industry, some shortcomings appear to arise (O'Grady and Young, 1991): difficulties in effective management of the team, team member's limited knowledge, and the cost of maintaining a team. As more sophisticated computer tools emerge constantly, the team-based approach is being enhanced by the computer-based approach in which the concurrent engineering philosophy is woven into the internal logic operations, enabling design justification or optimization with respect to the entire aspects of a product's life-cycle.

A constraint programming language has been developed for designers to take into account all life-cycle implications related to the product (Bahler *et al.*, 1990; Bowen *et al.*, 1990; O'Grady and Young, 1992; Bowen and Bahler, 1992). Non-directional constraint networks were used to represent the mutually constraining influences of the life-cycle perspectives as well as the intended functionality. A constraint monitoring technique instead of constraint satisfaction was used as the form of constraint processing.

Recognizing that a concurrent engineering approach for sophisticated products or mechanical systems may cause a large scale design project which is difficult to manage as a whole, Kusiak and Park (1990) proposed a methodology for clustering numerous design tasks into groups that can allow effective organization of resources required in the design process. Cluster analysis was used to provide an underlying theory for solving the grouping problem, and a knowledge-based approach was employed for scheduling and managing the design activities.

Lu (1990, 1992) discussed knowledge processing technology as a unified foundation for software techniques that can be used in CE. The rationale behind the discussion is that as our society developed and technological tasks in CIM environment became more complicated, we have started to experience the deficiencies of data processing technology which has served well, as the foundation for manufacturing automation over the past several decades. That is, we are facing a challenge of converting massive amounts of data, often referred to as data explosion crises, into structured useful knowledge so that we can utilize it effectively to improve decision-making productivity in CE.

1.3 COMPUTER-BASED CE ENVIRONMENT

The underlying prerequisite to the computer-based approach is systematically to acquire, represent, integrate, and coordinate the requisite concurrent engineering knowledge with which computers can perform required analyses. A large number of computer-aided design (CAD) tools will be also required during the design stage to examine the influence of the design on the product's life-cycle. Consequently, the expectation is to see an integrated design environment in which all the CAD tools interact and cooperate to find a globally optimized or compromised design.

Figure 1.2 shows a conceptual model of the approach [Jo *et al.*, 1990]. In the figure, the outer layer of the 'concurrent engineering wheel', product modelers are advanced which can provide designers with the capability to invoke any tools in the inner layer to evaluate or optimize their designs. The core of the wheel is the control logic which involves steering of various CAD tools to provide a variety of services, helping to find a globally satisfied

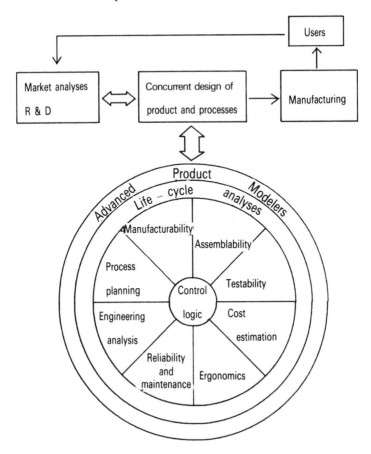

Fig. 1.2 Product development cycle employing 'concurrent engineering wheel'.

design. It must be emphasized that it is the core of the wheel for which scientific theories of the design process are badly needed. Between the outer layer and the core is the functional layer that comprises various life-cycle analysis tools. It should be noticed that many other aspects of a product's life-cycle may be inserted into this layer. They may include market analysis, disposability, packing/shipping, social impacts, and so on. Brief descriptions of a few of the functions within the wheel follow.

1.3.1 Product modelers

It has been widely recognized that the integration of a manufacturing system as a continuum is a must to meet today's highly competitive market demands. The integrated system that does not require any human intervention may yield significant advantages to reduce the product realization time, increase product quality, and reduce the cost. In other words, human interface, if required, often introduces errors and inconsistencies, reduces the extent of automation, and increases product development time and cost. However, there has been an apparent gap between design and manufacturing.

In bridging the gap, the development of more versatile product modeling software should play the major role. In an attempt to overcome the gap between design and process planning, the concept of feature has been introduced, because the process planning is feature-based. For an ideal computer integrated manufacturing system, one of the very basic requirements is that the part design feature as well as its dimensions and orientation must be readily available to a set of analysis software for justification of the design. The definition of feature is application specific so that it does not have a general definition agreed upon by every researcher. One simple definition from a geometric view may be stated as a collection of surfaces that needs to be machined and closely matches the way process planners view a subset of the geometry on an engineering drawing. Unger and Ray (1988) introduced some other definitions given by major researchers in the area. The Automated Manufacturing Research Facilities (AMRF) process planning team at the National Institute of Standards and Technology (NIST) proposes a more abstract and thorough definition (Unger and Ray, 1988): 'A feature is a higher lever grouping of geometrical, topological, and functional primitives into an entity more suitable for use in design, analysis, or manufacture'. A feature has its own geometry and some associated attributes such as dimensions, tolerances, manufacturing information depending on the application, leading to different definitions accordingly.

Recall that the feature closely matches the way process planners view the design. Since the most current CAD systems do not store features and their attributes explicitly, a generic methodology for automatic feature recognition and extraction is required to make the manufacturing system an integrated continuum. The extracted information is then sent either to a set of advisor modules for the evaluation of the design or a computer aided process planning (CAPP) system if the design has been justified. Such an environment would enable designers to justify design decisions from the standpoint of downstream processes, simultaneously satisfying structural and functional requirements of the part, processing constraints, and the capital budget. Because of the important characteristics that features may contain, several algorithms have been proposed to recognize and extract features from geometric models.

Choi *et al.* (1984) proposed an algorithmic procedure to identify the features directly from its 3D-boundary representation model and Perng *et al.* (1990) introduced a method of extracting the feature from 3D constructive solid geometry (CSG) model. Pande and Prabhu (1990) reported an expert system for automatic feature extraction.

A few feature recognition/extraction approaches are reviewed. However, those attempts would be unnecessary if a methodology could be developed, in which all the feature information were retained in the geometric modeler. This idea led to the development of a modeling scheme called 'feature-based design', where the product is represented by features from the outset. Features are stored in a database as entities and the part functionalities can be represented by combining the related features. In this environment, the designer is provided with certain desired higher level abstractions of the model that relate directly to certain design and manufacturing characteristics (Chung *et al.*, 1988). Yet most feature-based designs are concerned about supporting single-point cutting processes, while a few support casting and assembly (Whitney, 1990).

Cunningham and Dixon (1988) explain, however, that different manufacturing processes (casting, forging, metal cutting, assembly, etc.) may require different features for their various activities (fitting, manufacturability evaluation, manufacturing cost analysis, tool/die design, and so on). This implies that the geometric model designed with features needs to be refined to meet a process-activity set's requirements. The refinement may involve some feature extraction to obtain more abstract features or breaking down certain features into sub-features. This is the subject of a geometric reasoning area that might be indispensable in bridging the gap between design and manufacturing.

A more ambitious project has been carried out completely to define a product for all applications over its whole life cycle. A huge project, Product Data Exchange Specification (PDES), at NIST is being developed to include the following data in a single database (Ssemakula and Satsangi, 1989): administrative and control data, geometry, topology, tolerances, form features, feature attributes, material properties, part assembly, and so on. The challenge of the project, however, appears formidable, because it requires the development of such a thing as the 'Standard for the Exchange of Product Model Data' (STEP) that can be used to describe a product in a thorough way that supports all facets of its life-cycle. STEP, being undertaken by the International Organization for Standardization (ISO), is aimed at providing a consistent and computer interpretable definition of the physical and functional primitives of each unit of a product through its life-cycle (Davies, 1991). The PDES project is now the United States' contribution to STEP and stands for Product Data Exchange using STEP. If the projects are successfully completed, then we may say that we have come one major step closer to the real computer integrated manufacturing system that links CAD to CAM without human intervention.

1.3.2 Computer-aided process planning (CAPP)

As previously mentioned, the philosophy of CE entails that product design and process plans be developed simultaneously. This is best assured by having the designers work in manufacturing modes, rendering the freedom to add features to the design or manipulate the process plan itself (Cutkosky and Tenenbaum, 1990). The conventional operation of product design and process planning proceeds sequentially, coupled with annotated engineering drawings. The problems with this sequential practice are well known in terms of quality, cost, and time-to-market. One of the most promising methodologies to solve the problems is the concurrent design of product and process which is a part of concurrent engineering. In such an environment, it is even necessary to evaluate alternate manufacturing processes such as casting, forging, and machining according to the preliminary product design using expert systems.

Ahluwalia and Ji (1990) proposed a structure of a CAPP system in a CE environment. The major three components of the structure are: interface to CAD and CAM databases, process selection advisor, and process planning advisor.

An interesting finding described by Lu (1990) is that CAPP functions must be modularized and distributed throughout the product and process design phases so that they can be called upon at any stage during the design phase.

An integrated design procedure is proposed by Yoshimura *et al.* (1989), in which an integrated optimization problem is formulated. The problem is formulated as a multiobjective optimization problem which produces a set of Pareto optimum solutions, and the heuristic solution algorithm is presented. The procedure is implemented and applied to design a cylindrical coordinate robot to show the significance of the simultaneous product and process design.

1.3.3 Design for manufacturability

In concurrent engineering environments, the time to develop a new product can be significantly reduced by avoiding design errors and features difficult to machine. The basic idea of design for manufacturability (DFM) addresses this time-to-market issue as well as those of quality and cost. Many noticeable results of DFM practices have been announced (Edosomwan, 1988; Owens, 1988; Wade and Colton, 1990). For instance, Drasfield (1990) described that the product introduction time had decreased by 75% in two years and design modification could be reduced by 74% over four years. It should be emphasized, however, that one of the major prerequisites to the successful practice of DFM concept is to develop and use *quantifiable metrics* for the evaluation of manufacturability. Only with the quantifiable measures may the designer be able to appropriately evaluate his/her design's manufacturability (Priest, 1990).

Jansson *et al.* (1990) described a set of manufacturability indices that can provide a comprehensive definition of manufacturability and be used for evaluating the design early in the design process. Six indices were defined to represent the major aspects of manufacturability: compatibility, bug index, availability, complexity, standardization, and efficiency. Although the assessment of manufacturability based on these indices may be inappropriate or depend on diverse manufacturing environments, it may provide designers with an opportunity to understand the design trade-off analyses, which is a very important step to develop intelligent CAD systems.

A group of people propose an approach, known as the rapid prototyping protocol (RPP), which uses a database of flexible manufacturing cell capabilities in order to analyze a design for manufacturability (Anand *et al.*, 1988a, b; Anjanappa *et al.*, 1989; Hudson *et al.*, 1990). The RPP covers a wide variety of manufacturing constraints, providing the system with the ability to reject the design the cell protocol is unable to produce, and suggest a design that may be more manufacturable to the user.

Lu and Subramanyam (1988) describe another structured design environment based on knowledge-based systems to design part and process simultaneously. It consists of a features database, user interface, and manufacturability advisor.

Lim (1987) also proposed an intelligent knowledge-based system, called Holdex, for integrating component design to tool engineering. It was stated that although Holdex initially started as an expert system for designing modular fixtures, its functions and purposes were enlarged to allow the designers to evaluate their component designs from the perspectives of cost and manufacturability.

Sutherland *et al.* (1988) described a methodology for an aspect of concurrent engineering which incorporates machining process modeling and the design of experiments to find robust product/process design in terms of the factors under study. The factors included, in the face milling model case, were part geometry, part material, number of inserts and their spacing, insert geometry, insert runout conditions, amount of surface error left by the preceding machining pass, and machining conditions including feedrate and depth of cut.

1.3.4 Design for assembly

Assembly is often the most labor intensive operation and accounts for a major portion of the total cost. Assembly robots have been used in industry that manufacture simple products in large volumes. However, most products are still not well suited for robotic assembly. In order fully to utilize robotic assembly the designer must modify some of the current design practices.

The concept of design for assembly (DFA) arises in the hope that all discrete component parts will be designed so that they are easily assembled and the cost is significantly reduced.

A DFA system was developed at the University of Massachusetts (Boothroyd, 1982) for rating the efficiency of assemblability based on two factors: the easiness of handling and assembling of component parts and the number of parts used in the product. The system evaluates scientifically the parts feeding and orientation problems and sets guidelines for rating product design producibility. If the rating is low, then redesign is recommended. Use of the handbook version of the system may significantly reduce the assembly cost but it requires significant human intervention: a large amount of component-coding and form filling.

Holbrook and Sackett (1988) studied existing DFA methodologies and classified them into four groups: specific assembly operation theories, unstructured rules and concepts, procedural methods of applying rules, and expert system/knowledge-based systems. The previously described handbook approach is an example of the procedural methods of applying rules.

Runciman and Swift (1985) developed an expert system that estimates the costs of handling and assembly equipment for a given part design in two dimensions. The system informs the designer of any difficulties that the part may present for automatic handling during manufacture and suggestions for improving the design. The deduction sequence used by human experts is emulated and the four major stages are selected: feeding, orientation, presentation problems of component parts and assembly costing problem.

As a methodology for implementing the concept of design for assemblability, Steiner and Derby (1989a, b, c) described an approach for estimating robot assembly cycle time during product design. An assembly sequence is determined using standard sequence constructs and part mating features of the design. A velocity profile for the sequence is then developed and synthesized using a velocity profile shrinking algorithm. Cycle time is finally estimated based upon the motion distance and synthesis of motion velocity profiles.

As they discuss a strategic approach to product design, Nevins and Whitney (1988) assert that the assembly sequency itself should be considered as part of product design, because the sequence affects so many design and manufacturing decisions that are to be made early in the design process. Whitney *et al.* (1990) describe algorithms for automatic generation and evaluation of the sequence.

1.3.5 Design for reliability and maintainability

While quality control activities are dealing with conformance to the design specifications, reliability is an inherent attribute of a product resulting from its design (Kapur, 1984). Unfortunately, conventional reliability and maintainability (R&M) analyses have been typically done very late in the design cycle, being ineffective in changing the design. As a part of regaining the manufacturing competitive edge, it is necessary for today's R&M analyses to serve as a design advisor in the initial design phase. This issue exactly complies with the philosophy of concurrent engineering in which all 'ilities' characteristics are captured as early and quickly as possible in the design process.

Grassman and Rodriguez (1989) briefly describe a system, the Supportability Design Evaluation System (SDES), which embeds analysis software to test design for thermal stress, reliability, maintainability, and other 'ilities' directly into the designer's workstation. The system, developed at Hughes Aircraft Company, was intended to provide the designer with immediate feedback information as to the impact of replacing or moving components of the design on the failure rate, the consequence of upgrading the reliability factor of components on failure rate, and so on.

R&M analyses should no longer be considered as a task performed by R&M organizations, because the R&M activities must start early in the product development cycle. The designers must be considered as the persons who are responsible for the analyses. But, we need to provide the designers with some tools that they can use effectively. One tool may rely on a formalized design review and evaluation procedure such as failure mode and effects analysis.

The quality function deployment (QFD) process can also be extended to help the development of the R&M analyses plan in the design process. Vannoy (1989) describes an approach with each step of the extended QFD process that can be used for the purpose. It is believed that by using this approach as early as possible in the design process, the failure governing factors can be identified long before the reliability test is performed. Notice that the reliability test takes the nature of confirming the adequacy of the design rather than of exploring the design alternatives. It should be emphasized, however, that the whole procedure should be accomplished as quickly as possible in a concurrent engineering environment so that the designer can meet the time-to-market pressure.

1.3.6 Design for engineering analysis

Suri and Shimizu (1989) proposed a design strategy based on the concept of 'design for analysis'. With the aims of reduced design times, shorter manufacturing lead times, superior quality, and lower cost, the concept states that designers should work with only those designs which can be analyzed easily and quickly by simple tools. The reason for the proposed strategy is that there are practical difficulties in going through the analysis process mainly due to the time required to perform the analysis. To resolve these difficulties, it may require two simultaneous approaches. One is to provide designers with good tools which are easy to use, provide results quickly, and answer key design questions. The other is actively to change the design itself so that tools can be used effectively. This concept has a significant implication in the design process models where the designs are refined iteratively.

1.3.7 Cost estimation

Concurrent engineering must also address questions about how product costs can be estimated at the design stage so that the product's profitability can be assessed. If the projected cost of the product being designed exceeds the cost limit, then cost estimation software may suggest discontinuing the further development or redesigning the product. The estimation module should be developed in such a way that it collects data from all related modules in a CE environment and evaluates the design based upon the predicted costs of material, machining, overhead, testing, assembly, and other related drivers. It should be emphasized that the reliable data from all those cost drivers are crucial for the accuracy of the estimates.

The most important step of cost estimation is the construction of cost models that can derive meaningful cost estimates based on the collected data. The model required to track costs in a CE environment will be different from the usual models for estimating product costs, because the traditional cost estimating systems are not structured adequately to support CE. An integrated cost estimating system is required to adhere to and enhance the CE philosophy. Because a surprisingly large portion of the product cost is influenced by product design, it is important to obtain the cost estimates as early and quickly as possible during the design phase. However, various levels of uncertainty exist for the cost estimation through-out the design process. These uncertainty factors, such as in material, geometry, and manufacturing processes, certainly limit the choice of estima-ting methodologies. Therefore, as a design project progresses from concep-tual stage to detailed final stage, different methods should be brought to bear on the cost estimation problem (Ulrich and Fine, 1990).

Creese and Moore (1990) briefly describe a few methods that can be used at each of the three stages of the design process: conceptual, preliminary, and detailed design stages. Wierda (1988) emphasizes that the designer should do the job, typically that has been done by the cost estimator, with the tools that can give the designer a quick and accurate estimate of the product cost. He briefly describes a survey of possible cost estimation techniques and research efforts mainly encountered in Europe. A procedural cost estimation technique called predictive cost modeling is introduced by Huthwaite and Spence (1989) who believe it enables companies to predict the total life-cycle costs for many design alternatives. Their technique appears to be based on the spirit of teamwork.

Schmitz and Desa (1989) describe a design methodology based on the cost estimation, called the producibility evaluation package (PEP), for precision planner stamped products. A basic idea in implementing the methodology is that they model the entire stamping process in terms of parameters so that the designer can control both the product functionality and most of the manufacturing costs. It is stated that PEP contains all the cause–effect relations from the design specifications to the set of manufacturing cost factor values so that the designer can identify, during the design process, the design specifications which cause the value to be high.

Moore and Creese (1990) briefly describe the new accounting philosophy for the model that may be used in a CE environment, and Wong *et al.* (1991)

proposed a totally integrated manufacturing cost estimating system (TIM-CES), which can be easily integrated with the other engineering software systems. The input to the proposed TIMCES will be provided by several databases including CAD, CAPP, material cost, labor cost, etc.

1.3.8 Control logic

As mentioned before, a large number of CAD tools will be required during the design stage to examine the effects of the design upon all aspects of its life-cycle from conception to disposal. As a consequence, an integrated design environment will be needed to accommodate the various tools which interact and cooperate to find a globally optimized or compromised design. The integrated design environment may consist of sets of various CAD tools and an underlying CAD framework.

A CAD framework is 'a software infrastructure that supports the application environment, providing a common operating environment for CAD tools' (Graham, 1991). It is a logical extension to the general computing environment that provides the functions and services demanded by the design community. The framework is intended to be a well-organized software infrastructure, which properly guides designers through the entire design process so that they can reach the design of maximum quality which can be produced at the lowest cost and in the shortest time. A CAD framework is actually layers of code between the operating system and application environment, which provides a variety of functions and services to designers. A notable improvement in design productivity can be accomplished, because designers using the framework can concentrate on their main design tasks rather than on other tasks such as the user interface.

It is believed that development of sound CAD frameworks to support fully concurrent engineering practices will not be trivial and will require some fundamental redesign of framework architectures (Smith and Cavalli, 1990). A CAD framework should contain several facilities which coordinate the entire design process that complies with the CE philosophy. Since today's technological milieu changes rapidly, these frameworks should be constantly updated. To ease the burden from the CAD system integrators or designers, frameworks are intended to be open architectures that make them more domain independent, easier to use and update, easier to integrate the tools, and more capable of supporting a large population of tools and corresponding various data models (Daniel and Director, 1989). A methodology for implementing *soft* integrated systems that facilitate the adaptation to changing environments is described by Weston *et al.* (1991).

At the heart of a CAD framework is the task of invoking and controlling numerous CAD tools, which remains as an error-prone and a largely unsolved problem (Fiduk *et al.*, 1990). The CAD tool communication and control mechanism is perhaps the most important technological part of the framework. It provides the communication backbone for a large amount of design data, and controls various design events that occur throughout the entire design process. Invoking a large number of tools manually is a difficult or at best tedious and time-consuming task. The issue of CAD tool control is critical for solving many problems that have plagued CAD frameworks.

1.4 FUTURE TRENDS

Concurrent engineering has a key role to play in the computer integrated manufacturing systems (CIMS) and the extent to which the CE principles are applied will be the major deciding factor of the competitiveness of a nation's industries. However, full implementation of CE is still far from reality.

The implementation of CE principles will utilize new hardware as it becomes available. For instance, sensor-equipped intelligent robots will emerge and be used for fully automated assembly. This will be able to relax the constraints imposed on the product design from the standpoint of assemblability, enabling the product to be made with more power and sophistication in a CE environment.

As mentioned before, the product data representation schemes currently available are far from satisfactory and the subject is considered one of the bottlenecks to integrate the manufacturing system as a continuum. In some business situations, product drawings and order quantities will be transmitted from the buyer's CAD system to the vendor's CAD system. The vendor will in turn transmit any potential revisions, product costs, and shipping schedules. The feature recognition and extraction schemes will be required in this case for the vendor to manufacture the product based on the transmitted CAD data. It appears that the product representation schemes and feature recognition/extraction methodologies will be a topic for further development effort.

It is not difficult to predict that there will be an effective combination of human and artificial intelligence, with the human intelligence creating a design and the artificial intelligence analyzing every designing phase and suggesting the possible improvements based on a previously constructed knowledge base.

Many managers and directors as well as manufacturing engineers are frustrated with many different databases created for a single product. They are constantly seeking a single source for all information on the product. This problem might be resolved by using the concept of single database (SDB) that includes all the data from design, analysis, process planning, tooling, quality/inspection plan as well as BOM, MRP, production scheduling, and post-process information, etc. (Alting and Zhang, 1989).

1.5 CONCLUSIONS

A survey of concurrent engineering has shown that practicing the concurrent engineering philosophy is a huge project spanning all phases of a product's life cycle. Furthermore, concurrent engineering is essential to attain a computer optimized manufacturing (COM) system that may be the next generation beyond computer integrated manufacturing (CIM) systems. It will be a challenging task to implement this philosophy.

If the 'Concurrent Engineering Wheel' is carefully designed, constructed, and implemented in industry, then it may become the *Wheel of Fortune*.

The Computer Integrated Manufacturing (CIM) milieu in the twenty-first century is conceptually described (Goldhar and Jelinek, 1990) as a situation called 'the automation of custom manufacture' where the customer participates at least to some degree in the design of the product he/she wants and decides how much 'service' in terms of customer requirements and speed of delivery he/she is willing to pay for. The gap between this concept and reality will diminish if CE is understood and becomes a part of our industrial practice.

1.6 REFERENCES

Achterberg, R.C. (1974) Product design for automatic assembly, *ASME Paper*, No. AD 74–435.

Ahluwalia, R.S. and Ji, P. (1990) Process planning in concurrent engineering environment, *1990 International Industrial Engineering Conference Proceedings*, pp. 535–40.

Allen, C. W. (ed.) (1990) *Simultaneous Engineering: integrating manufacturing and design*, Society of Manufacturing Engineers, Dearborn, MI.

Alting, L. and Zhang, H. (1989), *International Journal of Production Research*, **27** (4) 553–85.

Anand, D.K., Kirk, J.A., Anjanappa, M. and Chen, S. (1988) Cell control structure of FMP for rapid prototyping, *Advances in Manufacturing Systems Engineering*, ASME (Publication) PED, **31**, 89–99. Presented at the ASME Winter Annual Meeting, Chicago, IL.

Anand, D.K., Kirk, J.A., Anjanappa, M., Nau, D. and Magrab, E. (1988) Protocol for flexible manufacturing automation with heuristics and intelligence, *Proceedings of the Manufacturing International '88*, Atlanta, Ga, May, 209–17.

Anjanappa, M., Courtright, M.J., Anand, D.K. and Kirk, J.A. (1989) *Manufacturability Analysis for a Flexible Manufacturing Cell*, ASME Design Engineering Division (Publication) DE, **19–1**, 303–8.

Bahler, D., Bowen, J., O'Grady, P. and Young, R.E. (1990) in, *Issues in Design/Manufacturing Integration*, (ed. A. Shron) presented at the Winter Annual Meeting of the ASME, Dallas, TX, November, 25–30, 59–67.

Boothroyd, G. (1982) *Assembly Engineering*, March, 42–5.

Boothroyd, G. (1988) *American Machinist*, **132**, August, 54–7.

Bowen, J. and Bahler, D. (1992) Constraint nets for life cycle engineering: negotiation among multiple perspectives in an interactive design advisor, *Proceedings of the 1992 NSF Design and Manufacturing Systems Conference*, Atlanta, GA, January, 8–10, 661–4.

Bowen, J., O'Grady, P. and Smith, L. (1990) *Artificial Intelligence in Engineering*, **5** (4), 206–20.

Burling, W.A., Lax, B.J.B, O'Neill, L.A. and Pennino, T.P. (1987) *AT&T Technical Journal*, **66**, (5), 21–38.

Charles, D.P.S.T. (1989) Simultaneous engineering, *AUTOFACT '89*, Conference Proceedings, Detroit, Michigan, October 30–Nov. 2, 14.1–14.6.

Chisholm, A.W.J. (1973) *Annals of the CIRP*, **22**, (2), 243–47.

Choi, B.K., Barash, M.M. and Anderson, D.C. (1984) *Computer Aided Design*, **16**, 81–6.

Chung, M.J. and Kim, S. (1990) An object-oriented VHDL environment, *Proceedings of the 27th ACM/IEEE Design Automation Conference*, Orlando, FL, June, 24–28, 431–6.

Conradson, S.A., Barford, L.A., Fisher, W.D., Weinstein, M.J. and Wilker, J.D. (1988) Manufacturability tools: an engineer's use and needs, *1988 Proceedings, IEEE/CHMT 5th International Electronic Manufacturing Technology Symposium*, Lake Buena Vista, FL, October, 10–12, 155–8.

Correia, M. and Freeman III, W.J. (1988) Building the bridge between design and manufacture, *1988 Proceedings, IEEE/CHMT 5th International Electronic Manufacturing Technology Symposium*, Lake Buena Vista, FL, October, 10–12, 45–50.

Creese, R.C. and Moore, L.T. (1990) *Cost Engineering*, **32**, (6) 23–6.

Cunningham, J.J. and Dixon, J.R. (1988) Designing with features: the origin of features, *Computers in Engineering 1988: Proceedings of the 1988 ASME International Computers in Engineering Conference and Exhibition*, San Francisco, California, July 31–August, 4, **1**, 237–43.

Cutkosky, M.R. and Tenenbaum, J.M. (1990) *Mechanism and Machine Theory*, **25**, (3), 365–81.

Daniel, J. and Director, S.W. (1989) An object oriented approach to CAD tool control within a design framework, *Proceedings of the 26th ACM/IEEE Design Automation Conference*, Las Vegas, June, 25–29, 197–202.

Datsko, J. (1978) *International Journal of Production Research*, **16**, (3), 215–20.

Davies, M. (1991) The status and future of product data exchange, *Proceedings of the International Conference on Computer Integrated Manufacturing*, Singapore, October, 2–4, 259–62.

Dixon, J.R. and Duffey, M.R. (1988) Quality is not accidental – it's designed, *The New York Times*, June, 26.

Dixon, J.R. and Duffey, M.R. (1990) *California Management Review*, Winter, 9–23.

Donovan, J.R. (1989) *Proceedings of the Institution of Mechanical Engineers, Part D*, **203**, (1), 47–53.

Dransfield, J.D. (1990) *Proceedings of Manufacturing International '90*, **5**, Atlanta, Ga, March, 25–28, 27–32.

Dwivedi, S.N. and Klein, B.R. (1986) *CIM Review*, **2** (3), 53–9.

Edosomwan, J.A. (1988) Implementing the concept of early manufacturing involvement (EMI) in the design of a new technology, Technology Management I, *Proceedings International Conference on Technology Management*, Miami, FL, February, 17–19, 444–50.

Evans, B. (1988) *Mechanical Engineering*, February, 38–9.

Fiduk, K.W., Kleinfeldt, S., Kosarchyn, M. and Perez, E.B. (1990) Design methodology management – a CAD framework initiative perspective, *Proceedings of the ACM/IEEE Design Automation Conference*, Orlando, FL, June, 24–28, 278–83.

Gatenby, David A. and Foo, George (1990) *AT&T Technical Journal* **69**, (3) 2–13.

Gladman, C.A. (1968) *Annals of the CIRP*, **16**, 3–10.

Goldhar, J.D. and Jelinek, M. (1990) *Computers in Industry*, **14**, 225–45.

Graham, A. (1991) *IEEE Design and Test of Computers*, September, 12–15.

Grassman, N. and Rodriguez, E. (1989) Less reliability and maintainability...not more...is better, *1989 Proceedings of Annual Reliability and Maintainability Symposium*, Atlanta, Ga, USA, January, 24–26, 210–15.

Holbrook, A.E.K. and Sackett, P.J. (1988) Positive design advice for high precision, robotically assembled product, *Developments in Assembly Automation, incorporating Proceedings 9th International Conference*, March, 181–90.

Holt, D.J. (1991), *Automotive Engineering*, November, 26–41.

Hudson, D.C., Pang, D.-C., Kirk, J.A., Anand, D.K. and Anjanappa, M. (1990) Integrated Computer Aided Manufacturing for prototype machining *Proceedings of Manufacturing International '90. Part 2: Advances in Manufacturing Systems*, Atlanta, GA, March, 25–28, 55–62.

Hutchison, K. and Hoffman, D.R. (1990) *High Performance*, April, 40–3.

Huthwaite, B. (1987) Product design for manufacture and assembly: the five fundamentals, *Proceedings of the 2nd International Conference on Product Design for Manufacture and Assembly*, Newport, Rhode Island, April, 6–8.

Huthwaite, B. and Spence, G. (1989) *National Productivity Review*, **8**, (3), 239–48.

Inglesby, T. (1989) *Manufacturing Systems*, **7**, (4), 26–32.

Jansson, D.G., Shankar, S.R. and Polisetty, F.S.K. (1990) in *Design Theory and Methodology – DTM '90*, (ed. James R. Rinderle) 85–96.

Jo, H.H., Parsaei, H.R. and Wong, J.P. (1990) *Computers and Industrial Engineering*, **21**, 35–9.

Kapur, K.C. (1984) in *Handbook of Industrial Engineering* (Salvandy, G., ed.), John Wiley & Sons, New York, 8.5.1–8.5.34.

Kusiak, Andrew and Park, Kwangho (1990) *International Journal of Production Research*, **28**, (10) 1883–900.

Lim, B.S. (1987) *Expert Systems*, **4**, (4), 252–67.

Lu, S. C.-Y. (1990), *Robotics and Computer-Integrated Manufacturing*, **7**, (3/4), 263–77.

Lu, S. C.-Y. (1992) Knowledge processing tools to support concurrent engineering tasks, *Proceedings of the 1992 NSF Design and Manufacturing Systems Conference*, Atlanta, GA, January 8–10, 373–9.

Lu, S. C.-Y. and Subramanyam, S. (1988) A computer-based environment for simultaneous product and process design *Advances in Manufacturing System Engineering – 1988*, ASME Production Engineering Division (Publication) PED, **31**, 35–46. Presented at the Winter Annual Meeting of the ASME, Chicago, IL, November 27–December, 2.

Moore, L.T. and Creese, R.C. (1990) Cost prediction in concurrent engineering, *1990 International Industrial Engineering Conference Proceedings*, 547–52.

Nevins, J.L. and Whitney, D.E. (1989) *Concurrent Design of Products and Processes*, McGraw-Hill Publishing Co., New York, New York.

O'Grady, P. and Young, R.E. (1991) *Journal of Design and Manufacturing*, **1**, 27–34.

O'Grady, P. and Young, R.E. (1992) Constraint nets for life cycle engineering: concurrent engineering, *Proceedings of the 1992 NSF Design and Manufacturing Systems Conference*, Atlanta, GA, January, 8–10, 743–8.

Oliver, G.D. (1990) in *Issues in Design/Manufacturing Integration* (ed. A. Shron), presented at the Winter Annual Meeting of the ASME, Dallas, TX, November, 25–30, 37–42.

Owens, N.L. (1988) Design for manufacturability in hermetic product die attach, *1988 proceedings, IEEE/CHMT 5th International Electronic Manufacturing Technology Symposium*, Lake Buena Vista, FL, October 10–12, 57–61.

Pande, S.S. and Prabhu, B.S. (1990) *Computer-Aided Engineering Journal*, **7**, (4) August, 99–103.

Pennel, James P. and Winner, Robert I. (1989) *IEEE Global Telecommunications Conference and Exhibition (GLOBECOM '89)*. Part 1 (of 3), Dallas, TX, November, 27–30, 647–55.

Pennell, J.P., Winner, R.I., Bertrand, H.E. and Slusarczuk, M.M.G. (1989) Concurrent Engineering: an overview for Autotestcon, *AUTOTESTCON Proceedings '89: The System Readiness Technology Conference*, Philadelphia, PA, September, 25–28, 88–99.

Perng, D.-B, Chen, Z. and Li, R.-K. (1990) *Computer-Aided Design*, **22**, (5) June, 285–95.

Priest, J.W. (1990) State of the art review of measurement procedures in product design for manufacturing *Proceedings of Manufacturing International '90*, **5**, Atlanta, Ga, March, 25–28, 3–8.

Purssell, R.A. (1988) The product release interface: environment and concerns, *1988 Proceedings, IEEE/CHMT 5th International Electronic Manufacturing Technology Symposium*, Lake Buena Vista, FL, October, 10–12, 51–6.

Runciman, C. and Swift, K. (1985) *Assembly Automation*, **5**, (3), 17–50.

Schehr, L. (1987) *Manufacturing Systems*, **5**, (12), 24–8.

Schmitz, J.M. and Desa, S. (1989) The development and implementation of a design for producibility method for precision planar stamped products, *Computer-Aided and Computational Design: Advances in Design Automation – 1989*, ASME Design Engineering Division (Publication) DE, **19**, (1) 295–302. Presented at the ASME Design Technical Conferences – 15th Design Automation Conference, Montreal, Quebec, Canada, September, 17–21.

Siegal, B. (1991) *Industrial Engineering*, **23**, (12), 15–19.

Smith, R. and Cavalli, A. (1990) *High Performance*, June, 63–7.

Ssemakula, Mukasa E. and Satsangi, A. (1989) *Computers and Industrial Engineering*, **17**, 234–9.

Steiner, M. and Derby S. (1988) Estimating robot assembly cycle time during product design – Part 1: algorithms, *Trends and Developments in Mechanisms, Machines, and Robotics – 1988*. ASME Design Engineering Division (Publication) DE, **15**–**3**, 525–33. Presented at the 1988 ASME Design Technology Conferences – 20th Biennial Mechanisms Conference, Kissimmee, FL, September, 25–28.

Steiner, M. and Derby, S. (1988) Estimating robot assembly cycle time during product design – Part 2: case studies, *Trends and Developments in Mechanisms, Machines, and Robotics – 1988*. ASME Design Engineering Division (Publication) DE, **15**–**3**, 535–41. Presented at the 1988 ASME Design Technology Conferences – 20th Biennial Mechanisms Conference, Kissimmee, FL, September, 25–28.

Steiner, M. and Derby, S. (1988) An information architecture and methodology for estimating robot assembly cycle time for producibility analysis, *Computers in Engineering, 1988: Proceedings of the 1988 ASME International Computers in Engineering Conference and Exhibition*, San Francisco, CA, July 31–August, 4, 383–95.

Suh, N.P. (1990) *The Principles of Design*, Oxford University Press, New York, 40–2.

Suri, R. and Shimizu, M. (1989), *Research in Engineering Design*, **1**, 105–20.

Sutherland, J.W., DeVor, R.E., Kapoor, S.G. and Ferreira, P.M. (1988) *SAE Transactions*, **97**, 215–26.

Tomiyama, T. (1990) in *Design Theory and Methodology – DTM '90*, (ed. James R. Rinderle), pp. 219–23.

Ulrich, K.T. and Fine, C.H. (1990) *Proceedings of Manufacturing International '90*, **5**, Atlanta, GA, March, 25–28, 19–25.

Unger, M.B. and Ray, S.R. (1988) *Computers in Engineering 1988: Proceedings of the 1988 ASME International Computers in Engineering Conference and Exhibition*, San Francisco, California, July 31 – August, 4, **1**, 563–9.

Vannoy, E.H. (1989) *1989 IIE Integrated Systems Conference and Society for Integrated Manufacturing conference proceedings*, 33–8.

Wade, J. and Colton, J.S. (1990) The development of a design for manufacture expert system, *Proceedings of Manufacturing International '90. Part 2: Advances in Manufacturing Systems*, Atlanta, GA, March, 25–28, 69–75.

Weston, R.H., Hodgson, A., Coutts, I.A. and Murgatroyd, I.S. (1991) *Journal of Design and Manufacturing*, **1**, 47–56.

Whitney, D.E. (1990) *Research in Engineering Design*, **2**, 3–13.

Whitney, D.E., De Fazio, T.L., Gustavson, R.E., Graves, S.C., Amblard, G.P., Abell, T.E., Baldwin, D.F., Cooprider, C., Pappu, S. and Lui, M.-C. (1990) *Proceedings of NSF Design and Manufacturing Systems Conference*, **1** (of 2), Tempe, Arizona, January, 8–12, 17–26.

Wierda, L.S. (1988) *Engineering Costs and Production Economics*, **13**, 189–98.

Wong, J.P., Parsaei, H.R., Imam, I.N. and Kamrani, A.K. (1991) An integrated cost estimating system for concurrent engineering environment, *Proceedings of the 13th Annual Conference on Computers and Industrial Engineering*, Orlando, FL, Mar. 11–13.

Yoshimura, M., Itani, K. and Hitomi, K. (1989) *International Journal of Production Research*, **27**, (8), 1241–56.

CHAPTER 2

Concurrent engineering's roots in the World War II era

M. Carl Ziemke and Mary S. Spann

2.1 INTRODUCTION

A review of the literature reveals that little or nothing was written about 'concurrent engineering' or 'simultaneous engineering' before 1980. It would appear that concurrent engineering, as well as its subsets, design for manufacture (DFM) and design for assembly (DFA), are recent concepts in both management and engineering. Yet, few would doubt that these ideas have been a part of US engineering design philosophy for many decades. In the present rush to emphasize concurrent engineering, DFM, and DFA, there has been a strong tendency to reinvent practices that were common during World War II and earlier. Neglecting the lessons of history has two major disadvantages: time is wasted in developing procedures and methods that are a matter of record and effort is wasted in selling these concepts on the basis of their anticipated benefits when overwhelming historic evidence of their worth exists.

One could argue that concurrent engineering could not have been practiced during World War II because the process did not have a distinctive name at that time. In fact, as will be shown, emphasis on concurrent engineering, DFM, and DFA in manufacturing was the norm during this time of war; therefore, no specific terminology was required. In fact, wartime manufacturing and design practices are traceable to periods before 1939–45 and had their roots in the practices of some of America's industrial giants whose work began much earlier.

2.2 BEHIND THE ARSENAL OF DEMOCRACY

President Franklin D. Roosevelt referred to the gigantic military/industrial complex organized to arm America and its World War II allies as the

'Arsenal of democracy'. This was no idle boast. During the war, the US produced more ships, tanks, guns and aircraft than all of its enemies and allies combined. This massive outpouring of equipment included 296 000 planes, 12 000 ships, 64 000 landing craft, 86 000 tanks, 15 million guns and 40 billion bullets (Fallows, 1991).

This arsenal of democracy was a civilian, not a military one. Prior to World War II, the US standing army of only 100 000 men ranked fourteenth among the nations of the world. A small standing army was a goal of this nation's founding fathers who saw a large peacetime civilian military establishment as a threat to civilian government. America had no huge private arms manufacturers such as Krupp in Germany, Vickers in England, Hotchkiss in France and Skoda in Czechoslovakia. What few military arsenals that did exist were too small for major weapons production and even lacked staffs capable of providing all the basic designs for modern weapons. Thus, US weapons would have to be designed and produced by expanding existing civilian industries. Naturally, America's largest civilian industry, automobile manufacturing, would play a dominant role in war production.

2.2.1 Contribution of Henry Ford

During World War II, the Ford Motor Company produced jeeps, trucks, tank engines, full-tracked vehicles, B-24 bombers and other military hardware. The peacetime engineering philosophy, a legacy of Henry Ford, of using small, integrated, multidisciplinary teams similar to modern concurrent engineering practices was extended to Ford's military production. Ford, too old to take a major role in the wartime operation, had his grandson, Henry Ford II, released from the Navy in 1943 to take over at Ford Motor Company.

While active in his company's design projects, Henry Ford directly supervised small teams which included his top manufacturing people. After he had been told it was impossible, Ford led such a group to produce the world's first low-cost, integral-block, V-8 engine in 1931–2 (Lacey, 1986). Much earlier, heading an even smaller group of engineers, Ford designed the Model T automobile which embodied several technical advances, used new manufacturing techniques and was designed to target the market it would create, the lower middle-class automobile buyer (Ziemke and Spann, 1991). Thus, the concurrent engineering concept was well-established at Ford Motor Company when it was ordered to cease all civilian goods production and convert to weapons of war after December 7, 1941.

2.2.2 Contributions of Walter Chrysler

Together with Ford and General Motors, Chrysler Corporation was a major producer of World War II armaments. Chrysler produced more tanks than any other manufacturer, and today's excellent M-1A2 main battle tank is the design of Dr Phillip Lett, formerly of Chrysler. Chrysler built over

42 000 M-4 Sherman tanks. The design of the M-4 Sherman tank reflected the engineering philosophy of Chrysler's founder and namesake, Walter P. Chrysler.

Like Henry Ford, Chrysler was a mechanic and machinist as well as a self-educated engineer. After purchasing his first automobile, he disassembled and reassembled it to learn how it worked and how it went together (Ziemke and McCollum, 1990). As a machinist who made his own tools, Chrysler was very interested in how parts were made. Later, as head of the Chrysler Corporation, he led a design team of three other persons in developing the 1924 Chrysler, an instant success due to its outstanding performance and unique design features. The direct involvement of the head of a corporation in a design effort naturally brings a financial and marketing perspective to a primarily technical group. The inclusion of financial, marketing, and manufacturing perspectives, as in modern concurrent engineering design teams, was provided by Chrysler himself in the 1924 Chrysler design effort.

When Chrysler inherited the original M-4 tank design from the Army in 1941, much work was needed to make it easy to produce and yet competitive with enemy tanks. The peacetime practices of small, integrated, production-oriented design groups developed by Walter Chrysler proved very useful in this effort.

2.3 CONCURRENT ENGINEERING AND WARTIME DESIGN ACCOMPLISHMENTS

The US entered World War II with a great variety of weapon designs from several sources. The famous M-1 rifle was designed by John Garand, a US arsenal employee, whereas the M-1 carbine was a Winchester design based on a concept patented by David Marshall Williams. No military arsenal was involved in aircraft design so these designs were the products of private industries. Most of the trucks and later models of the tanks were the products of civilian design groups.

One of the distinctive characteristics of wartime weapon designs by US civilian teams was speed. For example, fighter planes were designed and flown routinely in six to twelve months. A heavy bomber design such as the B-29 Superfortress was more complex and required two years from placement of the contract until the first flight (Gunston and Bridgman, 1946). Of course, additional time was needed to tool up and then mass-produce the weaponry. As will be shown subsequently, DFA and DFM played major roles in this process.

2.3.1 The design of the P-51 Mustang fighter

If concurrent engineering's aim is to provide superior, easily manufactured products in record time, then this concept must have flowered at North

American Aviation Corporation on May 29, 1940. At that time, the British Air Purchasing Commission placed an order for 320 NA-73 fighter planes that were later to be famous as US P-51 Mustangs. The condition of this order was that the first prototype, NA-73X, had to be ready for testing 120 days after receipt of contract (Gruenhagen, 1976).

All that the North American Mustang design team had to start with was a set of British specifications and a three-view drawing of the airplane previously done by engineer Edgar Schmued. Given the almost impossible schedule for designing, developing, building and testing the Mustang, one might assume that the ninety-seven engineers and technicians designated to create this aircraft would have chosen only proven, conservative design features. Instead the reverse was true. Some of the novel features of the NA-73X Mustang included the first use of the National Advisory Committee on Aeronautics (NACA) laminar flow airfoils and the introduction of a combined radiator housing-ejector nozzle that provided 300 pounds of jet thrust, instead of the usual radiator air drag (Gruenhagen, 1976).

This remarkable aircraft was designed and built in 102 days. During that time, 2 800 drawings representing 600 000 hours of effort were produced. The first models were called Mustang I by the British who later produced an experimental Mustang X version by installing a Rolls Royce Merlin 61 engine in place of the original US Allison power plant. The Merlin 61 engine provided several hundred extra horsepower above 20 000 feet altitude, and the true potential of the Mustang was seen at once. Compared to a contemporary Spitfire with the same engine, the Mustang climbed faster, had 50 mph greater top speed and had a much longer range, despite being 1 600 pounds heavier. The Mustang X was rushed into production as the P-51B (Gruenhagen, 1976).

If the research and development team that designed and built the Mustang fighter plane in 102 days did not use concurrent engineering, it would be well to discover what system was used and copy it! In retrospect, it now seems that critical success factors in such wartime design/development teams were their small size and their broadly experienced leadership.

2.3.2 The Skunk Works paradigm

Today the famous Lockheed Aircraft Corporation's Skunk Works (Advanced Development Projects) is sometimes selected as the paradigm for the use of concurrent engineering in product design. There is considerable historical basis for this choice. The Lockheed Skunk Works was established by Kelly Johnson, a leading design engineer, in 1943 (Rich, 1991). The location in Burbank, California was so shrouded in secrecy that it was likened to the mysterious, never-seen skunkworks of the Li'l Abner comic strip. The secrecy was necessary because Lockheed was about to develop the first mass-produced US jet fighter plane, later named the P-80 Shooting Star.

To form his small, integrated, multidisciplinary design/development team, Kelly Johnson had his pick of top company specialists. He chose twenty-three engineers and 103 shop mechanics (Rich, 1991). This team had to design and assemble the aircraft. The project was complicated by the aerodynamic problems involved with producing the fastest plane ever built by Lockheed and the need to incorporate its first jet propulsion unit in the design. Nevertheless, the first prototype was designed and assembled in just 143 days, thirty-seven days ahead of schedule. According to Ben Rich (1991) who succeeded Kelly Johnson as head of the Skunk Works, one of the success factors in such an operation is restricting the number of people connected with the project. In the early years of the Skunk Works, this restriction was possible; and several outstanding aircraft projects were developed in record time. These include the F-104 Starfighter and U-2 spy plane. The latter was conceived, built and flown in only eight months while staying within the project budget (Rich, 1991).

The success of the F-104 and U-2 programs was derived largely from the integrated design team concept developed during World War II to produce the P-80. Again, this concept featured a small, select team of design and manufacturing personnel who worked within concrete specifications and an adequate budget with minimal interference from outside the group. The similarity to the current concept of an effective concurrent engineering team is striking.

2.4 DFM AND DFA IN WARTIME PRODUCTION

It is sometimes assumed that the wartime 'miracles' of weapons production are rather easily achieved because cost is no object and little or no marketing is involved. This is a highly inaccurate view. During any major, long-term war, there are severe shortages of materials and labor. Transportation facilities are strained and the necessity for secrecy impedes information exchange. Despite these constraints, America's gross national product doubled during World War II and her aviation industry grew ten-fold (Gordon, 1991). Besides extensive use of integrated weapons design teams as in the concurrent engineering concept, production rates were enhanced by extensive use of DFM and DFA concepts and practices. This was entirely logical because of the wartime shortages of both tooling and labor. Of course, overuse of DFM/DFA practices could result in weapons that were plentiful but sub-optimal in performance. This proved not to be the case.

2.4.1 US aircraft design experience

Earlier, the outstanding performance of the P-51 Mustang fighter plane was discussed. One should not assume that an advanced aircraft could not be designed for ease of manufacture as well. To facilitate ease of manufacture of the P-51 Mustang, stabilizers, fin and wing tips were squared off. The fuselage assembly was configured in three major sections that separated at the firewall and the station behind the radiator air scoop fairing. The complete airplane was designed with plumbing and cable disconnects to facilitate assembly and also disassembly for packing and shipping.

The P-51 Mustang was unusually aerodynamically clean. Its low air resistance contributed to its high speed and exceptional range. However, it is interesting to consider the performance of a later US fighter designed with more DFM constraints on its envelope, the Grumman F-6F Hellcat. This aircraft used fewer compound curves in its external design. It was intended to be produced in very large quantities to overcome the shortcomings of its predecessor, the Grumman F-4F Wildcat, in combating the faster, more maneuverable Japanese Zero Naval fighter.

As a consequence of the DFM influence on design, the F-6F Hellcat was rather an ugly aircraft with slab sides and severely clipped wing tips as compared with its nearest competitor, the Chance Vought F-4U Corsair, an elegant pre-war (1938) design. Nevertheless, the F-6F proved to be easy to learn to fly and more than a match for Japanese fighters. The F-6F and the sleeker F-4U shared the same engine and weighed about the same. The only penalty on performance imposed by the DFM design of the F-6F was that it was about 20 mph (32 kph) slower than the F-4U (Heritage, 1980).

Dedication to DFA was common to most US aircraft and was evident when they were prepared for shipment by boat or transport by land. Outer wing panels and even tail assemblies were removed and then replaced at the destination point. This feature also permitted a considerable amount of major repair and/or cannibalization at the front-line air bases.

The rapid wartime enlargement of the US aviation industry meant that the majority of the production workforce in many plants consisted of women with no previous industrial experience. This fact was taken into consideration in aircraft designed after 1941 and could be seen in the modular approach to assembly and sub-assembly.

2.4.2 Spitfire versus Messerschmidt: two views of DFM/DFA

During the early years of World War II, classic aerial duels occurred between the elegant British Supermarine Spitfire fighters and the deadly ME-109s of the German Luftwaffe. The differences in design philosophy in the development of these two aircraft were striking.

The Spitfire was the product of the brilliant English designer of Schneider Cup winning racing seaplanes, R.J. Mitchell. During the early 1930s, his seaplanes exceeded 400 mph at sea level and were each hand built by master craftsmen at the rate of only one or two copies per model. Some were powered with Rolls Royce racing engines that produced over 2 000 horsepower. When asked in 1935 to design a low-winged monoplane to replace England's obsolete biplane fighters, Mitchell employed many of the features of his successful seaplane racers including a V-12 Rolls Royce engine, the then new 1 000 horsepower Merlin. Unfortunately, the Spitfire was designed for optimum aerodynamic performance and structural efficiency, with almost no consideration for DFM or DFA. The beautiful elliptical wings, chosen for their aerodynamic excellence, were a nightmare to mass produce. Overall, the Spitfire required 13 000 labor hours to produce (Evans, 1992).

As fate would have it, the difficulty in mass producing the early Spitfire models did not prevent them from becoming a major factor in the Battle of Britain air campaign in 1940–1. The reason was that top priority had been given to Spitfire construction over all other types of aircraft. Also, the limiting factor in the air defense of Britain was a shortage of pilots. The Spitfire was a highly maneuverable aircraft but was difficult to learn to fly. Over 1 000 student pilots were killed during wartime training. Thus, the desperate Royal Air Force recruited volunteer foreign pilots, mostly Poles and men from the Commonwealth nations. These foreign volunteers actually shot down 25% of the German aircraft lost to aerial combat during the Battle of Britain.

At almost the same time that R.J. Mitchell was designing the legendary Spitfire, a similar fighter plane was being developed by Mitchell's opposite number in Nazi Germany, Herr Dr Professor Willi Messerschmidt. Dr Messerschmidt designed his ME-109 fighter (German designation BF-109) with classic Teutonic efficiency. He knew the plane would have to be mass-produced with tooling and manpower already straining to overcome the effect of the Versailles Treaty provisions for the prevention of German rearmament. This same treaty forced Messerschmidt to design the aircraft around an underpowered engine, a Daimler-Benz DB-600 of 690 horsepower (Gunston and Bridgman, 1946). Thus, the ME-109 was made especially small and light with a very cramped cockpit as compared to that of the Spitfire.

The ME-109 design was a DFM/DFA classic. A minimum of compound curves was used in its exterior. Rudder and horizontal stabilizers were squared off for production efficiency. The cockpit canopy was made almost entirely of flat glass. The 20 mm cannon fired through the hollow propeller shaft to overcome the complexities of wing installation and to eliminate sighting parallax. The aircraft was used in the Spanish Civil War between 1936 and 1939. During this period, it received the new Daimler-Benx 1 000 horsepower engine, the DB-601 (Gunston and Bridgman, 1946). With this major power increase, together with its light weight and compact size, the ME-109 was as fast as the more aerodynamic Spitfire although not as maneuverable.

In 1940, this easily produced Messerschmidt fighter plane outnumbered the available Spitfires at least three to one. This fact would have been more significant except for one feature of the ME-109 design attributable to the DFM philosophy. Early freezing of dimensions for production efficiency did not permit enlarging the fuel tanks to match the later more powerful engines. During the Battle of Britain, these planes had only enough fuel reserves for 30 minutes of aerial combat over South England. British Spitfires and Hurricanes entered combat with nearly full tanks and could return to base with minimum fuel reserves. This circumstance helped the British to achieve a two-to-one kill ratio over their Luftwaffe opponents despite the German advantage in the number of aircraft.

Throughout the war, the ME-109 continued to receive more powerful engines such as the DP-605D of 1 800 horsepower to keep it competitive with the Spitfire in speed and altitude. The German preoccupation with DFM and DFA permitted the production of more fighters during 1944 than any other year of the war. One reason that this was possible was that the ME-109 required only 4 000 labor hours to produce compared to 13 000 for the British Spitfire (Evans, 1992). However, shortages of well trained pilots and later also of fuel prevented the full use of this production miracle, achieved despite massive Allied bombing.

Several German ME-109 aces obtained over 100 victories with the aircraft, a feat not matched by any Spitfire pilot. The last combat use of the ME-109 came during the Arab-Israeli War of 1948. At that time, former RAF pilots in the Israeli Air Force flying ME-109s cleared the skies of Spitfires flown by Egyptian pilots. Thus, while air war historians still designate the Spitfire as the better aircraft, the long and successful career of the ME-109 amply demonstrates that major attention to DFM and DFA need not produce an ineffective aircraft design.

2.4.3 The Japanese Zero and beyond

The Japanese counterpart of the German ME-109 was the Mitsubishi Model AGM5, Naval Designation Type 'O', Model 52. This famous fighter aircraft was called either the 'Zero' or 'Zeke' by the US and its allies.

The Japanese Zero design reflected the realities of that nation's production capabilities under wartime conditions. For their purposes, the primary design constraints were to build an aircraft with excellent maneuverability, good firepower and extreme range around a radial engine in the 1 000 horsepower class. This was achieved with a rather clean, lightweight design that also conserved very scarce aluminum supplies. The Zero was better streamlined compared to its initial US counterparts, the Curtiss P-40 and Grumman F4F Wildcat. Unlike them, the Zero had a teardrop shaped canopy with 360 degree visibility and landing gear covered by fairings when retracted. The loaded weight of the Zero was only 5 750 pounds, compared with 7 412 for the F4F and 8 720 for the P-40 (Gunston and Bridgman, 1946).

The wing areas and engine power of all three planes were similar so the Zero had a great advantage both in turning radius and climb rate. The agility of the Zero and its experienced pilots caused high losses among Allied fighter aircraft in 1942. But aircraft design soon felt the DFM/DFA constraints of an overstressed Japanese economy. The Zero and its Army counterpart, the Nakajima Ki-43 Oscar, lacked armor plating and self-sealing fuel tanks. The weight limitations also precluded the use of rugged construction. Thus, whereas these nimble aircraft were hard to hit, when they did sustain damage, they tended to lose pilots, burn or disintegrate.

When faster American aircraft such as the Lockheed P-38 and Grumman F6F Hellcat appeared in numbers during 1943, they began to overwhelm the slower, more vulnerable Zeros and Oscars. Yet the Japanese continued to build these easily manufactured aircraft until the end of the war, despite the availability of much more potent designs like the Mitsubishi J2M2 Jack. The Jack had an 1 850 horsepower engine, armor plating and self-sealing fuel tanks (Gunston and Bridgman, 1946). It was reputed to be faster than the US P-51D Mustang at some altitudes. As was the case with the excellent German FW-190 fighter, production of the Mitsubishi Jack was curtailed in favor of older, more easily produced models.

The Japanese aircraft design teams were formed around a single chief designer, much as was the case with the Kurt Tank (FW-190) and Kelly Johnson (P-38). Also, as in the US and Germany, Japanese aircraft design teams tended to be small and highly integrated. This situation continued after the war, although the teams grew in size. The designer of the Zero and Jack fighters, Jiro Horikoshi, also led the design of the Mitsubishi twin turboprop airliner imported into the US about 25 years ago (Heritage, 1980).

The Japanese spirit of design team cooperation called 'wa' provides a natural basis for continuation of the wartime concurrent engineering function. Japanese devotion to quality aircraft design is seen in their recent agreement with Boeing to share development of the B-777 airline scheduled to fly in 1995.

2.4.4 DFM/DFA in German small arms

The long-established German tradition of craftsmanship and handiwork in weapon design and manufacture yielded slowly to considerations of DFM and DFA. The world's first military semiautomatic pistol, the Mauser model 1896 'broom handle' gun, had a barrel machined integral with the forging that contained the receiver! The later 'Luger' pistol, officially the Pistole Model 1908 or P-08, had a screwed-in barrel but was still too expensive for adequate volume production during wartime. It was succeeded by the P-38 which has a quick removable barrel and other main parts. This gun design used low-cost production methods including stampings. Yet it is a double-action weapon that would be thoroughly modern today if it had been equipped with a double-column fifteen cartridge magazine like the present US 9 mm service pistol.

More dramatic was the redesign of the MG-34 light machine gun for DFM/DFA considerations. The MG-34 was probably the world's finest light machine gun when introduced in 1934. However, following the traditional German practice of small arms craftsmanship, the MG-34 employed expensive forged parts with close tolerances. By 1942, the German war machine was heavily involved in fighting on two fronts while its industrial plants were being bombed. Thus, the excellent MG-34 was replaced with the MG-42, a weapon built with heavy use of DFM and DFA. Stampings replaced forgings and ease of assembly/disassembly was provided. One outstanding feature was the quick-change barrel that allowed gun crews to replace overheated barrels easily under combat conditions, a convenience not available on US light machine guns. The MG-42 had a cylic rate of fire twice that of the comparable US infantry weapons. The distinctive sound of the MG-42 caused American troops to name it the 'burp' gun.

Again, recourse to DFM and DFA did not produce an inferior weapon. The present US M-60 light machine gun looks like the MG-42 because it uses many of the features of the German gun. Today, the modern German army still uses the MG-42, now converted to fire the 7.62 mm NATO cartridge.

2.4.5 The tale of the tommygun

Before World War II, a few US Army and Marine units were equipped with the M-1928 Thompson submachine gun manufactured by Auto Ordnance Corporation. This was the famous tommygun seen in the gangster movies of the 1930s. It was provided to the original British commandos in 1940 and appeared on their shoulder patches.

The Thompson M-1928 model was designed during the pre-DFM/DFA weapons design era after World War I. It featured a long, finned barrel designed for continuous firing (rarely required of submachine guns). The end of the barrel was fitted with a Cutts compensator to control recoil, hardly a problem with an 11 lb weapon firing forty-five ACP caliber pistol cartridges. Ammunition was carried in a fifty-round 'L' model drum magazine that was operated with clock-spring activated follower. There was even a 100-round 'C' magazine available. The gun designer, General John T. Thompson, hoped that the military would adopt the M-1928 as a light machine gun, a plan that fortunately failed to be accepted.

Despite its excessive size, weight and complex mechanism, the Thompson M-1928 proved to be a reliable and effective weapon. A wartime model, the M-1942, was redesigned to reduce production costs. It featured a shorter, unfinned barrel without the Cutts device, eliminated the front pistol grip and was issued with the lighter twenty- or thirty-round straight magazines.

Although many Thompson Model 1942s were manufactured, no way was available to simplify the mechanism and eliminate the excess weight. Several competitive designs including the Reising were considered but the ultimate solution was the M-3 'grease gun'. This simple, compact weapon looked like the hand-held tools used to grease automobile chassis. The US M-3 submachine gun was lightweight, simple to assemble/disassemble and was very easy to manufacture. In short, it was a DFM/DFA accomplishment. Yet the M-3 gun was effective and reliable, providing unique features such as an ejection port dust cover that also served as a safety. It compared favorably with the British replacement for the tommygun, the simple, all-metal 9 mm Sten submachine gun.

2.4.6 Saving an obsolete tank design through DFM/DFA

Military historians agree that the main US battle tank of World War II, the M-4 Sherman, was inferior to most equivalent German, Russian and even some British tanks. Compared to its later competitors, the M-4 lacked both effective armor and a powerful cannon. It also had the highest profile of contemporary medium tanks. Although the shortcomings of the M-4 were known by the end of 1942, it was decided to freeze the design to achieve maximum production. While the British had a superior tank of their own design, they could not produce enough of them and, thus, depended heavily on US tanks.

The story behind the development of the M-4 Sherman tank is one of American Army bureaucracy versus the ingenuity of civilian automotive engineers. Between the world wars, the US Army budget for new tank design and development was minuscule. Consequently, the Army looked to British tanks being produced in significant numbers for export. One of these models, the Vickers 6 ton type A light tank, was purchased by the US and redesigned as the T-3 light tank (Ellis and Chamberlin, 1972). This US tank used an improved form of the Vickers 'bogie' type track suspension that featured small road wheels in pairs, pivoting about a common axis. This type of track suspension became the standard for all US tracked vehicles until 1944 despite the fact that it was inferior to that developed by American automotive engineer, J. Walter Christie. Meanwhile, Germany, Russia and Great Britain adopted the Christie suspension for their best tanks.

In addition to an obsolete suspension, the M-4 was plagued by a lack of a purpose-built tank engine. The British had the same problem, powering their early World War II tanks with bus engines and World War I Liberty aircraft engines. Thus, early M-4s and their predecessors, the M-3s, were powered by de-rated radial aircraft engines. For the M-3 and M-4 medium tanks, this was the 450 horsepower Wright R-760. Although they were light, powerful and available, these were not ideal tank engines. They caught fire regularly when started and required a higher grade of gasoline than ordinary Army trucks and automobiles.

The wartime automotive engineers initially ignored the tank engine problem and concentrated on mass-production through DFM and DFA techniques. The original M-3 tanks had been of riveted construction but the M-4 was designed to use a cast turret and hull. Because of the lack of foundry capacity, some hulls were of welded construction. The bogie suspension was further improved by replacing the leaf springs with horizontal volute springs for a smoother ride. These suspension units were designed for easy removal and repair in the field by relatively unskilled crews. Tracks were also easily removed and replaced by tank crews.

The original M-4 main gun was a modernized version of the World War I French 75 mm field piece. In North Africa, in 1942, it outclassed most British and German tank guns. The gun was made more potent by using a gyro-stabilizer that permitted firing while the tank was in motion (Ellis and Chamberlin, 1972).

The lack of a suitable M-4 tank engine was eventually satisfied by Ford Motor Company. Ford produced the Model GAA 500 horsepower V-8 liquid-cooled engine that was used in the M4-A3 and subsequent models. This engine was less fire prone than the radial types but not as fire safe as the diesel engines especially designed for the German and Russian tanks.

In 1942, M-4 tanks faced mainly German Pzkpfw IV medium tanks with 25 mm frontal armor thickness and medium velocity kwk L/24 75 mm main guns (Bauer, 1972). However, by 1943, Germany was producing large numbers of Panther tanks (Pzkpfw Vs) with 75 mm high velocity guns and a maximum frontal armor thickness of 120 mm. The Russian response was to produce new tank models but Detroit's automotive engineers had to meet the threat of the Panther tanks without restricting the huge output of M-4 vehicles from the tank plants.

Obviously the basic 75 mm Sherman main gun could not pierce the frontal armor of the German Panther and Tiger tanks. There was no time to develop a new, high-powered gun. However, the US Navy had developed a 3 inch (76 mm) high-velocity gun for use on small auxiliary vessels. Production was shifted to the Army, and a new, larger M-4 turret was cast to accept the larger 76 mm gun. These guns appeared on some M4-A3 models.

There was still a problem with the inadequate armor of the M-4 which was 75 mm on the turret and 12 mm over the engine. The full frontal armor was only 50 mm thick. Hull side armor could be pierced by German 50 mm high velocity guns. The DFM solution to this problem was to weld extra armor onto the tank in critical areas. This led to the production of the M-4A3E2 'Jumbo' tank with 150 mm turret armor and 100 mm hull armor (Ellis and Chamberlin, 1972).

Compared to the best German, Russian and British medium tanks, the M-4 tanks tracks were too narrow for good traction in mud, sand and snow. It was not practical to redesign the M-4 track system, but engineers devised bolt-on 'grousers' to increase track width. Also, the M-4 tank was adapted to many special purposes. There was a duplex drive (DD) swimming version, a bridge-laying version, a rocket-firing 'calliope' tank and a tank that cleared mine fields with forward-mounted chain flails. In all, about 42 000 Sherman tanks were built. Despite the fact that the basic Sherman M-4 was obsolete after 1942, it was kept reasonably up-to-date by DFM/DFA design practices that did not interfere with volume production.

2.4.7 The miracle of the Liberty ships

During the decade of the Great Depression that preceded World War II, the US Smoot-Hawley Tariff Act of 1931 went into action and foreign trade stagnated. Few US cargo ships were built, and no new shipyards were constructed. The US Merchant Marine shrank to a fraction of its former size. Suddenly, with the outbreak of war in September 1939, there was a demand for merchant ships and crews. Crew training was accelerated, but ship construction was another matter. To build cargo ships, three major items were essential: high pressure marine boilers, steam turbines and shipyards. All three were in short supply because of the demands of naval construction. Yet, the free world's merchant fleet was rapidly being depleted by submarine attacks. The solution to this problem appeared remote until the Liberty ship concept appeared.

The Liberty ship was a DFM/DFA dreamboat. It was designed to use low-pressure boilers which could be built in quantity by any number of firms. These boilers powered old-fashioned, triple-expansion piston engines that were also easy to manufacture. In other words, the Liberty ship had a 1918 propulsion design which provided a speed of about 12 knots (22 km per hour). This slow speed was not the disadvantage that it might first appear. Many wartime ships convoys consisted of a variety of merchant vessels. Convoy top speed was that of the slowest ship, often 12 knots or less.

A third problem remained: a shortage of shipyards. At this point, West Coast industrialist Henry Kaiser entered the picture. Kaiser proposed to assemble the Liberty ships at his shipyards using all-welded construction and a new method of modular assembly never before attempted on large vessels. To understand the implications of Kaiser's proposal, it is well to consider the conventional method of constructing a large ship. On a slipway, the heavy steel keel, the backbone of the ship, is laid. To the keel, ribs and frames are attached, piece by piece. Heavy steel plates are attached to the keel and ribs while internal bulkheads and decks are built. This type of construction requires considerable scaffolding and the availability of giant shipyard cranes. A merchant ship might require a year to build and an ocean liner, three years.

Kaiser sought to eliminate most of the construction at scarce shipyard space and have it done in modules at other, less critical sites sometimes quite distant from the slipway. Thus large sections, including complete deckhouses, were propped in place and welded to the ship. The time savings were enormous. Ships were started and launched in as little as four days (Chiles, 1988).

The Liberty ships were modular in other ways besides construction. Ships like general cargo ships and tankers were built on the same hull. More than 1 200 of all types were built before production ceased in 1944. By that time, the capacity to produce modern marine propulsion units had caught up with demand. Shipyards shifted to production of Victory ships, fast, turbine-powered vessels that could outrun submarines and, therefore, needed no naval escort.

2.5 LOSING THE ENGINEERING LESSONS OF WORLD WAR II

As detailed above, the use of small, integrated, interdisciplinary design teams, identical in principle to the engineering design concept we now call concurrent engineering, were used in the design of outstanding weapons in World War II. DFM and DFA, similarly viewed as recent engineering concepts, were also used extensively in World War II in the national effort to produce the huge quantities of armaments needed to supply America and its allies. Despite the success achieved through the use of wartime concurrent engineering teams and a concentration on DFM and DFA techniques to maximize production volume, the concurrent engineering design team concept virtually disappeared until about 1979 and the emphasis on DFM and DFA was greatly reduced.

As early as 1949, Henry Ford II found that his 1949 Ford was costing $100 too much to manufacture because of inadequate DFM (Abodaher, 1982). Similarly about eight years ago, when the new plant for production of Ford Taurus/Mercury Sable cars in Atlanta was designed, Ford engineers had to re-invent DFA by asking production workers how cars should be designed for easy assembly. Comparable situations existed at Chrysler and General Motors. Yet, all of the automotive industry's 'Big Three' had an extensive heritage of concurrent engineering, DFM and DFA from World War II and before.

2.6 REINVENTING CONCURRENT ENGINEERING AT FORD

In the late 1970s, the Ford Motor Company was in serious financial condition. By the late 1980s, Ford had become the most profitable automaker and a leader in both productivity and quality in the American auto industry. Despite a weak economy, tight credit, and higher fuel prices in the early 1990s, Ford maintained a five-year average return on equity of 25.5% and a 12.8% return on sales through 1990 (Flint, 1991). The turnaround at Ford took more than five years and required more than $5 billion in investment as well as a revolution in Ford's corporate culture. Ford bet the company and won. This remarkable transformation is identified with a single product: the Ford Taurus.

In 1979, Ford set out to make a truly new American car with the quality of the best of Japanese and European imports. The sweeping changes at Ford began with the design of the Taurus and its twin, the Mercury Sable. Ford brought together a cohesive, multidisciplinary team from product planning, marketing research, design engineering, manufacturing, logistics, finance – anyone and everyone who could help bring a winning new car to the market. In 1979, Ford reinvented concurrent engineering with the design of the Taurus. Design engineers at Ford did another remarkable thing, they went to the factory floor and involved production workers in the design of the Taurus, thus, rediscovering DFM/DFA about the same time. Ford made one more radical departure from business-as-usual in the auto industry, they involved customers and suppliers in the planning for the Taurus (Doody and Bingaman, 1988). Those with short memories heralded a new philosophy of car making in Detroit.

Just as concurrent engineering and DFM/DFA produced superior weapons in World War II, these same engineering philosophies produced superior cars at Ford. *Car and Driver* named the Taurus as one of the world's ten best cars in 1986, 1987, and 1988, the first three years of its existence. In 1988, *Road and Track* named the 1988 Taurus the best sedan in its class and price range. The Taurus was popular with consumers as well; it was the best selling mid-sized car in 1986 and 1987. Taurus owners were highly satisfied with their purchases; in 1988, 91% of Taurus owners reported that they would buy another Taurus and 94% another Ford.

The rewards of concurrent engineering, DFM/DFA and the corresponding change of corporate culture went beyond the awards and the popularity of the Taurus. In 1986, the Mercury Sable was runner-up to Taurus as *Motor Trend*'s 'Car of the Year'. The following year, the 1987 Ford Thunderbird won the award. For the years 1982–7, the Ford Escort with 6.5 million sales in sixty countries was the best-selling car of any type in the world. In the first quarter of 1988, Ford had five to ten best-selling cars in the US: Taurus (first), Escort (second) and Tempo (fifth) (Doody and Bingaman, 1988).

2.7 INSTITUTIONAL AMNESIA

It took the Ford Motor Company about thirty years to reinvent an engineering philosophy pioneered by founder Henry Ford and demonstrated so capably during the World War II. As discussed earlier, Ford managers and engineers forgot their concurrent engineering and DFM/DFA heritage and competencies fairly quickly and rather completely until desperation brought them back to the founder's knee.

How do organizations forget? How do organizations lose competencies? We don't really know, but looking at some organizational theory as well as auto-making and automakers from the 1950s to the 1980s may yield some clues.

2.7.1 Bureaucracy in theory

To the pioneering organizational theorist, Max Weber, the bureaucracy was designed for productivity and efficiency. Later theorists, like Merton, Selznick and Goulder, recognized the dysfunctional consequences of the bureaucratic organization (March and Simon, 1978). Merton recognized that bureaucracies are controlled by standard operating procedures that can lead to overly rigid systems and behaviors. Selznick pointed out that departmentalization and specialization flourish in bureaucratic organizations.

As an organization grows larger, it becomes necessary to divide it into subunits, usually on the basis of function or product. In general, the problems created by departmentalization and specialization grow as the organization grows. Members of subunits become more and more specialized and develop values consistent with the subunit. Marketing department personnel may value unique product design features whereas accounting departments personnel may value cost control. Departments also differ in their time orientation. Production departments may have a very short time frame whereas R&D departments may have a much longer one. Based on their values and time orientation, departments develop their own goals, which may or may not be consistent with overall organizational goals.

As organizations grow, organization structures grow taller, response time slows down and bottlenecks occur. Interdepartmental communication becomes more difficult and barriers arise. Departments build and protect their own turf. Rivalries develop between departments which become especially bitter whenever resources are scarce. Organizational goals get subverted by department goals (Ziemke and Spann, 1991).

2.7.2 Bureaucracies in the auto industry

From 1950 to 1963, the worldwide market for cars and trucks doubled and then doubled again by 1973 (Gilmour, 1988). By the late 1980s, even after some downsizing, Ford had about 350 thousand and General Motors had about 900 thousand employees (Bateman and Zeithamel, 1990). At a typical General Motors plant in the 1970s, there were six layers of management between the plant manager and the production worker (Bateman and Zeithamel, 1990); add to that several more layers at the division and corporate level to get some idea about the height of a Big Three automaker's organization. As automakers grew, managers and their departments became more and more specialized as did union workers. By the late 1970s, autoworkers at General Motors could work in 183 separate job classifications (Bateman and Zeithamel, 1990).

Before the new team concept emerged at Ford starting in 1979, Ford managers were characterized as autocrats, blazing stars, swashbuckling personalities with attending cults and cliques. Each superstar ruled his own fiefdom. The organization was rife with professional jealousies, interdepartmental rivalries, territorialism, poor communication and buck passing (Doody and Bingaman, 1988). As the Big Three automakers grew, their organizations grew into large bureaucracies plagued with all of the dysfunctional behaviors predicted by organizational theorists. Some industry analysts believe that crippling bureaucracies remain a major problem for Detroit automakers (Woodruff and Levine, 1991).

The typical pattern of new car design and development that emerged after World War II is best described as sequential. Development of a new car was shuffled from design engineering departments to specification engineering departments to manufacturing engineering departments to production departments to marketing departments. New car development was carried out in a series of ordered, unrelated, uncoordinated steps by people isolated from each other. To make matters worse, each specialization was autonomous, and no set of experts meddled in the affairs of other experts (Doody and Bingaman, 1988). How far Ford engineering management had strayed from Henry Ford's philosophy! No wonder, the team concept embodied in concurrent engineering seemed so new.

2.8 SUMMARY AND CONCLUSIONS

Concurrent engineering and its subsets, DFM and DFA, were part of the engineering design philosophy, especially among automakers, prior to World War II and were used extensively in the design and manufacture of the weapons of that war. The terms, concurrent or simultaneous engineering, were not used in this period because small, unified, multidisciplinary design teams were the norm rather than the exception. Fairly early after World War II, this engineering design philosophy got lost in the growth and prosperity of the post-World War II period in America. Firms like Ford Motor Company grew big and, perhaps, proud and forgetting the engineering lessons of the war, designed automobiles and other products in a sequential process using a multitude of specialists who refused to talk to one another. In 1979, Ford rediscovered its engineering roots and heralded a new age in auto design. Other firms got on the concurrent engineering bandwagon. In the 1980s, concurrent engineering re-entered engineering management philosophy and practice.

2.9 REFERENCES

Abodaher, D. (1982) *Iacocca*, Kensington Publishing, New York.
Bateman, T.S. and Zeithamel, C.P. (1990) *Management: Function and Strategy*, Irwin, Homewood, IL.
Bauer, E. (1972) *The Illustrated Encyclopedia of World War II*, Marshall Cavendish, New York.
Chiles, J.R. (1988) *The American Heritage of Invention and Technology*, **3**(2), 22–9.
Doody, A.E. and Bingaman, R. (1988) *Reinventing the Wheels: Ford's Spectacular Comeback*, Ballinger, Cambridge, MA.
Ellis, C. and Chamberlin, P. (1972) *Fighting Vehicles*, Hamlyn Publishing Group, London.

Evans, D. (1992) America in '92 is like Britain in '45, *Chicago Tribune*, January 6.

Fallows, J. (1991) *US News and World Report*, **3**(23), 54–5.

Flint, J. (1991) *Forbes*, **147**(1), 130–6.

Gilmour, A.D. (1988) *The Academy of Management Executive*, **2**(1), 23–7.

Gordon, J.S. (1991) *The American Heritage*, **42**(8), 16–18.

Gruenhagen, R.W. (1976) *Mustang, Revised Edition*, Arco Publishing, New York.

Gunston, B. and Bridgman, L. (1946) *Jane's Fighting Aircraft of World War II*, Jane's Publishing, London.

Heritage, J. (1980) *The Wonderful World of Aircraft*, Octopus Books, London.

Lacey, R. (1986) *Ford: The Man and the Machine*, Little, Brown and Company, Brown.

March, J.G. and Simon, H.A. (1978), in *Classics of Organization Theory*, (eds. J.M. Shafritz and P.H. Whitbeck), Moore Publishing, Oak Park, IL, pp. 110–16.

Rich, B.R. (1991) *Product and Process Innovation*, **1**(2), 28–34.

Woodruff, D. and Levine, J.B. (1991) *Business Week: Special 1991 Bonus Issue*, 70–3.

Ziemke, M.C. and McCollum, J.T. (1990) *Business Forum*, Winter, 14–18.

Ziemke, M.C. and Spann, M.S. (1991) *Industrial Engineering*, **23**(2), 45–9.

CHAPTER 3

Implementation: common failure modes and success factors

Stephen Evans

3.1 USING THIS CHAPTER

In implementing concurrent engineering (CE) there are many things you have to get right. Most companies have a track record of successful change and can be expected adequately to deal with certain elements of concurrent engineering implementations (such as process analysis, budgeting, systems analysis and design). CE offers a specific challenge to management by demanding radical change in the way we develop products; a challenge that managements' previous experience is unlikely to have prepared them for. Here we will concentrate on things that have commonly gone wrong and learn how to identify them and how to avoid them. Putting together the techniques needed to avoid common failings allows us to form a scheme for increasing the success of any CE implementation. The scheme is not a comprehensive plan for implementing CE – indeed it ignores those implementation activities that are either of little overall impact, or of interest only to a small part of the total population, or those that are commonly conducted successfully – but it does identify the most likely weaknesses of a CE implementation plan and should be used to support internal experience and to increase the success of CE.

The common failure modes and success factors presented here are the combined and distilled knowledge of many companies operating a CE product development process. This is based on the author's research into CE implementation and experience in CE implementation consultancy. To avoid any issues of interpretation by the researchers, however, the prime source is the direct output of CE user forums. Attended by the best practitioners of CE, their often painful experiences in gaining success have a remarkable degree of similarity. Clearly the reasons for failing to implement CE successfully are repeated in most companies. The intention of this work is to help potential practitioners of CE identify and avoid these same failings. Though the emphasis here is the negative (and avoiding it) it is not necessary for companies who are embarking on the CE journey to create detailed plans to remove all the identified common failure modes. Substantially positive results have been obtained by many companies with poor CE implementations. This demonstrates the robustness of CE and creates the context for this work – practitioners should not delay any existing CE implementation to incorporate this work as they will never recapture the delayed benefits *but* practitioners should review their plans to identify opportunities to increase the benefits. The best learning may come from hard won experience but it is slow and can be expensive; the efficient learner will incorporate others' learning to rapidly increase knowledge and benefits while reducing costs.

3.2 IMPLEMENTATION AND SUCCESS

Implementation refers to the planning, organization and delivery of change. In CE implementation we are concerned with changing the process used to deliver new products; a process which begins with a new idea or a company product plan and ends with transfer of the new product into the standard company delivery process (in different industries this may be upon customer receipt of first product, when product is launched to the market or when the product generates profit equal to development cost).

Within the product development process the tools and methods of CE have delivered increased performance. Much has been written about the faster development of products but CE can equally be used to generate better products, and at lower cost, with each company balancing their priorities differently. CE implementation takes a company from one standard of quality, cost and delivery performance to a higher standard. By using the information on common failure modes companies can achieve a yet higher standard.

3.2.1 Concurrency and implementation quality

The achievement of superior quality, cost and delivery is an after-the-fact measure; we can obtain proof after the action of implementation. To convince ourselves and others of the advantages of a specific technique, it is useful to have a more immediate yardstick; and to find one we must look to the very heart of concurrent engineering.

Concurrent engineering delivers better, cheaper, faster products via the continuous consideration of all constraints. In a perfect CE world all constraints are considered at every decision point – be they manufacturability, maintainability, reliability, testability, performance, etc. In traditional product development we have dealt with each constraint separately as a matter of policy. The success of CE, in general or in any single CE tool, is achieved by increasing the number of constraints we consider at each decision point. Increasing the number of constraints under consideration can therefore be the simple yardstick to measure the appropriateness, or otherwise, of our CE implementation plan.

Typical tools to achieve consideration of multiple constraints are described throughout this book. For example, design for manufacturability brings manufacturing constraints into those design phases normally reserved for product performance constraints. Multifunction teams use individuals to represent different constraints and ensure consensus decision making. Advanced information technology tools are often based on constraint satisfaction techniques.

Though the only true test of success for CE implementations is the delivery of better, cheaper and faster products, we will use the test of increasing or decreasing concurrency as an indicator. Common failure modes will be seen to be missed opportunities to increase concurrency – particularly in the early phases of product development where the leverage of concurrency is high (the incidence of concurrency in the very final phases of product development is high and is often a pull-all-the-stops-out, break-the-rules, emergency activity).

3.2.2 Concurrent engineering – where are we now?

A similar product development process across many companies and industries is the result of common pressures and common history. Henry Ford and Frederick Taylor demonstrated the massive advantages of dividing labor into small, specialist tasks. It was only after fifty years of continuous development (which brought unprecedented prosperity) that we began to reach its limits of performance. The division of labor into smaller and smaller specialist skills created two problems – coordination and communication. Specialists created their own languages and were only interested in their own goals (marketing, software design, test, etc.) making communication between disciplines difficult. This was exacerbated by the measurement system – specialists were measured only by criteria they had control over (e.g. test productivity, test quality) and had little motivation to communicate or cooperate with other specialists.

Coordination of non-cooperating, non-communicating specialists is *tough* and the mechanism we created to develop new products was a serial process by which one specialism worked on an activity till complete and documented, only then passing this information on. Coordination was by management who would identify tasks, allocate them to individuals and monitor progress; staff concentrated on delivery of tasks to management plans.

The weaknesses of a serial product development process are well documented yet it remains by far the most common method of organizing product development. Most companies now recognize the potential of operating it differently and are actively investigating CE as a route forward. Some companies have implemented pilot projects, typically with a multidisciplined team as a first move, while some companies have 'bought into' CE via advanced CAD. Very few companies have mature CE implementations where a high degree of concurrency is the norm. This chapter addresses the majority position – companies that are traditionally functional in organization, using a largely serial product development process which provides competent performance in cost, quality and speed, and who are actively investigating the potential of CE.

3.3 COMMON FAILURE MODES AND TECHNIQUES TO INCREASE SUCCESS

The experience of many companies indicates that the quality of the CE implementation plan is more important than any other factor. The vast array of tools and methods available can have the appearance of a catalogue selection process – choose the right ones and install into your company. In practice the implementation of each individual tool can vary enormously in the benefits achieved. This variability is more significant than the tool selection itself (Fig. 3.1). The dashed area indicates the volume of elective benefit available and it is significantly higher than the benefits available from any one tool. It is clear that the decision to implement one more tool into the CE plan is of less total benefit than increasing the quality of the plan itself, which can increase the benefits gained from *all* the tools used.

Studies of CE implementations demonstrate that the occurrence of specific implementation techniques is related to the scale of benefits achieved. In most cases the techniques used to increase success improved the usage of most or all tools in operation, thus achieving a wide impact. Conversely, the selection of tools shows little relation to scale of benefits achieved, indicating that the marginal benefit of adding any one tool to a CE implementation plan is restricted. The only exception to this is the use of multidisciplined teams, which consistently achieves significant benefits and increases the scale of benefits generated by other tools. This can be understood by recognizing that any technique for increasing the concurrent consideration of all constraints will itself be enhanced by the information-sharing and expertise-sharing team environment.

The selection of CE tools is a concern for most companies but, with the exception of teamworking, it is not a significant cause of low or high achievement in product development. The ability of most companies to deal with this technical issue is standardly high and extra effort is input on a Pareto basis. At the other end of the Pareto scale are the organizational and behaviorial issues; traditionally weak in these, most companies can significantly improve product development performance with little cost. Yet we

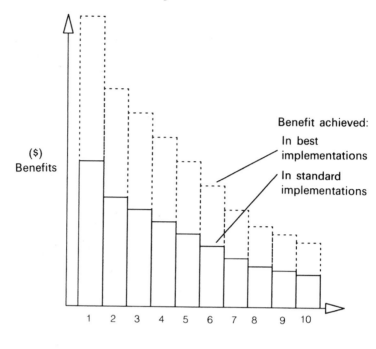

Concurrent engineering tools

Fig. 3.1 Variability in benefits achieved from CE tools under best and norm conditions.

continue to underperform in these areas and companies continue to experience common failings.

Understanding the common failure modes (CFMs) and developing techniques to overcome them will move companies rapidly upward from the standard performance. The CFMs presented here are not a comprehensive list of all possible failings in any CE implementation – they concentrate on activities that companies regularly do badly and which have significant negative impact on the ultimate success of CE. The CFMs are presented to allow readers to understand and learn from others' experience. Though the way in which you may use this learning will and should be different for each company.

3.3.1 Where and when are the common failure modes?

To describe the common failure modes it is necessary to know at what point(s) the CFM happens. The implementation itself can be divided into three time phases encompassing all parts of the company organization hierarchy (Fig. 3.2). The earliest implementation phase is 'Getting started,' it begins at the first recognition of desire radically to improve the product development process and ends when the company decides to investigate and/or implement CE. The 'Preparing/planning' phase includes information

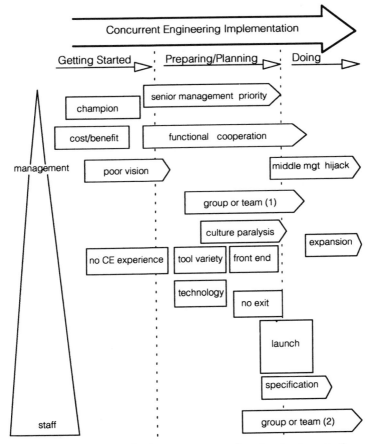

Fig. 3.2 Common failure modes in concurrent engineering implementation.

collection and analysis and ends when a full CE plan is presented and approved. The final phase, 'Doing' starts with the first process changes (which may be training); this phase never ends.

Common failure modes occur throughout the organization, from senior management to staff. Over the implementation time frame, the majority of activity passes from management to staff and the population of CFMs shows this, with early failings mainly occurring within management. The pattern of CFMs illustrates that the early phases contain many failings, proving that right-first-time design is equally relevant whether designing a CE implementation or designing a product. The following sections tackle each CFM individually.

3.3.2 Cost/benefit

Before commitment to CE is sought individuals and groups within the company must believe there is a benefit to be gained. Too often individuals fail to capture the minds of others by concentrating on the wonderful tools of CE and failing to demonstrate an unarguable business need for CE. The claimed benefits for CE, even when demonstrated first-hand, can be difficult to believe. Previous experience is in achieving 5% to 10% performance improvements per year; CE can deliver 50% or more, so colleagues are often doubtful.

Traditional measures of performance in new product delivery are often poor, typically ignoring the high cost of disruption of late engineering changes for example. The CE emphasis on improving the whole product delivery process means that new measurement systems have to be used, such as break-even time. (This can appear to be a desire to create a new playing field rather than compete against current performance measures).

Costs are also difficult to predict. The on-going nature of a CE implementation means that costs continue *ad infinitum*. The return on investment of team training, awareness and management training are more difficult to calculate than of, say, a CAD system, and are prone to easy cost-cutting.

If a CE implementation is viewed as a once only activity with a calculable, low-risk return, the money may become available but permanently improved performance is unlikely. If the CE implementation is viewed as an infrastructure project without financial focus it may never be approved or, more likely, will never gain enough urgency to be implemented.

The solution is to combine the infrastructure and low-risk return arguments. The long-term need must be met by an implementation program that delivers specific financial returns; both objectives are necessary.

3.3.3 Champion

Middle management is the usual location of the first CE implementation activities. An individual or small group recognizes the need for change and the ability of CE to meet that need. Middle management can never support a full implementation. However, the cross-functional changes will require cooperation at the highest level if CE is not to remain an engineering project. A senior champion must be recruited early as there are a lot of senior hearts and minds to be won over. A champion must commit considerable personal time to learning, and to working with others to increase their understanding. These activities cannot be delegated.

3.3.4 Poor vision

The group that develops the CE implementation plan must be targeted. It is acceptable to motivate people by saying 'We want to be number one,' but it is unacceptable to use that as a direction-setter (the question is 'number one what?'). The group needs to be told their boundary – what they can and cannot change – in terms of market, product range, new product plans, structure and design methods. They also need performance targets – for example, best lead time in class on next product – and an early target date for implementation. It is useful to set a very short-term goal to the planning group by asking them to deliver a competitive benchmark and a set of target values. Without these targets, options for the CE implementation plan will be chosen on the basis of experience only. This is not in itself bad, but it is unlikely to be optimal.

3.3.5 No CE experience

In the early phases of CE implementation there is no CE experience within the organization. The size of the task ahead and the number of unknowns can make it impossible to generate sufficient confidence to progress. 'Only when we get there will we know enough to get us there'; external experience is valuable but still requires translation and consultants do not know the company. A typical symptom of this common failure mode is constant searching of external material – visits, conferences, etc. – without obvious progress.

The solution is two-fold. First, recognize the dilemma of being inexperienced; remember that your predecessors were in the same position, and move on with the planning activity. External reference sources can be helpful here but are not strictly necessary. Second, recognize how valuable your own experiences are going to be and plan to maximize their effect. Internal experience needs to be incorporated into the implementation plan quickly and effectively; this means a willingness to change the plan and the inclusion of a learning system within the plan (see Fig. 3.3). A learning

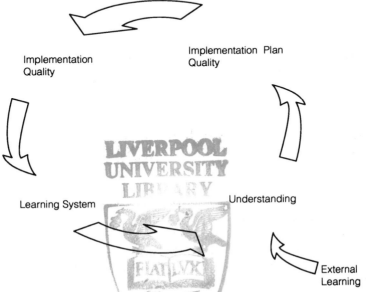

Implementation Quality

Implementation Plan Quality

Learning System

Understanding

External Learning

Fig. 3.3 Need for systematic learning in concurrent engineering implementation.

system will include review mechanisms and analysis activities to decide if and how to incorporate the learning; best practice indicates that the review and assessment should be primarily conducted by the staff (typically the multidiscipline team) in conjunction with the management. Specific training in review skills is recommended.

3.3.6 Senior management priority

Concurrent engineering demands that a different approach to product development be adopted. This is a major change for any company and is typically accompanied by calls for 'senior management commitment'. Unfortunately many companies have been able to claim that senior management commitment existed without gaining the benefits. Only when senior management act resolutely to support CE will the full benefits flow.

The company-wide impact of CE will alter how most departments act so senior management must take an active part in the planning work if they are to control where the company should be. Relegating the planning tasks appears to be efficient in the use of senior management time but results in a lesser understanding and lower commitment to the change. Often the planning task is passed to engineering with other departments represented; the true customer becomes the Head of Engineering and the final implementation is engineering centered due to lack of interest elsewhere. The benefits produced are dramatically reduced by this approach.

The real issue is senior management priority. If CE is truly important and company-wide then all senior management should be part of the planning and should ensure support from their functions is forthcoming. CE is not 'somebody else's problem'. To take part in the planning and actively to sponsor CE, a program of senior management education is vital.

Clarifying business benefits is part of the process, as is plotting the non-CE scenario – what may happen if we do not dramatically improve our product development performance. CE is often described as common sense and it is certainly not a revolutionary theory, but senior management will be aware of the risks in any change program of this scale. Recognizing, articulating and planning for the risks will demonstrate competence and generate some comfort. Finally to capture the imagination, however, it is likely that the CE champion will have to work one-on-one and complement discussion with real life examples. Use of external cases and visits to like companies are often the final motivator.

If the preparation phase is succesful we can expect to see:

- Senior management involved in planning;
- Departments planning internal complementary changes;
- CE planning group reporting to the CEO/Board; and
- Best personnel being released to the CE program.

These are the true tests of commitment, most senior managers are extremely busy with hundreds of short-term problems. Only if the priority of CE is high enough to compete with these will a company maximize the benefits.

Without first creating sufficient top management priority for CE it is unreasonable to move out of the getting started phase. Planning where to go and how to get there must follow from top management priority for CE or the plan itself will either not be workable (by asking for cooperation from departments not involved in the planning of CE) or will be sub-optimal (by restricting implementation to only those departments who have been involved in the planning).

The elevation of concurrent engineering to a priority item for senior management is a critical test that must be passed before moving into the planning phase.

3.3.7 Functional cooperation

Having a senior management team who know what CE is and who have a vision for what they want to do with it and how it will benefit the company usually leads to the creation of a CE planning group. This group is typically multidisciplined and tasked with developing a CE implementation plan. It is often unclear what the group can or cannot do, what management wants or how the group will know they have been successful.

To start the planning activity senior management should ensure representation from all the company functions, including finance, marketing and personnel. Where possible a work area should be created for the planning group and time commitment agreed and equal between functions (e.g. we all work here Monday to Wednesday and return to our functions Thursday and Friday). The initial group meeting should be held by senior management who present to the group:

- why CE is necessary for the company;
- what they think CE is;
- the boundaries of the planning task;
- the escalation and reporting procedures;
- what they can/cannot tell other employees; and
- how they will measure the success of the group.

If senior management understands the nature of CE it will be natural and correct not to set a task boundary, as it is best to consider everything the company does. A regular, organized escalation and reporting system should be used to keep senior management informed of ideas, problems and progress. Senior management should also be responsible for briefing the whole company on the activities of the group.

The most critical element is the setting of target measures. Having a common target is the single most important method for achieving cooperation between the functions. Target values for quality, cost and delivery time for new products should be set against a date along with a target date for delivery of an implementation plan, to include cost, benefit and risk analysis. Only if all the group members are measured against these common targets will functional cooperation happen; an essential prerequisite to gaining functional cooperation in the product development process.

3.3.8 Group or team

Multi-disciplined teams are the biggest aid to developing better, cheaper and faster products. Teams are the backbone of CE, providing a mechanism for all other CE tools to play their part. For these reasons most companies correctly introduce the multidisciplined team into their CE plan, but most fail to recognize what it takes to extract the maximum benefit from this new environment. Lack of experience in using multidiscipline teams leads most companies to implement multidiscipline groups, where the prime common factor for the members is that they work on the same product. This is not a team.

Group members typically attend meetings on the new product and represent their functions in the discussions; they will have many other duties and a boss that stresses performance at base. To avoid creating this situation the CE planning team must first recognize the difference between a group and a team: 'A team is a group which shares, and says it shares, a common purpose, and recognizes it needs the efforts of every one of its members to achieve this'.

Second, it must recognize what environmental factors are needed to enable teamworking. A team should have full-time members and be colocated to create a physical unity. Common purpose can only be created with a common performance measure – the whole team is responsible for all aspects of the product (therefore the maintainability engineer is not berated for poor product maintainability). Both failure and success should be attributed to the whole team. All interested parties should be equal team members, for example finance should not send performance data to calculate cash flows and returns, they should be part of the team and be present full-time.

The creation of a plan for the team's activities should be left to the team as a first action. It should not be imposed from outside. This is an example of letting the people who know the most about the product and process decide their own fate – a very powerful motivator. Another example is to allow the team members authority to make detailed decisions without constant referral to the functional manager. This is best achieved by informal discussion, by having all the functional managers join a reporting group and by agreeing a policy guideline which can be used by each team member to identify whether policy may be broken (and therefore whether it may need discussing with functional management). Of all the common failure modes this is greatest, both in terms of occurrence and impact.

Some of the CFMs already discussed will make implementation of an excellent team somewhat easier. Only a specific recognition of the differences between groups (who share a common name) and teams (who share a common purpose) will allow a planning group to deliver an excellent team environment.

Some elements of the environment – full-time membership and colocation for example – are seen as luxuries. Team environment therefore becomes an area of compromise. Those companies that have successfully navigated the teamworking and CE journey will offer simple advice – do not compromise because it does not pay.

3.3.9 Culture paralysis

Any company that has progressed thus far with concurrent engineering planning will have regularly met the terms 'Change Management,' 'Culture Change' and 'Team Culture'. These and many other terms are a foreign language to most manufacturing companies, a language which they feel they must learn and understand. This is an enormous task in itself and will cause massive delays to the CE implementation. Many companies recognize the validity of the problem but fear their own ability to understand this new topic and so paralysis ensues.

A similar paralysis occurs in planning the team leader role and in selection. Many advocates of CE suggest that the best, or only, way to have an effective team is to find a team leader with almost supernatural qualities. Technical, managerial, commercial and personal skills in equally high quantities must be present in the team leader to ensure the team works well together, meets its goals and does not conflict with management. Of course, finding such a person, if they exist, takes time and slows the implementation. Both forms of paralyses occur through a similar misunderstanding – a failure to understand why we need it (culture change or great leader) and therefore to understand how we might otherwise achieve the same end.

We need a culture change to make sure all individuals do the right thing in all circumstances – a massive undertaking if we attempt to organize it from the top down and inform everyone what is the right thing to do. The challenge is far less daunting if we tackle it bottom up; help individuals begin to change some of the things they do and grow from there. The sum of the changed individual behaviors eventually becomes a noticeable change in company culture. Changing individual behavior is primarily achieved through changing how the individual is measured and rewarded. Team-supporting behavior comes from a common goal and common measurement, and is complemented by lack of external (management) interference. The great team leader form of paralysis fails to recognize that we need a great leader to deal with the weaknesses in the system, such as separate measurements and management interference. Using the techniques discussed in this chapter, many of these weaknesses are removed and the role becomes less critical, but it does not become unimportant! The selection of a team leader can be reduced from a near-impossible task to a crucial, but feasible, choice.

3.3.10 Tool variety

The Institute for Defense Analysis lists over sixty CE tools. With other sources the total may be one hundred or more. This is bewildering to those companies who have decided to investigate CE fully and can cause considerable delay while each tool is investigated, understood, assessed for suitability and built into the CE implementation plan (or put aside). Obviously the delay is bad, but the risk of making an error is usually the greater concern and many companies generate extensive delays until their knowledge is sufficient to assess the risk of moving on.

It is possible to exit this stage of CE planning relatively quickly. Three factors make this feasible:

1. Delaying the implementation is expensive;
2. The best people to select a tool are the users; and
3. The most appropriate tools are obvious.

Relatively little research into CE is needed to learn about the capabilities of the main tools (such as teams, quality function deployment, 3-D CAD, design for manufacture). The less well-known tools are often built for specific product constraint problems (such as design for maintainability). If this was a problem area for your company you would be likely to identify it with little effort.

More importantly, the CE planning group should not attempt to prescribe how the product development process should take place. Assuming a team approach is planned, then it should be the team members' task to identify what tools may be used and how. The CE planning group should allow the CE implementation to move quickly to a team environment and then provide support to the team by delivering information on possible tools and, if selected, by installing the tools.

This approach is made possible and necessary by the use of teams who have both the capability and need to decide how they deliver to their goals.

3.3.11 Technology

A historical emphasis on using the very best technology to deliver improvements, both within our products and within our businesses, is continued in CE. Vendors claim tools make CE possible and are careful not to deny the power of the low-tech tools, like teamworking or QFD, while emphasizing the value of high-technology. The shared use of 3-D CAD models is claimed to go part-way to the creation of a 'virtual team'. Though the high-tech tools can and often do deliver improved performance, their return on investment is poorer than many of the chapter tools and their implementation is longer.

Evaluation and implementation of high-technology tools into a CE implementation should not be placed on the critical path. It should occur in parallel with other elements and will benefit from the extra knowledge gained during the implementation.

3.3.12 Front end

The majority of CE based product development projects start when approval is given to bring the product to market. Up to half the total lead time (from idea to customer) may have already disappeared before we bring any CE tools to bear; many opportunities to consider how different constraints may affect the product concept have also been missed. The assumption that the clock starts on project approval is passed onto the CE planning team and implementation plans tend to tackle only the latter parts of the product development process.

The solution is obvious and non-trivial; by including all of the product development process in the analysis of the current situation, high payback improvement techniques can be recognized and included early in the implementation plan.

3.3.13 No exit

Fear of the large change that CE will bring makes all CE implementation planners wary of starting the actual implementation. The desire to plan more carefully and in more detail, to win more support, etc. acts as a brake. This fear of the unknown can be overcome if senior management set specific and aggressive targets, acknowledge that the pace is deliberately high, that mistakes will be made and openly discussed and that any lessons learned will be incorporated.

Errors will occur and management must accept this and encourage discussion of errors to remove the fear of failure – which makes planning safer than doing. Measurement of errors should be on the basis of repetition and senior management should be rightly concerned with errors that recur.

3.3.14 Specification

This common failure mode is similar to that of front end. Most CE implementations are designed assuming a product design specification is in place. The implementation therefore fails to address the weaknesses already built in. Product specifications are typically lacking in sufficient content to guide design decisions or detailed to the point of describing how the product will work. By including the product specification in the analysis phase the CE planning team will inevitably tackle the weaknesses identified. For the first CE project this is even more critical as the product specification forms the major part of the teams goal. The team should develop its own product specification following the guidelines laid down by management.

The product specification should describe in customer terms what the customer wants in sufficient detail to identify trade offs (such as how much more price will be paid for 10% greater mean time between failure). These data are difficult to gather but using the team to collect them by speaking directly to customers is a positive technique to ensure the best targeting of product characteristics.

3.3.15 Launch

Most companies use pilot projects to introduce and learn about concurrent engineering. Team members are selected by management and brought together to work on a specific product. Typically, the team is brought together and given a product specification with target launch dates and costs, possibly given a presentation on CE and then left to tackle the problem.

The failure in this form of project launch is that of missed opportunity. The CE implementation process is new and stressful for everyone and the project launch is the best opportunity to bring all the stakeholders together and create a common understanding. Many of the subsequent weaknesses in a project can be reduced or removed if the launch is well-organized.

The launch is an event. The importance of the new product and of CE must be stated by the CEO, who should lead the event. The new belief system – that the team is major unit, that teams will have sufficient power and responsibility to deliver to the goals, that management will support the team method – should be expressly stated by the CEO.

Management from all functions must attend and visibly agree to support the team method by agreeing the team goals, team responsibilities and the problem escalation mechanism to be used. Both of the above actions have to be discussed and agreed with senior management well in advance of the project launch. This part of the launch is designed to clarify what the team can and cannot do; without it the team will have to discover the new boundaries of responsibility by trial and error – a slow and painful experience.

The second part of the project launch focuses on the team. The roles of team members and team leader must be presented and agreed. Note that all team members must have the same role. Subdivision so that manufacturing engineers tackle manufacturing issues, say, must be avoided; this is to ensure that all the team share responsibility for all issues. The team leader's role is often left unclear and the historical view of the project leader as chief decision-maker prevails. In a CE team the team leader's role is to communicate information, not to make decisions. The team leader must first escalate problems that the team cannot deal with – lack of resources or data or decisions which impact company policy, for example. Second he or she must organize meetings, formal and informal, to ensure communication between team members; and third, the team leader must collect data on progress against plan, escalate problems to management and keep the plan up-to-date. The team leader's objective is to create the most efficient environment for the team members and therefore acts more as a coach than a player – something that must be made clear to all.

The final element of the launch is to give the team a target. This should be set in terms of meeting agreed product performance goals, cost goals and time goals – this must be prepared in advance and the team members must be given an opportunity to analyze it before agreement. A team activity to produce a project plan is advocated as a first task; this gives the team a task with a short-term deliverable which they must cooperate on to deliver successfully – an excellent learning opportunity.

Goals should be aggressive and quantified, whenever possible, in terms of profit. The trade-off between cost, quality and speed should be presented and given a scale value. For example, if each day's delay of product launch is worth $30 000 then the team are likely to work harder to improve the speed of delivery. The goals must be expressed in product terms and avoid any association with functions (e.g. purchased parts cost, manufacturing cost). The measure of the project success must be in terms of the product, not specific elements, and a recognition/reward system should be presented which clearly rewards all team members equally, regardless of functional contribution. In the author's opinion the reward system should be recognition not cash; (e.g. send all team members plus families to sister plant in Italy to share CE experiences).

All of these messages can be delivered without a formal, high-profile launch. The launch is needed to ensure that everybody understands their own and everybody else's job; this is very difficult to achieve after the project has started.

3.3.16 Middle management hijack

When a multidisciplined team is being used the position of functional manager alters dramatically. Chief Stress Engineers, Production Engineering Managers, Sales Managers and the like, have released people to the project and attended presentations on CE but little real change has normally taken place. They are likely to recall the resource to solve short-term problems, to require the individual to report back on their work and to alter that work as they see fit. This removes real responsibility from the team and slows the process, as well as re-emphasizing the functional alignment of team members. The eventual result is failure or reduced ability to meet project and product goals.

The solution is to clarify and agree the roles and responsibilities of management and the team. This can only be achieved if the individuals understand and believe the principals of CE and education is a prerequisite for this. Management belief in CE is hard-won and should be based on fact not hope – case study material from other companies plus visits, may be the only way to win over the 'seen it all before' skeptic.

Middle management hijack is a common failure mode in all CE implementations – only the extent and impact varies. Dealing with the issue before and at project launch is best but constant vigilance by the CE steering group is required. Problems that do appear should be escalated to the CE steering group to allow senior management or the CEO the opportunity to remind middle management of their agreed role. In many situations the project process is affected so badly that replanning is necessary and a re-launch event should be used to create a focus. The relaunch should be organized on similar lines to the launch discussed above; in particular it should deal with the future and so avoid becoming a battleground for examining past errors.

3.3.17 Group or team

Team members allocated to a concurrent engineering team for the first time will find the new role disorientating. Even if management does not interfere, the individual team members will often revert to their previous way of working (such as waiting for drawings, raising engineering change requests and not contributing to concept design). In this situation the opportunity is being lost to bring all of the team resources together to ensure the product design progresses with all constraints being considered.

Most often this is caused by middle management hijack and/or unclear roles and responsibilities. The solution in these cases is the same and requires refocusing and a project relaunch.

In some cases the cause is poor team processes – too many, too long meetings is one symptom, team leaders centralizing decision-making is another. The solution is to use education as a reminder that the measures are of team performance and of the value of others' contributions, along with training in team problem-solving. The CE steering group must support and encourage the team leader in particular.

3.3.18 Expansion

The improvement of all product developments is the aim of CE implementation yet this chapter has focused on the stages of implementation up to and including the first pilot projects. This is still a learning phase and the whole company must adopt the lesson by changing permanently. Multidiscipline teams, QFD, etc. have to become the normal method of product development.

The hurdles are many – reward systems, home locations, career development, skills development, role of management and organization structure being most discussed. The common failure here is not making CE the norm. CE planning groups expend enormous energy identifying all possible problems and designing a solution which meets them. As each problem is analyzed then further, more detailed, concerns are raised which the solution must address. This is a form of paralysis as real change does not happen, and the longer that the delay continues the less likely change becomes.

The key is confidence. Companies must accept that it is impossible to design a perfect CE implementation and stop trying. Instead we must accept the ethos of continuous improvement and ensure that we implement a change program that can quickly identify problems and opportunities from the implementation and incorporate the lessons into the plan. Such a 'learning implementation' must have a system for learning to increase the plan quality (see Fig. 3.3) which includes the team members, CE steering group, stakeholder management and senior management. Each group should meet specifically to discuss and review the CE implementation – what is working, what is not working and what may be missing. The coordinated output of the meetings should be incorporated into subsequent stages of the CE expansion.

3.4 COMMON FAILURE MODES: CAUSES AND STRUCTURE

The effort in all concurrent engineering implementation is initially based in management and it is no surprise that the majority of early common failure modes are caused by management. The inability to state clearly where the company is going and why it should use CE is prevalent; it is not enough to agree that CE is good for you, no matter how true, for without targets we cannot aim our sights!

During the preparation and planning phase a lack of targeting is felt strongly and leads to confusion about what is acceptable and which tools to use. In part this is also due to the large amount of material published about the use of information technology in CE and to the extreme complexity of the 'soft team'/organization/culture issues.

This typically leads to a planning phase which concentrates on technology and avoids the non-technical issues. The power of multidiscipline teams cannot be ignored however and most CE implementations result in a pilot project with a team component. As with the earlier phases the tendency is to leave problems to be solved at some unspecified, later point and to divert energy to topics we are more comfortable with (such as IT). This obviously creates problems which have to be dealt with either after they appear or by not implementing CE at all! Neither option is sensible yet most implementations travel this difficult journey.

In analyzing the later CFMs it is clear that they are less likely to occur if earlier CFMs had been resolved (for example, middle management hijack is less likely if the project launch and functional cooperation common failure modes do not occur). The emphasis therefore seems to be in the earlier, management dominated implementation activities, when the understanding of CE is at its lowest. The risk can itself create uncertainty and delay implementation, so how can a company progress successfully?

First we must be absolutely clear that the implementation of concurrent engineering is a major challenge for management. Second, management must accept the challenge, no matter how daunting. The consequence of standing still in product development quality, cost and speed while others move ahead is dire.

Last, and most important, concurrent engineering is based on the principle of common sense; in retrospect much of concurrent engineering is obvious. This gives concurrent engineering implementations a robustness that is difficult to shake; even poor implementations can be an improvement on existing product development performance.

The challenge set out here by the identification of common failure modes is therefore a positive one – to increase the benefits generated by concurrent engineering and so improve competitiveness.

3.5 REFERENCES

Byrd, J. and Wood, R.T. (1991) *CALS Europe 91 Conference*, 549–60.
Evans, S. (1991) Implementing concurrent engineering: learning through action. *Computers in Manufacturing Conference 1991*, Birmingham, UK.
Garrett, R.W. (1990) *Manufacturing Engineering* November 1990, 41–7.
Manton, S.M. (1991) *CALS Europe 91 Conference*, 593–608.
Smith, P.G. and Reinertsen, D.G. (1991) *Developing Products in Half the Time*, Van Nostrand Reinhold, New York.

CHAPTER 4

Overcoming barriers to the implementation of concurrent engineering

Gary A. Maddux and William. E. Souder

4.1 INTRODUCTION: THE IMPLEMENTATION PROBLEM

As organizations struggle to become more competitive in a global market-place, concurrent engineering has surfaced as one of many concepts that promises major benefits for its practitioners. Along with total quality management (TQM), quality function deployment (QFD), Hoshin kanri, kaizen, kanban, and a growing list of similar terminologies, concurrent engineering has both captivated and bewildered the world with its simplistic yet radical philosophy (Ouchi, 1981; Akoa, 1991). Its adoption and adaptation by an organization can have a profound effect. Users of concurrent engineering boast of better designs, fewer engineering changes, improved quality, improved marketability of products and increased profits (Hauser and Clausing, 1988; Winner, 1988; Maskell, 1991; US Army Material Command, 1991; Hartley, 1992). Why hasn't everyone jumped on the concurrent engineering bandwagon?

The answer to this question may lie in the natural reluctance of an organization to change. Concurrent engineering is a non-traditional approach to the design process. While many of its concepts are logical, the implementation of these ideas in an organization steeped in the accustomed mindset of sequential product design may be perceived as radical change. To convert the design process from a sequential operation with clearly differentiated functions to one that involves the simultaneous involvement of several inter- and intraorganizational units can encounter enormous barriers.

The incidence and effects of such barriers were clearly dramatized in one recent study of approximately 300 randomly selected industrial new product developments (Souder, 1987). The effective use of concurrent engineering was made difficult by the unwillingness and psychological inabilities of the R&D, manufacturing and marketing personnel to collaborate. As Table 4.1 shows, nearly 60% of the projects had some incidence of disruptive disharmony that stood in the way of interdepartmental collaboration. As shown in Table 4.2, the lack of collaboration and concurrent engineering correlated with the failure of the projects.

Table 4.1   Incidence of harmony and disharmony

States	Percentage of projects experiencing this state
Mild disharmony	
Lack of interaction	7.6
Lack of communication	12.9
Subtotal	20.5
Severe disharmony	
Lack of appreciation	26.9
Distrust	11.8
Subtotal	38.7
Harmony (total)	40.8
Total	100%

Source: Souder, Wm.E. *Managing New Product Innovations*, p. 168.

Table 4.2   Distribution of project outcomes by harmony/disharmony states

| States | Project outcomes in percentages | | |
	Success	Partial success	Failure
Harmony	52	35	13
Mild disharmony	32	45	23
Severe disharmony	11	21	68
Totals	100%	100%	100%

Source: Souder, Wm.E., *Managing New Product Innovations*, p. 170.

This chapter examines the barriers to the successful adoption and implementation of concurrent engineering. Several actions are recommended that managers can take to overcome these barriers and achieve the full benefits of concurrent engineering.

4.2 TWO TYPES OF BARRIERS: ORGANIZATIONAL AND TECHNICAL

In any organization, two types of barriers are likely to exist that inhibit the successful implementation of concurrent engineering: organizational and technical. Organization barriers relate to management style, organizational policies, organization cultures, personnel behaviors, risk taking propensities and accustomed ways of doing things. Technical barriers involve a lack of supporting or facilitating technologies and know-how to implement concurrent engineering. For example, the lack of CAD/CAM facilities can be a significant barrier to the use of concurrent engineering.

Organizational barriers may be the most imposing to overcome, since they involve deep seated and well-entrenched ways of doing things and behaving. Moreover, organizational barriers are likely to be complex and interrelated. For example, a change in top management style can directly affect risk-taking propensities. However, as discussed below, technical barriers may also be formidable obstacles. They are also likely to be interrelated with the organizational barriers, e.g., a policy that restricts the use and availability of CAD/CAM facilities to some organizational members.

4.3 SEVEN COMMON ORGANIZATIONAL BARRIERS

4.3.1 Lack of top management support

Any attempt to improve quality (whether through design, manufacturing, or service improvements) must receive the full endorsement and support of top-level management if it is to succeed. One universal rule is that implementation begins in the boardroom. Without top management support, any attempt to affect positive change within an organization is doomed. The communication and cooperation required of the design engineers, production engineers, purchasing agents, and others must begin with the vice presidents of design, production, purchasing, etc. To quote Marion Wade, the founder of ServiceMaster, 'If you don't live it, you don't believe it' (Belohlav, 1990).

Dr James Harrington vividly describes the 'waterfall effect' of top management support that is necessary for organizational change (Harrington, 1987). The philosophical shift must be initiated at the top of the organization, then wash down to the next layer of management. This effectively touches all levels of management, with each layer being brought on board before the next is asked to sign on. This technique mandates that the executive level understand and embrace the concepts before attempting to sell them to subordinate levels.

Another requirement for top management is to allow sufficient time for the new philosophy to generate benefits. It takes time to get departments that have never been required to work together to cooperate on optimizing a design. For example, the establishment of supplier–customer partnerships is an activity that may take years to accomplish. Expecting concurrent engineering to transform the company overnight is a weakness of myopic management expectations, not a weakness of the concurrent philosophy.

4.3.2 Indequate organizational climates

There is a general wisdom that organizational climate affects the implementation of concurrent engineering. Bad climates are said to retard its implementation, and good climates to promote it. Climates are generally held to be the responsibility of top management (Souder, 1987; Souder and Padmanabhan, 1990). But just what is 'climate'? A generally acceptable definition is: 'an organization's prevailing attitude, atmosphere or orientation'. Climate is thus a reflection of the example that top management sets through its actions, policies and decisions (Souder, 1987; 1988).

In an empirical study of fifty-two firms, three climate factors controlled by top management were found to stimulate interfunctional cooperation, interdepartmental collaboration and the adoption of new techniques. These factors were: the degree of uncertainty in task assignments, the amount of role flexibility and the level of perceived openness and trust (Souder, 1987). When each individual's task assignments and responsibilities were clearly stated and known to all parties (low degree of uncertainty in task assignments), cooperation proceeded well and new ideas and techniques were readily implemented by all parties. The extent to which each individual had the capability to play multiple roles (role flexibility), e.g., the ability of a manufacturing person to perform some marketing tasks, the ability of a design engineer to perform some research tasks, etc., was also found to encourage the adoption of new ideas and techniques. The perception that the organization was open to new ideas and that each party trusted each other was another correlate. Thus, for the successful implementation of concurrent engineering, organization climate matters. And top management is responsible for organization climates.

4.3.3 Protective functional managers

One key to implementing successfully an organization culture compatible with the concurrent engineering philosophy lies in achieving cooperation between the various staff functions. The members of a number of departments, e.g., design, manufacturing, purchasing, and marketing, must be willing to share information and insights about the product. An impediment to this requirement is the functional manager who is overly protective of his area.

Managers who exhibit this characteristic often do so out of fear or insecurity, rather than a blatant disregard for the organization's objectives. This fear can be motivated by a lack of confidence in the individual's own talents, the talents of the staff, or the fear of being shown up by one's colleagues. Regardless of the stimulus, fear must be eliminated in order successfully to implement any organizational improvement philosophy. The open exchange of ideas and information must be conducted in a non-threatening environment. To expect that full cooperation and innovation can exist otherwise is a fallacy that will lead to project failure (Souder, 1987).

4.3.4 Inadequate reward systems

The traditional reward systems employed by many organizations can be a significant barrier to achieving the cooperation required for concurrent engineering implementation. In particular, reward systems based on departmental goals rather than organization-wide objectives can lead to a suboptimization of the organization's performance. When departments pursue a limited number of resources (bonuses) based on achieving some numerical goal, cooperation is not a likely outgrowth. The concurrent engineering philosophy is built on the optimization of the whole. Engineering must cooperate with production, production with purchasing, engineering with purchasing, etc. for the good of the entire organization. If a department head perceives himself in an adversarial role with his organizational counterparts, he is less likely to share the information needed for a successful product design.

An example of suboptimization at the expense of a product's design is a purchasing department rewarded for performance based on dollars saved on the purchase of manufacturing materials. If the company can purchase gears from a number of suppliers, all of which meet the specifications, most purchasing departments would normally select the lowest priced vendor. However, if the design team knows that Supplier A not only meets the specification but also achieves significant quality, then the team is likely to recommend Supplier A. The purchasing agent, being rewarded on a pay for performance incentive plan, may reject the team's recommendation in favor of the lowest priced Supplier B. Hence, the product's quality suffers because the system was suboptimized due to conflicting objectives. To quote Dr Deming:

> There has to be teamwork, but the annual system of rating destroys teamwork. How could someone in purchasing get a good rating for paying a higher price? Even if it saves 10 times as much in production, you do not get a good rating by paying a higher price. (Peters and Austin 1985)

The organization must ensure that its employee reward system is not contradictory to the cooperative efforts required of concurrent engineering.

4.3.5 Lack of customer involvement

Lack of customer involvement has historically led to widespread failures in design, from consumer products to computer systems. Sullivan contrasts the Japanese philosophy of design, where the voice of the customer drives the product design, with that of the United States. In the US, the voice of the executive or engineer has traditionally driven design (Sullivan, 1986). Concurrent engineering mandates involvement and design interaction from all facets of the product's life cycle. The most important factor determining the success or failure of a product is the user.

The role of the customer as a member of the design team can require a diplomatic approach for the organization. There must be the realization that customers vary in their degree of sophistication. Some customers know exactly what they need, and can explain those needs in the level of detail required for the design. Others, however, may supply specifications that are incomplete or incorrect. Designs based on these erroneous specifications are turned into unusable products. Bob King refers to this as 'redesign caused by the customer' (King, 1989). Therefore, the customer must be a member of the team, but the customer's specifications must be analyzed to determine their correctness.

Souder (1988) has proposed the model shown in Fig. 4.1 as an aid to the developer in determining when to expose a prototype to a customer, and how to manage that customer in demonstrating the prototype. The key dimensions for determining the level and intensity of developer–customer

Customer's level of sophistication

Developer's level of sophistication	Understands own needs and can translate them into product specifications	Understands own needs but cannot translate them into product specifications	Does not understand own needs
Understands the product specification and the technical means to develop new products	A	B	C
Understands the technical means but does not understand the product specification	D	E	F
Understands the product specification but does not understand the technical means to develop it	G	H	I
Does not understand either the technical means or the product specification	J	K	L

Source: Souder (1988) p. 15.

Fig. 4.1 Customer-developer conditions (CDC) model.

relations required are matters of the customer–developer sophistication. For example, if the conditions described in cell A of Fig. 4.1 characterize the situation, close developer–customer involvement is not necessary. The developer simply builds to the known customer specifications. The extreme situation is epitomized by the conditions in cell L of Fig. 4.1. This is a situation of complete uncertainty, in which it is essential that the developer and the customer jointly collaborate to define the user's needs and specifications, the nature of the product to be developed, the technologies to be employed and the technical means for developing the product. Intensive discussions, exchanges of perceptions, joint definitions of the user's needs and application requirements, collection of performance data on similar products, concept testing and prototype testing in the user's facilities are all activities that must occur in cell L. It seems clear that the appropriate vendors and suppliers must also be collaboratively involved in these activities.

Most situations will lie somewhere between the extremes of cells A and L, e.g., various degrees of developer–customer–supplier collaboration will be needed. For example, the conditions in cell E are often encountered in industrial products. In such cases, early involvement of the customer is necessary in order to help the developer set the performance attributes for the new product. Supplier involvement is then necessary to assist the developer in optimizing the technical means (materials, components, etc.). Later, when a prototype has been developed, customer inputs are again required to confirm its efficacy.

4.3.6 Lack of supplier involvement

In his book *Out of the Crisis* (Deming, 1986), considered by many to be the bible of today's quality movement, Dr W. Edwards Deming contrasted the traditional design methodology with that of concurrent engineering (although without using the term). With the traditional methods:

> Engineers formulate the design of a part or subassembly. Purchasing people let contracts for the parts. Some of the contracts went to the company's own allies, other contracts went to outside vendors. Difficulties in manufacture and faults in assemblies led to many engineering changes. Engineering changes led to increases in cost.

Dr Deming goes on to describe the components required for what is now known as concurrent engineering.

> Teams composed of experts from the supplier chosen for this material (part or component); plus your own design engineer, process engineer, manufacturing, sales, or any other knowledge that is needed.

The result: 'Better and better quality as time goes on, with lower and lower costs'.

The requirement of supplier involvement in the design process is also described by Giorgio Merli in what is referred to as co-makership (Merli, 1991). In Merli's partner–supplier relationship, the supplier is integrated in the operations of the client company. This requires a number of changes in the operating environment of the client company. First, the suppliers chosen should be kept to a minimum, and there must be open information exchange and cooperation between the client and supplier. These concepts are also widely promoted by Dr Deming's famous 'Fourteen point managerial philosophy'. Examples of the successful implementation of this practice include both Xerox and Ford (Dertouzos *et al.*, 1989).

The barriers to the supplier–partner arrangement are embedded in the culture of many organizations. For decades, clients have sought to lessen costs by continually requiring lower and lower prices from their suppliers. This adversarial relationship often resulted in low priced, inferior materials that destroyed the cooperation so necessary for the concurrent engineering philosophy. The creation of trust and open communications will be a significant barrier for some firms to overcome.

4.3.7 Fear of loss of creativity

One of the areas of resistance to concurrent engineering is that it imposes standardization on the design engineer. A common complaint is that standardization stifles creativity (Hall, 1987). While the design engineer is somewhat restricted to using proven techniques and accepting the advice of others, the benefits of improved design far exceed the restrictions on creativity.

When designing the product, is creativity constrained because the voice of the customer is required? Is creativity jeopardized because the supplier is requested to contribute knowledge of what he can provide? Creativity and innovation that lead to improved design is always promoted, and is not restricted by concurrent engineering. However, creativity that has little basis of improving the product, that lessens the value of the product by increasing its costs without improving its function or marketability – this type of creativity is justifiably limited by concurrent engineering.

4.4 TECHNICAL BARRIERS

The first technical requirement to support concurrent engineering is a proper computer-aided design/manufacturing (CAD/CAM) setting. According to John R. Hartley, 'Concurrent engineering is wasted without CAD/CAM' (Hartley, 1992). With the appropriate tools being utilized correctly, engineers can work simultaneously on the design, constantly communicating with one another on the same set of specifications and parameters. This common set of data does not relieve the engineers from the personal interaction and communication required of concurrent engineering; rather, it serves to enhance it.

A related technical kind of barrier results from the misconception that an organization can buy its way into concurrent engineering by purchasing the appropriate tools and software. While the hardware and software are necessary, their purchase will not ensure a successful concurrent engineering implementation. The tendency to buy technology, without first understanding how it is to be utilized, can sap an organization of its resources and turn off managerial support for an otherwise credible strategy. Companies must be careful first to identify the need before they buy the cure.

The technology that exists to support concurrent engineering is increasing at a rapid pace. A significant amount of research is being conducted by joint teams of government, academia, and industry. For example, the US Army Missile Command has concurrent efforts in effect with several university partners. Typical efforts include Purdue University's Quick Turnaround Cell, The University of Delaware's Manufacturing of Composite Materials, Ohio State University's Design for Injection Molding and Net Shape Manufacturing, and The University of Alabama's Intelligent Manufacturing Laboratory. As this illustrates, there are a number of areas where technology currently does not exist, but is in rapid development.

Much of the need for technology is in the areas of integrating existing know-how to form a more cohesive system. According to Dr Daniel Schrage of the Georgia Institute of Technology, one of the obstacles to the creation of a concurrent engineering design environment is 'the failure to integrate quality engineering techniques such as QFD and Taguchi's PDOM (parameter design optimization method) into the entire process' (US Army, 1991). Therefore, for a CAD/CAM system or a stereolithography rapid prototyping workcell to be completely effective, it must be integrated with design techniques and methods that are separate but equally important.

4.5 OVERCOMING THE BARRIERS: FIVE ACTIONS

The above barriers to the successful implementation of concurrent engineering can be overcome if managers focus their attention on five actions. These actions are: making the cultural transformation, effecting organization change, CE team building, providing adequate support technologies, and fostering role definition and interaction.

4.5.1 Making the cultural transformation

One of the more important realizations about concurrent engineering is that it is a culture, not a program. In fact, some organizations have recognized this in their definition of concurrent engineering. Concurrent engineering is a 'culture wherein a multifunctional design team conducts the simultaneous design of a product and the process required to produce it' (US Army, 1991).

The means of changing a culture is primarily through education. Starting with top management, all members of the organization must be made aware of the benefits of the concurrent engineering philosophy and what is required of them. Each level of the organization and each individual within that level must be made fully aware of what concurrent engineering is and is not as the first step in implementation. Implementation cannot and should not proceed until this first step is complete.

4.5.2 Effecting organizational change

The use of cross-functional teams is critical to concurrent engineering success. Management must be quick to recognize that cross-functional management serves as an excellent method of breaking down departmental barriers, especially in companies that have suffered from intense internal strife in the past (Imai, 1986). Cross-functional implementation teams are the best vehicle for implementing concurrent engineering.

To aid the process of multifunctional teams, it has been recommended that 'product planning, development and production be integrated together, reporting to one vice president' (Souder, 1987; Clausing, 1991). This structural change in the organization can enhance the cooperation between department heads by reinforcing the idea that the manufacture of a product should be the output of a single team, rather than a joint project of several teams.

4.5.3 CE team building

Though several approaches may be taken to build the concurrent engineering team, Rockwell International's approach has been very effective. This approach requires that:

1. Each functional department assign staff members to the design team;
2. Team members are co-located;
3. The team has one team leader with a strong technical background;
4. The team has total project budget and schedule control;
5. The team leader coordinates all efforts;
6. The team reports to the program manager; and
7. The team concurrently develops the product description and the manufacturing processes (US Army, 1991).

The members of the team must not only represent each of the departments required for the design process, but also must pass certain criteria. Among these are the ability to represent the knowledge of that function adequately, and to gain the commitment of their areas once a decision has been made

(Clausing, 1991). If the team member cannot sell the design to his own department once the design has been made, then cooperation will not exist and the design effort will be in vain.

Once the barriers to cooperation are overcome within the organization, suppliers and customers must be brought into the process. This may pose a number of issues, e.g. how to reduce the number of suppliers and how to involve these outside parties in competitively confidential activities (Harrington, 1987; Crosby, 1986). This poses obstacles to the development of the trust, openness and full information exchange necessary for an effective team.

Souder (1987) has shown that one effective approach to creating a concurrent engineering team is to focus on 'task-only' integration. This is a commonality of goals and consensus commitment for only the task at hand. It is a kind of detente for a particular task. Task-only integration relieves the parties of complete organizational cohabitation, and eliminates the fear that each will lose their specializations and their independent identities. Task-only integration works as follows. First, management must set the end objective and its boundaries, e.g., the development of a new widget that performs in such and such way, for client X. Second, top management must specify where joint action is mandatory, and set sanctions and penalties for non-compliance. Third, top management must be available for consultation, guidance and assistance in resolving disputes in a timely fashion. Fourth, top management must agree to assume the risk for some types of errors that the joint effort may make, and to define what these errors are. This approach has been found to be effective in several organizations (Souder, 1987).

4.5.4 Providing adequate support technologies

A number of technologies exist to support concurrent engineering: CAD/CAM, stereolithography, rapid prototyping, and other computer aided engineering techniques. There are also several methodologies that may be used to support concurrent engineering implementation and utilization, e.g. quality function deployment (QFD), Taguchi methods, and statistical process control (SPC). QFD helps 'hear the voice of the customer', and is useful for brainstorming sessions with the members of the design team to determine how best to deliver what the customer desires. Taguchi methods can help determine how to make improvements in the existing processes to minimize the variability of the production and thus maximize the quality of the product. SPC can be used to monitor the processes used during production, to ensure that the systems are in a state of control. It is management's responsibility to provide these supporting technologies, to encourage an awareness of their availability, and to provide the required training to use them intelligently.

4.5.5 Fostering role definition and interaction

A major barrier to the adoption of concurrent engineering lies in the improper role interactions that often occur between the various parties that need to be involved. In a series of empirical studies of the barriers to the adoption of advanced manufacturing techniques, Souder and Padmanabhan (1990) found that the *right* parties need to be involved at appropriate times, and they must play the *right* roles, otherwise the techniques are not adopted. Specifically, they found that top management, the chief manufacturing officer, the chief technical officer, the chief marketing officer, the first line supervisors, the key shop floor personnel, the key design person and the key development person each must play their proper roles. A lengthy set of prescriptions was developed to describe these proper roles (Souder and Padmanabhan, 1990). In general, the proper roles of the top managers and the chief officers involve setting visions and goals concerning what the implementation should accomplish. In effect, their proper roles are to define the benefits expected, and to develop metrics for measuring progress toward these benefits. The proper roles of the first line and key shop floor personnel involve applying these metrics and assuming the burden of implementing portions of the techniques within their responsibility areas. The proper roles of the design and development personnel are to serve as linchpins between the technology and the organization, to further assure that implementation is occurring at the lowest levels of the organization.

Souder and Padmanabhan's findings would appear to apply to the implementation of concurrent engineering. The notion that there are proper roles to be played by each party, at various times over the evolution of a plan to implement concurrent engineering, is an appealing notion. As Souder and Padmanabhan (1990) point out, each organization may have a slightly different prescription of roles. But there is a compelling argument that various responsibility roles n.ust be fulfilled in order to effect the successful implementation of any innovation, whether it be a new management technique or a new technology – and these roles must be in harmony.

4.6 SUMMARY

Many practitioners of concurrent engineering declare it to be a common-sense approach to manufacturing. It has also been said that nothing is so uncommon as common sense. The successful implementation of concurrent engineering faces many barriers within an organization. Some of these are self-imposed by the organization itself, others are the idiosyncrasies of the individuals within the organization, and still others are inherent in the nature of concurrent engineering. To achieve success, the organization must develop an environment where concurrent engineering can flourish. This may require changes in the organization's culture, and in many practices that have become embedded. This is a painstaking process, but one that must be accomplished before the technical aspects of concurrent engineering are considered. It is much easier to install a new piece of hardware than to establish a new culture, but that is exactly the experiences of the pioneers within the concurrent engineering community. There are bountiful rewards for those who are successful, but there may be miles to travel before they are attained.

4.7 REFERENCES

Akoa, Y. (Ed.) (1991) *Hoshin Kanri: policy deployment for successful TQM*, Productivity Press, Cambridge, MA.

Belohlav, J.A. (1990) *Championship Management: an action model for high performance*, Productivity Press, Cambridge, MA.

Clausing, D. (1991) *Concurrent Engineering*, Design Productivity International Conference, Honolulu, HI.

Crosby, P.B. (1986) *Running Things: the art of making things happen*, McGraw-Hill, New York, NY.

Deming, W. Edwards (1986) *Out of the Crisis*, Massachusetts Institute of Technology, Cambridge, MA.

Dertouzos, M.L., Lester, R.K. and Solow, R.M. (1989) *Made in America: regaining the productive edge*, Massachusetts Institute of Technology, Cambridge, MA.

Hall, R.W. (1987) *Attaining Manufacturing Excellence*, Dow Jones-Irwin, Homewood, IL.

Harrington, H.J. (1987) *The Improvement Process: how America's leading companies improve quality*, McGraw-Hill, New York, NY.

Hartley, J.R. (1922) *Concurrent Engineering: shortening lead times, raising quality, and lowering costs*, Productivity Press, Cambridge, MA.

Hauser, J.R. and Clausing D. (1988) *Harvard Business Review*, May–June, 63–73.

Imai, M. (1986) *Kaizen: the key to Japan's competitive success*, Random House, New York, NY.

King, B. (1989) *Better Designs in Half the Time*, GOAL/QPC, Methuen, MA.

Maskell, B.H. (1991) *Performance Measurement for World Class Manufacturing*, Productivity Press, Cambridge, MA.

Merli, G. (1991) *Co-makership: The new supply strategy for manufacturers*, Productivity Press, Cambridge, MA.

Ouchi, W.G. (1981) *Theory Z: how American business can meet the Japanese challenge*, Addison-Wesley, Reading, MA.

Peters, T. and Austin, N. (1985) *A Passion for Excellence*, Random House, New York, NY.

Sanno Management Development Research Center (1992) *Vision Management: translating strategy into action*, Productivity Press, Cambridge, MA.

Souder, W.E. (1987) *Managing New Product Innovations*, Macmillan, New York, NY.

Souder, W.E. (1988) *Journal of Product Innovation Management*, **5**(1), March, 6–19.

Souder, W.E. and Padmanabhan, V. (1990), in Parsaei *et al.* (eds) *Justification Methods for Computer Integrated Manufacturing Systems*, Elsevier, New York, NY, pp. 132–59.

Sullivan, L.P. (1986) *Quality Progress*, June, 12–15.

US Army Material Command (1991) *Concurrent Engineering*, Alexandria, VA.

US Army (1991) *US Army/National Science Foundation Joint Symposium for the Technology Transfer of Concurrent Engineering Tools and Methodologies*, Von Braun Civic Center, Huntsville, AL, June 4–5.
Winner, R.I. *et al.* (1988) *The Role of Concurrent Engineering in Weapons System Acquisition*, IDA Report R-338, Alexandria, VA.

CHAPTER 5

Improving interpersonal communications on multifunctional teams

Michael E. Fotta and Ray A. Daley

5.1 INTRODUCTION

In order to deal with the complexity inherent in modern product development there has been an increasing degree of specialization. Some engineers specialize in the design function, others in the manufacturing function, still others in reliability, etc. These specialists are then put together as a multifunctional team to develop a product. While there is an obvious advantage to having teams composed of well-trained, experienced specialists there can be interpersonal communication problems within such teams. This is due to the fact that the previous training and individual experience which each specialist brings to the team leads to terminology and the use of that terminology particular to the individual's speciality. In other words each specialist has their own viewpoint of product development.

In traditional product development the sequential interactions between different specialists have been constrained to occurring at specific times and only under well-established conditions (for example, a design engineer may not show a design to manufacturing until drafting has made a drawing). Thus, in a sequential development process most specialists, after some experience, could learn enough about another specialist's viewpoint to handle these interactions well.

In forming concurrent engineering (CE) teams we will be putting together specialists who have followed the sequential pattern in the past. Based on their past interactions many specialists have only a limited knowledge of other specialities. However, on CE teams we expect every specialist to be able to communicate in a meaningful fashion throughout the life cycle of the product with a greatly increased number of other specialists.

We could just let them flounder and learn on their own as they have in the past or we could look for ways to help this complex communication process. If we could put each specialist's viewpoint into a concrete communicable form, compare these viewpoints, and then show and explain all this to the team then we could enhance interpersonal communication. This chapter proposes a method to do just that.

The chapter is structured as follows. First, personal construct theory is described as the basis for providing concrete representations of viewpoints. Next a specific technique (entity-attribute grids) for developing these viewpoints is discussed. This is followed by an explanation of a communication classification scheme used to compare the viewpoints. Once this groundwork has been laid a specific method to apply to multifunctional teams is described. This is followed by an example of such an application.

5.2 PERSONAL CONSTRUCT THEORY

For over thirty-five years there has existed a way to put a person's viewpoint into a concrete form based on their vocabulary and how they use this vocabulary in a particular situation or domain. This is based on a theory (personal construct theory – Kelly, 1955) and a methodology for implementing that theory (repertory or entity-attribute grid methodology – see Hart, 1986). Personal construct theory (PCT) has been successfully applied to a variety of domains including: clinical psychology (Kelly, 1955), market research (Frost and Brain, 1967), knowledge acquisition for expert systems (Boose, 1984; Diedrich, *et al.*, 1987) and management team development (Kilcourse, 1984).

PCT is based on the premise that everyone develops constructs (ways of thinking) about a domain based on differences they perceive between entities (objects, people, design alternatives, etc.) in that domain. For example, a group of design alternatives may be seen as varying in cost. An individual would then have a construct concerning this variation in cost about the design alternatives. The alternatives are the entities in this case.

PCT further assumes that people can name these constructs and tend to use bipolar dimensions when they do so, (e.g., friendly–unfriendly, quality–no quality, fast–slow). In the example just given a likely name is 'high cost–low cost'. We refer to this as the terminology used to verbally explain a construct (see Fig. 5.1).

According to PCT, people develop sets of constructs (construct systems) for each of the domains of their experience. An individual's construct system for a particular domain establishes a personal model of that domain for the individual. This is basically a person's viewpoint of the domain. Based on this viewpoint one anticipates the domain and acts on the basis of these anticipations.

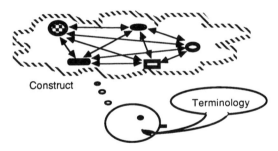

Fig. 5.1 Terminology and construct. A construct is an internalized way of thinking about a group of objects based on perceived differences between the objects. Constructs cannot directly be perceived by others, but can be communicated by verbal descriptions we call terminology.

Since a person's construct system is largely determined by past experience, the construct systems of team members with differing specialization will reflect their personal viewpoint. If we could determine these different construct systems and communicate them to others then they could be used as the basis for improving team members' understanding of one another's viewpoints.

5.3 ENTITY-ATTRIBUTE GRIDS

The entity-attribute[1] grid technique provides a way to elicit and view a person's construct system for a particular domain. The first step in using this technique is choosing a set of entities – items of interest in the domain being examined. They must all be understood by those whose constructs you wish to derive.

In order to illustrate this technique consider the viewpoint which one might derive when faced with the problem of 'buying a new car'. First, consider a number of entities in the domain which are being considered (Car 1, Car 2, Car 3, Car 4 and Car 5). Now one way in which a person might compare these entities is related to a car's 'beauty' – a construct. One could think of the entities being compared as falling along a dimension extending from the positive aspect of this construct to the negative aspect of this construct as shown in Fig. 5.2. The name assigned to this construct would be something like 'attractive–unattractive'.

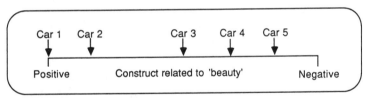

Fig. 5.2 Comparing entities along a construct of 'beauty'.

Furthermore, one could quantify this construct by letting one end of the dimensions be a low number and the other end a higher number. For example, let the positive aspect, attractive, be a 1 and the negative aspect, unattractive, be a 7. The entities (cars) could then be rated along the scale of attractive–unattractive as shown in Fig. 5.3.

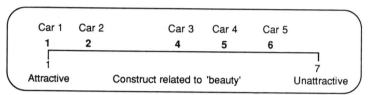

Fig. 5.3 Rating entities along a dimension of attractive–unattractive.

Now if a person continues to think of different ways to compare cars, names a scale for each comparison and rates the cars along the scales we would end up with a personal construct system (viewpoint) of buying a car as shown in Fig. 5.4. The entities are shown at the top, the positive aspect of a construct on the left and its corresponding negative aspect on the right side of the same row. If each construct is represented by a scale ranging from 1 to 7 then the numbers in each row represent the rating of the entities along the construct on that row. For example, on the row with the terminology 'Fits 5–Does Not Fit 5,' Car 2 received a rating of 1 which indicates Car 2 easily accommodates five people. The rating of 6 under the column for Car 5 on this row indicates Car 5 will not easily fit five people.

	Car 1	Car 2	Car 3	Car 4	Car 5	
Attractive	1	2	4	5	6	Unattractive
Fits 5	2	1	4	5	6	Does Not Fit 5
Low Cost	7	6	5	2	4	High Cost
4WD	4	5	1	1	3	Rear Drive
Not Dealer	6	6	3	4	5	Dealer

Fig. 5.4 Entity-attribute grid showing a viewpoint in considering a set of five cars.

Figure 5.4 is an example of an entity-attribute grid. The words along each row are the terminology used to describe a construct. Each pattern along a row gives a glimpse of how the person thinks in terms of comparing cars along the construct. An individual's entity-attribute grid (the patterns of ratings over all the constructs and the terminology used) provides a construct system or personal viewpoint of this domain.

Thus entity-attribute grids supply a method to quantify viewpoints. We then need a framework within which to compare these viewpoints. A communication classification scheme proposed by Shaw and Gaines (1989) supplies this framework.

5.4 CATEGORIZING COMMUNICATION

Shaw and Gaines proposed that viewpoints of experts in the same domain can be categorized in at least four ways – consensus, correspondence, conflict and contrast. Their explanation is slightly modified here to fit communication between product development specialists on multifunctional teams.

Figure 5.5 illustrates these four categories. First, consensus exists if two specialists use the same terminology to describe the same concept. There is no communication problem here. The specialists are thinking about the same thing and using the same terms. Second, if two specialists use different terminology to describe the same concept correspondence occurs. There is a communication problem here because they use different words to mean the same thing. Since the specialists have an underlying understanding of the same concept the problem should not be too severe if it is identified and communicated to the specialists.

Communication problems become severe when the specialists have different concepts but are using the same terminology to describe these. This

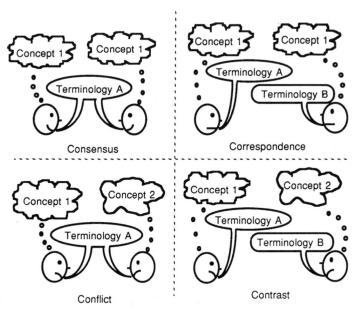

Fig. 5.5 Shaw and Gaines communication classification scheme (after Shaw and Gaines, 1989).

is referred to as conflict. Finally, a potentially severe communication problem exists when specialists have different concepts and use different terminology. This is referred to as contrast and often indicates different areas of expertise.

Developing entity-attribute grids for each specialist on a team provides the terminology used by each specialist to describe a construct. The pattern of ratings over a construct represents quantitative information about a concept. Thus entity-attribute grids can provide terminology and concept information. Using the Shaw and Gaines classification scheme we can then compare the grids in order to identify probable communication problems between specialists on multifunctional teams. The details of this comparison are explained below.

These potential communication problems are then used as the basis for team discussions which improve the understanding of each other's viewpoint. The next section details a methodology which elicits specialists' entity-attribute grids, compares these grids to identify problems and then uses team discussions to reduce communication problems on multifunctional teams.

5.5 CONCURRENT ENGINEERING METHODOLOGY FOR ENCHANCING TEAMS (CEMET)

The CEMET methodology is divided into five phases. Phase 1 is 'problem identification and evaluation'. In this phase the opportunity for the application of CEMET is first identified and the appropriateness of using CEMET is considered. Phase 2 discusses how to develop data collection materials for the problem at hand.

Phase 3 explains how to develop team members' viewpoints. This phase discusses how to use CEMET to enabl.e each team member to communicate their viewpoint of the problem. This is done by interviewing each team member to produce an individual's entity-attribute grid. This phase includes a feedback loop so that each team member can inspect and refine their viewpoint before they are compared.

Phase 4 uses the communication classification scheme to compare team members' viewpoints. As this scheme classifies viewpoints into only four categories a parsimonious basis for team discussions is provided.

Phase 5 is 'discuss comparison of viewpoints'. During team meetings the team progresses from identifying areas of existing team consensus, through discussing viewpoints which differ slightly and finally to resolving differences in viewpoints which directly conflict. The discussion will also help specialists to identify areas of other specialities about which they should learn more.

These phases are explained in more detail below. It is assumed that there is one person, referred to here as the team performance consultant (TPC) who is responsible for implementing CEMET.

5.5.1 Phase I: problem identification and evaluation

Opportunities for applying CEMET may be identified directly by the TPC working with a team or may come as requests to the TPC (e.g., from project managers, an organization's upper management, etc.). Once an opportunity is identified the TPC needs to gather information to evaluate whether CEMET should be applied. The information sources include documentation about the product under development, the project manager and/or team leader and the team itself. The TPC must meet with the team to explain what is being attempted, to get their buy-in, and to ascertain whether the first problem identified is appropriate. During these meetings the TPC may find a more appropriate opportunity for applying CEMET.

Next the TPC evaluates the possible use of CEMET using questions such as listed in Table 5.1. If all answers to these questions are 'yes', CEMET can be used. If the answers to any are 'no' then some judgment of the trade-offs is called for by the TPC.

Table 5.1   Evaluation questions for CEMET

- Are there sufficient differences between specialists on the team that the individual use of terminology is likely to be a problem?
- Are there sufficient differences between specialists on the team that important concepts understood by some members are not understood by others?
- Are all team members available for individual 1 to 2 hr interviews?
- Has the team agreed on a common problem/domain?
- Will the team agree on a set of entities within this domain quickly?
- Is there full management commitment for using CEMET?
- Can the TPC allocate a sufficient amount of time and effort?
- Does the TPC have an understanding of the method and its application?
- Does the TPC have previous experience in handling teams?
- Are there sufficient resources available in terms of software, meeting rooms (for interviews and team meetings), clerical support, etc.?
- If the TPC is not a usual member of the team, then is there a person on the team assigned as contact/coordinator for the TPC?
- Will the time and effort spent on CEMET detract from meeting deliverables?

5.5.2 Phase II: develop data collection materials

In other to collect the data to form entity-attribute grids the TPC has two basic choices:

1. Collect the data by hand, or
2. Use a software tool.

There are several different entity-attribute grid tools on the market (e.g., RepGrid[2]). Although an automated grid tool would assist the process, one is not necessary. If the TPC collects the data by hand then data sheets should clearly show the entities and leave plenty of room for descriptions of the constructs and the ratings. Figure 5.6 shows an example of a data collection sheet.

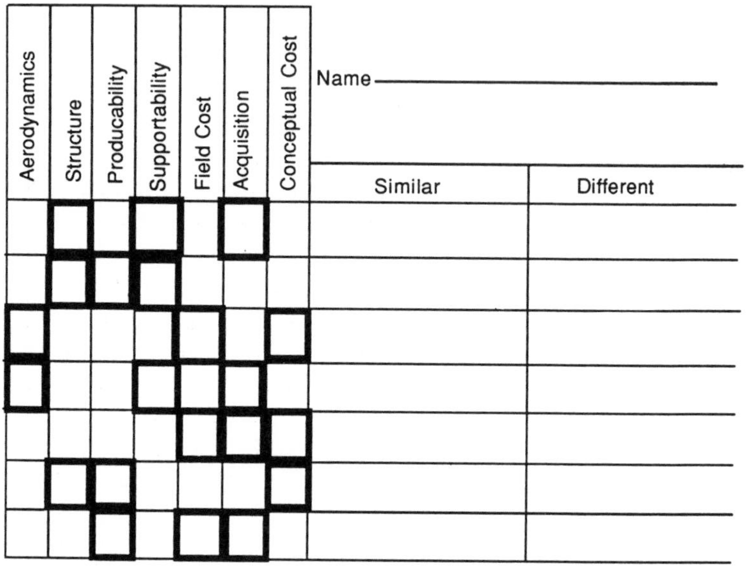

Fig. 5.6 Entity-attribute data collection sheet.

The crucial procedure in this phase is the development of a set of entities. The entities will be a short list of items drawn from a domain in which the entire team is working. The TPC may have enough information from the initial identification and evaluation phase to develop a working list of entities. These should be presented to the team in order to obtain agreement on the definition of each entity. Each person on the team will work with the same entities.

Entities must be drawn from one common sub-area of the domain. The sub-area may be high level or a specific problem. Examples of high level entities include: team roles, major inputs, major outputs, major requirements and project goals. Entities from detailed problems depend on the specific problem. Figure 5.6 shows the entities (vertically) at the top left-hand corner of the data collection sheet for criteria used in choosing a preliminary design for a turbojet engine. For more information on entities see the discussions by Fransella and Bannister (1977) on elements.

Once the entities are chosen triads must be formed. A triad is a group of three entities. In Fig. 5.6 the bold squares on each row identified a triad composed of the entities in the columns above the bold squares. For example, the first triad is 'Structure, supportability and acquisition'; the second is 'Structure, producability and support'. As many different triads of entities as the interviewer thinks appropriate are presented.

Each interviewee uses the same set of triads. The objective is to try to elicit different constructs for each triad by asking each person to describe how two of them are similar and the other different. This procedure is explained in the next phase.

5.5.3 Phase III: develop team members' viewpoints

Once the data collection materials are developed the TPC can proceed to interview team members to elicit constructs and ratings for the entities chosen. Each team member should be interviewed individually. In order to elicit the constructs the TPC asks the team member to compare the first triad of elements and describe briefly (a few words or phrase) how two of them are similar and the third is different from these two. The words used should form opposite ends of a logical scale or dimension to the individual.

Only the entities in the triad are compared at this point; however, after all triads are compared all entities will be rated along the dimensions described by the terminology for the constructs. Therefore, the data sheet needs to show which two entities in a triad were similar and which one different. This is often done by placing one number (e.g. 1) in the bold boxes for the two similar entities and a different number (e.g. 7) in the different entity box. These numbers then form endpoints of a scale described by the terminology for the construct elicited by the triad. An example will show this more clearly.

The TPC asks the person being interviewed first to consider the three entities (structure, supportability and acquisition) above the bold boxes in row one of the grid in Fig. 5.6. Figure 5.7 shows the terminology for the construct elicited directly under the 'Similar' and 'Different' heading to the right of row one. For this comparison the subject said that 'supportability and acquisition are the same in that both have high customer awareness, while structure is different from these two in that the customer is not aware of this factor (low customer awareness)'.

Note that the terms 'High customer awareness' and 'Low customer awareness' are placed to the right of this row. A '1' is placed in the boxes under the 'Supportability' and 'Acquisition' entities while a '7' is placed in the box under 'Structure'. The interview proceeds with the interviewee comparing each triad, deriving a constructs' terminology (describing how two are similar and one entity different) and placing a '1' for the similar entities and a '7' for the different one. Figure 5.7 shows the grid once all triads have been compared and constructs elicited. At this point the person doing the comparison has essentially formed a set of scales of interest to that person. The words used to describe the scales represent terminology of importance to the person in thinking about these entities.

After constructs are elicited for each triad the interviewee is asked to rate each entity on each construct. For example, for the construct 'high customer awareness – low customer awareness' Field Cost was rated a 2 meaning

Aerodynamics	Structure	Producability	Supportability	Field Cost	Acquisition	Conceptual Cost	Name	Dave Designer	
							Similar	**Different**	
	7		1		1		High customer awareness	Low customer awareness	
	7	1	1				Long term concern for system	Short term concern for system	
7				1		1	Cost items	Not cost items	
7			1		1		High customer awareness	Low customer awareness	
				1	1	7	High customer cost	Low customer cost	
	1	1				7	High customer rqmts	Low customer rqmts	
			1		1	7	Long term items	Short term items	

Fig. 5.7 All constructs elicited for one specialist. All triads have been compared.

customers are very aware of this, but not quite as aware as for supportability and acquisition. Aerodynamics, on the other hand, was rated as 6 indicating the interviewee believed customers are not very aware of this entity. Figure 5.8 shows a grid with all ratings filled in. This is the entity-attribute grid for this interviewee.

The entity-attribute grid forms the basis for constructing an individual's viewpoint. However, to clarify this viewpoint further and remove some redundancy these results would first be analyzed to group constructs which the individual is using in a similar fashion. There are a variety of analysis techniques (see Fransella and Bannister, 1977); however, the most common involve a correlational or similarity analysis. Either of these analyses compares the degree of similarity of the ratings of the entities along one construct with entity ratings along each of the other constructs. Constructs with a high degree of similarity are grouped together as a functional group.

A functional group of constructs shows all the terminology which the specialist is using in a similar fashion to compare the entities – or in terms of the classification scheme to 'express the same concept'. Table 5.2 shows a functional grouping for the individual's constructs from Fig. 5.8. The solid line separates functional groups. Following this grouping the resultant grid is discussed with the team member to see if any changes or refinements are desired. Any redundant constructs (those with the same terminology and similar patterns of ratings) are collapsed into one construct. Any changes in terminology or entity ratings can be made at this point. These 'fine-tuned' grids are then used for the comparisons in the next phase.

Aerodynamics	Structure	Producability	Supportability	Field Cost	Acquisition	Conceptual Cost	Name _____ Dave Designer _____	
							Similar	Different
6	7	3	1	2	1	3	High customer awareness	Low customer awareness
6	7	1	1	3	2	3	Long term concern for system	Short term concern for system
7	6	3	2	1	2	1	Cost items	Not cost items
7	6	3	1	1	1	2	High customer awareness	Low customer awareness
6	5	4	5	1	1	7	High customer cost	Low customer cost
5	1	1	3	2	6	7	High customer rqmts	Low customer rqmts
2	2	1	2	1	7	7	Long term items	Short term items

Fig. 5.8 Completed entity-attribute grid for a specialist. All entities have been rated using all constructs.

Table 5.2 Functional groups for constructs in Fig. 5.8

High customer awareness	Low customer awareness
Long-term concern for system	Short-term concern for system
Cost items	Not cost items
High customer awareness	Low customer awareness
High customer requirements	Low customer requirements
Long-term items	Short-term items

5.5.4 Phase IV: compare team members' viewpoints

Having developed individual viewpoints we need to compare these. Although we wish to use the Shaw and Gaines classification scheme, their specific method for achieving this is practical only for small groups. Since product development CE teams often become quite large some modification of the Shaw and Gaines method is needed for multifunctional CE teams.

We propose a method which compares the response of all team members to the same triad. The reasoning behind using each triad for this analysis is as follows. In considering the same triad each specialist has started with exactly the same stimulus. Everyone views the same triple of entities and responds in their own personal way. This response then represents a 'sample of each specialist's viewpoint' of the same stimulus. Comparisons of these samples should provide us with insight into classifying interpersonal communication.

The analysis for data on a triad proceeds in much the same fashion as for an individual; except here instead of comparing the degree of similarity between constructs within one specialist, the similarity of constructs between all specialists is considered. The analysis classifies team members' constructs into the four categories discussed above. Those constructs which have the same terminology and high similarity of ratings are in consensus. Constructs with a high similarity rating, but using different terminology, are in correspondence. When the terminology is very similar, but the ratings over the entities are not then the two constructs are in conflict. Finally, if both the terminology and the pattern of ratings are not similar than the two constructs are in contrast. Figure 5.9 shows, for one triad, the constructs of eight specialists classified in this manner. Only the terminology on the 'similar' side of the grid is included here. A detailed explanation of this figure follows the discussion of the methodology.

Similarity of ratings	**Terminology: Same**	**Terminology: Different**
High	**Consensus** Related to performance. Concerned w/ performance Related to how it works.	**Correspondence** Life cycle analysis Physical properties Material property analysis
Low	**Conflict** Related to cost More cost related	**Contrast** Material property analysis More cost related

Fig. 5.9 Terminology of eight specialists categorized by Shaw and Gaines communication classification scheme (after Shaw and Gaines, 1989).

This analysis and categorization forms the basis for conducting the team discussions in Phase 5. The TPC will most likely have a great deal of resulting information to use. Based on the time constraints the TPC is under, decisions will have to be made on which information to present during the team discussions. Some constructs falling in all categories should be discussed and constructs from every team member should be included. Constructs which are generally agreed upon by all team members (consensus) should be included as a starting point to show the team has a foundation of commonality. Also, constructs about which a large number of team members disagree (those in correspondence or conflict) should be included as these may well be a major source of the interpersonal communication problems on the team.

5.5.5 Phase V: discuss comparison of viewpoints

The first meeting to compare viewpoints should be done with the entire team. Follow-up meetings may focus on certain groups or even pairs of individuals. Each of the following steps should be repeated for as many triads as possible. If there is a severe time constraint, or a very large team, the TPC could focus on the first three to five triads. Usually a person's key constructs are most often elicited by the first five triads.

This phase of the CEMET methodology is largely subjective. CEMET uses the communication classification scheme to evaluate team member constructs in order to identify the possible communication problems. It is then up to the TPC to use these to focus team discussion. The length of time spent discussing each of the four categories must be decided by the TPC for the particular situation. The TPC should have skills in interpersonal communication and public speaking and experience in team processes, interviewing techniques, basic statistics and facilitating group discussion. The latter is particularly important in this phase. Some brief guidelines for the TPC to follow are presented here. Future work in CEMET will attempt to specify this phase in more detail, e.g. by specifying methods to resolve conflict.

Begin the discussion by verifying obvious consensus. This will be an easy task to start with and will help to get the discussion going. The TPC may even find that some 'obvious' consensual constructs are not actually in consensus. For example in Fig. 5.9 'Related to performance' could possibly be considered different terminology than 'Related to how it works' by the two specialists who generated these terms. In this case the TPC should re-classify these as in correspondence and draw out an explanation from the specialists as to why they do not believe they are the same terminology.

Next the TPC will want to move the discussion to the least severe of the communication problem areas – correspondence (different terminology, but similar ratings of entities). The TPC should first find out if even though the terminology for the constructs appear different (e.g. 'Life cycle analysis' and 'Physical properties' in Fig. 5.9) they are actually viewed as equivalent terminology by the team members involved. In this case the constructs must be re-classified as in consensus.

For those constructs which really use different terminology to express the same concept, the TPC must get the team members to give more detail on how they use the terminology in their job. Start with explanations about the specific entities used, extend this to similar entities in the domain and then perhaps to the entire product development domain. In this way team members will build up an understanding of what each other means by the terminology they use.

Next the TPC will usually want to move to resolving the constructs in conflict. The use of similar terminology with a dissimilar pattern of entity ratings indicates individuals are using the same words to describe different concepts. The use of such terminology on a team can be particularly devastating as it can lead to the team believing they all 'agree' when the key words used in those agreements mean something different to different team members. Such 'agreements' fall apart with team members making statements like: 'That's not what I agreed to at all! I thought I was agreeing to …' Resolving conflicting terminology will help to prevent such problematic 'agreement'.

As with consensus the TPC must first make sure the terminology which has been classified as similar is seen as similar by the team. If not then the constructs must be re-classified as in contrast. If the terminology is the same then the team members need to discuss what they mean by them. Ask team members to explain why they rated the entities in the way they did for that construct. Get the team members to consider other entities in the domain and then the domain in general when explaining their personal use of that terminology. The team should try to compare these explanations looking for points of agreement and the greatest points of disagreement in the use of the terminology in question.

Finally the TPC should lead the discussion of constructs in contrast. This dissimilar terminology with dissimilar entity ratings indicates different areas of expertise. Any use of CEMET should uncover contrast on multifunctional teams since different speciality areas are represented. The constructs elicited by entity-attribute methods usually reflect the most important terminology and concepts the individual applies to the domain.

Thus contrast may indicate some of the critical terminology and concepts which a specialist will use on a team. While it is impossible for every specialist to know as much as every other specialist, the constructs identified as in contrast can provide some important insight into other speciality areas. Team members should make an effort to explain or even make presentations (where appropriate) on constructs they use which are in contrast to most of the rest of the team. Team members should also make an effort to understand the constructs identified as in contrast – especially the constructs they do not understand at all.

The TPC may not be able to get through all four categories in one meeting. In any case it is unlikely, if CEMET is applied correctly, that one team discussion of the comparison of viewpoints will suffice. The number of team discussions will depend on various constraints, the demeanor of the team and how effective the discussions are. The TPC should quit team meetings when it is judged that further comparisons would yield little useful team interaction.

5.6 A SIMPLIFIED EXAMPLE USING THIS METHOD

As an example of the application of CEMET, consider an opportunity which has been identified (Phase I) because a team is having a problem making a choice of a preliminary turbojet engine design. Discussions with the project leader and team indicate that the different viewpoints of team members are an obstacle to coming to an agreement. Evaluating the use of CEMET (i.e., applying the questions in Table 5.1 plus general knowledge and experience) leads the TPC to conclude that CEMET could be applied to this team.

Moving into Phase II the TPC meets with the team to decide on the entities. It is found that although over seventy criteria are being used to make a decision, these can be grouped under seven categories. The team agrees to use these seven as the entities since team members agree on a basic definition of each, the entities are important to all team members and the number of entities can be handled in a reasonably short interview. The TPC chooses triads and prepares the data collection sheet shown in Fig. 5.6. The terms shown at the top of the columns in Fig. 5.6 are the set of entities.

Moving into Phase III the TPC interviews each team member. Each interview produces an entity-attribute grid such as that shown in Fig. 5.8. Each grid would then be analyzed to produce functional groups of constructs which the individual is using in a similar fashion (i.e., there is a high similarity of ratings over the entities). For the team members' constructs shown in Fig. 5.8 there are two functional groups as shown in Table 5.2. The solid line separates functional groups.

Each specialist's functional group analysis is then discussed with the specialist. Based on this analysis each specialist is asked for feedback on his grid (e.g. do the groupings make sense to the individual? why are some constructs with similar wording not grouped togther?). The specialist is then asked to clarify his constructs and modify his ratings as so desired. This fine-tuned grid is then used for the comparison in Phase IV.

In Phase IV the specialist's viewpoints are compared based on the terminology used and entity ratings elicited on each triad. The constructs for all specialists are divided into functional groups for each triad. As an example consider Table 5.3.

This shows the constructs which have been elicited from eight specialists using the triad with entities 'Structure, supportability and acquisition cost'

Table 5.3 Constructs of eight specialists on one triad

Specialist	Construct	
1	Related to performance	Related to administration
2	Concerned w/performance	Not concerned w/performance
3	Related to how it works	Not related to how it works
4	Life cycle analysis	Not life cycle analysis
5	Material property analysis	No material property analysis
6	Physical properties	Cost properties
7	Related to cost	Not strong relation to cost
8	More cost related	Less cost related

(the first triad in Fig. 5.8). Using the Shaw and Gaines classification scheme the TPC then proceeds to categorize the relationships between these constructs as follows. Those constructs with basically the same terminology in a functional group are classified as in consensus. Those constructs with different terminology but in the same functional group are in correspondence. Constructs with the same terminology but in different functional groups are classified as in conflict. Finally constructs with different terminology and in different functional groups are in contrast. Figure 5.9 shows examples of the classification scheme used with the constructs from Table 5.3. Only the positive aspect of each construct is shown in Fig. 5.9.

Once the constructs in each triad are classified the TPC convenes a team meeting to discuss the comparison of viewpoints (Phase V). First, the TPC discusses the constructs in consensus such as 'Related to performance,' 'Concerned with performance' and 'Related to how it works' (see Fig. 5.9). The three specialists producing these constructs are asked whether they do see this terminology as the same. If not they then discuss why. If they are the same then the TPC draws the rest of the team into a discussion to discover whether they agree or disagree.

Next, the constructs in correspondence such as 'Involves life cycle analysis', 'Physical properties' and 'Material property analysis' are discussed (see Fig. 5.9). The three specialists who produced these constructs appear to be using different terminology to describe the same concept. The TPC first asks these specialists if they agree that this terminology is different or has the TPC made an incorrect inference? If they are really the same does the rest of the team understand this? If, on the other hand, they are using different terminology to describe the same point of view can they explain in more detail what each means by the terminology they used? After getting explanations from the individuals the TPC involves the entire team in the discussion.

Next, constructs in conflict such as 'Related to cost' and 'More cost related' are discussed. First, the specialists producing the constructs discuss how they use these terms when they think about or apply them to the entities in question. They may want to explain how they use the terminology with other entities in the domain. They are asked to try to recall what they were thinking about when they described this construct in their interview. The team is then drawn into a discussion on how they use this terminology.

Finally, constructs in contrast such as 'Material property analysis' and 'More cost related' are considered. Recall that this may indicate different areas of expertise between the two specialists. A discussion here centers around each specialist attempting to explain to the other specialists why these terms are used in this way in this speciality area. If time does not allow or the explanation is a bit complex the specialist may want to prepare a brief presentation for a later meeting. In any event the team members should note the constructs in contrast and make an effort to learn about these after the meeting.

At the end of the meeting the TPC reviews what was learned about team members' viewpoints and the terminology used to communicate these viewpoints. Constructs in correspondence or conflict which were not resolved should be noted for future meetings or discussions between a few team members if appropriate. Constructs in contrast should be listed and all team members charged with working on understanding these.

5.7 SUMMARY

The example just discussed is a simplified case invented to convey the basic approach in a tutorial fashion. It is based on an application with twenty-two specialists on a turbojet engine exhaust nozzle development team. The application was not reported here since, due to an abrupt cancellation of funds, the team was disbanded before Phase 4 of CEMET could be completed. It can be reported, however, that the project manager and a company facilitator on the team were very pleased with how CEMET was proceeding and thought it well worth the time and effort. Also, in unsolicited comments some team members stated that simply seeing the entity-attribute grid of another team member improved their understanding of that member.

One crucial area in which CEMET could be applied is in helping each team member better to understand how other members view the goals of the team. Actually such an application will also likely help each team member to better grasp their own understanding of the goals. For this case the goals form the entities and the rest of the methodology is then followed.

Most applications would be more complex than the example shown above – involving teams much larger than eight, more complex terminology, more complex problems and/or repeated meetings and interviews to establish the entity-attribute grids. As the method is applied to increasingly detailed levels of development more constructs will be elicited from each specialist similar to knowledge acquisition from experts. Those applying CEMET will have to make trade-offs between the number of specialists, the detail desired, time, the potential for generating more data than can be reasonably handled (and communicated back to the specialists), etc. Such decisions, however, must always be made in any attempt to improve team communication.

The CEMET methodology proposed here can help each specialist to get a better understanding of the viewpoint of other specialists on a multifunctional team. Furthermore, use of CEMET can help to identify and work through the source of much dissension of these teams – viewpoints which are not communicated and terminology that is used in different ways by different team members.

5.8 ACKNOWLEDGMENT

This work has been sponsored by the Defense Advanced Research Agency (DARPA), under Contract No. MDA-972-91-J-1022 for the DARPA Initiative in Concurrent Engineering (DICE).

5.9 NOTES

[1] The term 'attribute' has been introduced in the personal construct area as fairly synonymous with the term 'construct'. We have chosen to stay with the term construct so as not to confuse the reader with an unnecessary additional term here.

[2] RepGrid is a copyrighted program commercially available from the Centre for Person Computer Studies, University of Calgary, Calgary, Alberta.

5.10 REFERENCES

Boose, J.H. (1984) *Proceedings of AAAI-1984*, American Association of Artificial Intelligence, Menlo Park, pp. 27–33.

Diedrich, J., Ruhman, I., and May, M. (1987) *International Journal of Man-Machine Studies*. **26**, 29–40.

Fransella, F. and Bannister, D. (1977) *A Manual for Repertory Grid Technique*, Academic Press, London.

Frost, W.A. and Brain, R.L. (1967) *Commentary*, **9**, 161–75.

Hart, A. (1986) *Knowledge Acquisition for Expert Systems*, McGraw Hill, New York.

Kelly, George A. (1955) *The Psychology of Personal Constructs*, Norton, New York.

Kilcourse, T. (1984) *Journal of European Industrial Training*, **8**(2), 3–38.

Shaw, M.L.G. and Gaines, B.R. (1989) *Knowledge Acquisition*, **1**, 341–63.

CHAPTER 6

Scheduling of concurrent manufacturing projects

Adedeji B. Badiru

6.1 INTRODUCTION

Manufacturing project management is the process of managing, allocating, and timing resources to achieve a production goal in an efficient and expedient manner. The objectives that constitute the specified goal may be in terms of time, costs, or performance. A project can be simple, such as cooking dinner, or very complex, such as launching a space shuttle. Project management techniques are used widely in many enterprises including construction, banking, manufacturing, marketing, health care services, and public services (Badiru, 1988a; Gill and Whitman, 1991). Project management uses a combination of analytical, managerial, and computer tools to address the following:

- Performance specifications;
- Schedule requirements; and
- Cost limitations.

This chapter presents the application of precedence diagramming to the scheduling of concurrent manufacturing projects. The major issues involved in scheduling concurrent manufacturing projects are addressed. The topics covered include resource allocation, heuristic scheduling, priority assignment, and task coordination.

6.2 CONCURRENT PROJECTS

Scheduling is the channel through which project goals are accomplished. A schedule evolves from a time-based allocation of resources to projects. The basis for scheduling a project is the analysis of the network of tasks making up the project. Basic project network analysis is typically implemented in three phases: planning, scheduling, and control. In concurrent projects, one of the main goals is to compress the project schedule while maintaining certain performance criteria. Schedule compression can be accomplished in a number of ways including reducing task times, eliminating tasks, and relaxing precedence constraints. Even though precedence relaxation has many advantages, it is not adequately exploited in practice. The precedence diagramming method (PDM) is one technique that offers an effective mechanism for relaxing precedence constraints. But the method is in limited use because of the complexity of implementing it for large projects (Crandall, 1973; Wiest, 1981). A computer-based implementation tool of PDM can facilitate its use in practical project scenarios. For example, the use of expert systems (Badiru, 1992) for implementing PDM should be investigated.

The concept of concurrent engineering (CE), also referred to as simultaneous engineering, requires that product and process designs be developed concurrently. This has the objective of facilitating shorter development cycles, higher productivity, more flexibility, better resource utilization, and higher product quality. There has been a heightened interest in concurrent engineering and manufacturing in recent years. The large quantity of recent literature on the subject attests to the growing interest (Apt, 1985; Evans, 1988; Lu and Subramanyam, 1988; Winner *et al.*, 1988; Foreman, 1989; Gopalakrishnan, 1989; Hollingum, 1989; Lu *et al.*, 1989; Madsen, 1989; Nevins and Whitney, 1989; St Charles, 1989; Eitzinger, 1990; Ben-Arieh and Miron, 1991; Cleland, 1991; Gunasingh and Lashkari, 1991; Shina, 1991).

In a case study presented by Eitzinger (1990), concurrent engineering was credited with up to 70% compression in development cycles in a manufacturing environment. Concurrent engineering differs from the traditional sequential product development by involving product and machine design, manufacturing, maintenance, purchasing, accounting, sales, distribution, and even outside suppliers at the start of the product development cycle. Concurrent engineering permits succeeding portions of a manufacturing process to be developed while the product is being designed rather than afterwards. A concurrent engineering project permits a global view of all the relevant components of an engineering process. A project management approach facilitates the coordination of technical groups within an organization to improve the interface between design, engineering, and manufacturing. Lu *et al.* (1989) provide details of numerous research efforts still needed fully to realize the potentials of concurrent engineering. Research on the development of an integrated approach to concurrent scheduling is expected to have significant potential for boosting engineering and manufacturing productivity.

6.3 SCHEDULING FOR CONCURRENT MANUFACTURING

Concurrent scheduling is accomplished through the conventional project scheduling techniques. However, the specific procedures for implementing the techniques are unique due to the fact that concurrent projects have special characteristics. The classical techniques of critical path method (CPM) and program evaluation and review technique (PERT) have been successfully adopted for different types of projects.

6.3.1 Critical path method

The primary goal of a CPM analysis of a project is the determination of the 'critical path' which is the sequence of tasks that determines the minimum completion time for a project. CPM analysis involves forward pass and backward pass procedures. The forward pass determines the earliest start time and the earliest completion time for each task. The backward pass determines the latest starting time and the latest completion time for each task. During the forward pass, it is assumed that each task will begin at the earliest possible time. A task can begin as soon as the last of its predecessors is finished.

6.3.2 Program evaluation and review technique

PERT differs from CPM in that it takes into account the variabilities in task times. PERT uses three time estimates of optimistic time, most likely time, and pessimistic time. The procedure calculates the expected value and variance of task time as a weighted average of the three estimates. With the task time variance and the use of the Central Limit Theorem, PERT can compute the probability of completing a project within a specific time. CPM and PERT have been documented extensively in the literature (Badiru, 1988b; Moder *et al.*, 1983; Wiest and Levy, 1977).

6.3.3 Precedence diagramming method (PDM)

Many extensions of PERT and CPM have been developed over the years to satisfy specific or unusual project scenarios. One such extension is the precedence diagramming method (Crandall, 1973; Moder *et al.*, 1983; Wiest, 1981). PDM offers some advantages over CPM and PERT. The advantages have been documented in the literature (Fondahl, 1962; Wiest and Levy 1977). PDM has some peculiar characteristics that are hardly recognized and inadequately documented (Harhalakis, 1990). These peculiar characteristics have not received enough attention needed to exploit the potential advantages. Only few significantly documented studies of PDM have been published over the past several years (Crandall, 1973; Wiest, 1981; Harhalakis, 1990). The most significant characteristic of PDM is the use of lead-lag factors which introduce additional and flexible precedence relationships in project networks. In CPM and PERT, there is only one type of precedence relationship: Finish-to-Start relationship. In PDM, this relationship is extended to additional relationships such as Finish-to-Finish,

Start-to-Start, Start-to-Finish, and so on. Figure 6.1 shows the graphical representation of the additional precedence relationships between two tasks, A and B. The relationships are summarized below:

SS (Start-to-Start):   This specifies that task B cannot start until task A has been in progress for at least SS time units.

FF (Finish-to-Finish): This specifies that task B cannot finish until at least FF time units after the completion of task A.

FS (Finish-to-Start):   This specifies that task B cannot start until at least FS time units after the completion of task A.

SF (Start-to-Finish):   This specifies that there must be at least SF time units between the start of task A and the completion of task B.

The lead or lag relationships may, alternately, be expressed in percentages rather than time units. For example, we may specify that 25% of the work

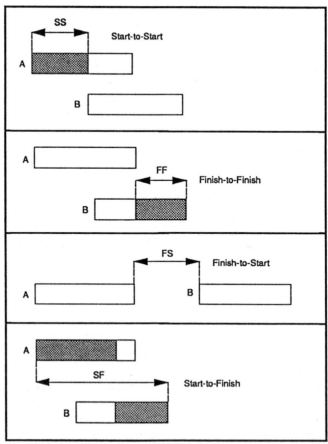

Fig. 6.1 Precedence relationships in PDM.

content of task A must be completed before task B can start. The lead-lag factors may also be expressed in terms of 'at most' instead of 'at least' constraints. With the additional relaxed precedence and increased flexibility, it may be possible to perform mutually dependent tasks partially in parallel instead of serially. Tasks that are permitted to overlap create opportunities for schedule compression (Hollingum, 1989). If PDM is used properly, significant gains can be obtained in terms of project duration, cost, and performance.

6.4 PRECEDENCE RELAXATION

There are many issues involved in implementing concurrent scheduling. This section focuses on task precedence relaxation and schedule development aspects of concurrent scheduling. Task precedence constraints can be classified into three types:

- Technical precedence constraint;
- Resource-imposed precedence constraint; and
- Procedural precedence constraint.

Technical constraints are those required because of the technical nature of the relationships between tasks. Technical constraints are difficult to overcome without having to redesign the processes involved. Resource-imposed constraints, such as manpower shortage, can be overcome by allocating additional resources. Procedural constraints are the easiest to overcome. This is where the greatest amount of gain can be obtained. Most of the task precedence relationships in engineering and manufacturing systems are of a procedural nature. Such inefficient procedures evolved from the 'good old times' when there was little pressure to maximize system performance. Precedence relaxation is governed by scheduling heuristics based on the type of precedence constraint, the experience of the project analyst, and the prevailing performance requirements. The relaxation attempt will require a compromise between time, resource availability, and performance specification as represented in Fig. 6.2. Such compromise analysis is difficult to model and solve mathematically.

6.4.1 PDM scheduling example

The procedure of PDM can be applied to multiple projects in the same way that it is applied to a collection of tasks in a single project. Figure 6.3 presents a simple multiple project network example adapted from Badiru (1991c). The example consists of three projects. The projects are to be performed serially and each has an expected duration of ten months. The expected durations are computed from three time estimates of *a*, *m*, and *b* used in PERT. The conventional forward and backward calculations

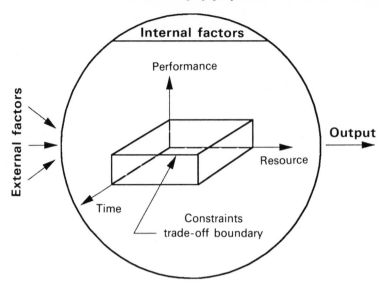

Fig. 6.2 Time–cost–performance compromise boundary.

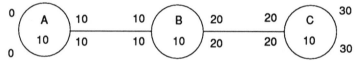

Fig. 6.3 Network of three serial projects.

indicate that the duration of the network is thirty months. The earliest times and the latest times are as shown in the figure. The Gantt chart for the network is shown in Fig. 6.4.

For a comparison, Fig. 6.5 shows the same network but with some lead-lag constraints. For example, there is an SS constraint of two months

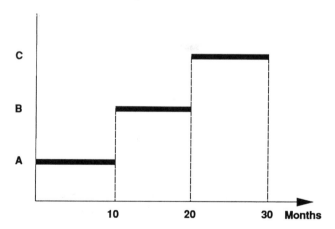

Fig. 6.4 Gantt chart of serial projects.

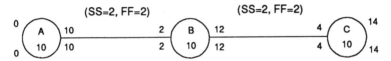

Fig. 6.5 Example of PDM network.

and an FF constraint of two months between projects A and B. Thus, project B can start as early as two months after project A starts but it cannot finish until two months after the completion of A. In order words, *at least* two months must be between the starting times of A and B. Likewise, *at least* two months must separate the finishing time of A and the finishing time of B. A similar precedence relationship exists between project B and project C. The earliest and latest times obtained by considering the lag constraints are indicated in the figure.

The calculations show that by being able to start B just two months after starting A, it can be completed as early as twelve months as opposed to the twenty months obtained in the case of conventional PERT/CPM. Similarly, project C is completed at the end of fourteen months, which is considerably less than the thirty months calculated by conventional PERT/CPM. The lead-lag constraints allow us to compress the overall schedule by overlapping projects. Depending on the nature of the projects involved, a project does not have to wait until its predecessor finishes before it can start. Figure 6.6 shows the Gantt chart for the example incorporating the lead-lag constraints. It is seen that a portion of a succeeding project can be performed simultaneously with a portion of the preceding project. A portion of a project that overlaps with a portion of another project may be viewed as a distinct portion of the required work. Thus, partial completion of a task or project may be evaluated.

Figure 6.7 shows how each of the three projects is partitioned into contiguous phases. Even though there is no physical break or termination of

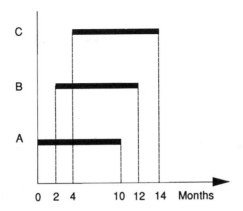

Fig. 6.6 Gantt chart for compressed PDM network.

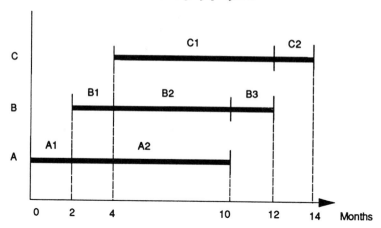

Fig. 6.7   Partitioning of projects into phases.

work in each project, the distinct phases (beginning and ending) can still be identified. The distinct phases are determined on the basis of the amount of work that must be completed before or after another project as dictated by the lead-lag relationships.

In Fig. 6.7, project A is partitioned into the phases A_1 and A_2. The duration of A_1 is two months because there is an SS$=2$ relationship between project A and project B. Since the original duration of A is ten months, the duration of A_2 is calculated to be $10-2=8$ months. Likewise, project B is partitioned into the phases B_1, B_2, and B_3. The duration of B_1 is two months because there is an SS$=2$ relationship between project B and project C. The duration of B_3 is also two months because there is an FF$=2$ relationship between projects A and B. Since the original duration of B is ten months, the duration of B_2 is calculated to be $10-(2+2)=6$ months. In a similar fashion, project C is partitioned into C_1 and C_2. The duration of C_2 is two months because there is an FF$=2$ relationship between project B and project C. Since the original duration of C is ten months, the duration of C_1 is then calculated to be $10-2=8$ months. Figure 6.8 shows a conventional project network drawn for the three projects after they are partitioned into distinct phases. The conventional forward and backward passes reveal that all the project phases are on the critical path. This makes sense since the original three projects are performed serially and no physical splitting of projects has been performed. There are three critical paths in Fig. 6.8, each with a length of fourteen months. It should be noted that the distinct phases of each project are performed contiguously.

Figure 6.9 shows an alternate example of three serial projects. The conventional PERT/CPM analysis shows that the duration of the network is thirty months. When lead-lag constraints are introduced into the network as shown in Fig. 6.10, the network duration is compressed to eighteen months. In the forward pass computations in Fig. 6.10, the earliest completion time of B is month 11 because there is an FF$=1$ restriction between

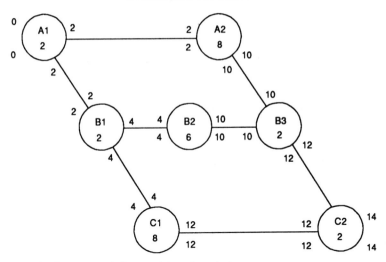

Fig. 6.8 Project network for partitioned projects.

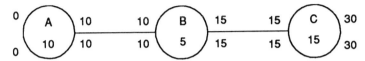

Fig. 6.9 Project network for Example 2.

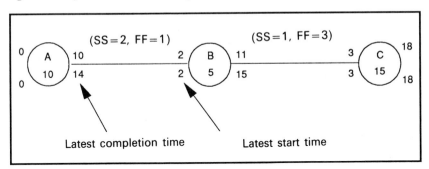

Fig. 6.10 PDM network for Example 2.

project A and project B. Since A finishes at month 10, B cannot finish until at least month 11. Even though the earliest starting time of B is month 2 and its duration is five months, its earliest completion time cannot be earlier than month 11. It is noted that C can start as early as month 3 because there is an SS = 1 relationship between B and C. Thus, given a duration of fifteen months for C, the earliest completion time of the network is $3 + 15 = 18$ months. The difference between the earliest completion time of C and the earliest completion time of B is $18 - 11 = 7$ months, which satisfies the FF = 3 relationship between B and C.

In the backward pass, the latest completion time of B is 15 (i.e. $18 - 3 = 15$) since there is an FF = 3 relationship between project B and project C. The

Scheduling of projects

latest start time for B is month 2 (i.e. $3-1=2$) since there is an $SS=1$ relationship between project B and project C. If we are not vigilant, we might erroneously set the latest completion time of B to 10 (i.e. $15-5=10$). But that would violate the $SS=1$ restriction between B and C. By a careful analysis, the latest completion time of A is found to be 14 (i.e. $15-1=14$) since there is an $FF=1$ relationship between A and B. All the earliest times and latest times at each node must be explicitly evaluated to ensure that they conform to all the lead-lag constraints. This is why a general mathematical modeling of PDM can be cumbersome (Crandall, 1973; Wiest, 1981). Manual evaluations of the lead-lag precedence network analysis can become very tedious and intractable particularly for large networks. Scheduling heuristic and computer implementation offer a convenient alternative for implementing PDM.

Using a procedure similar to the one explained for the previous example, the expanded project network in Fig. 6.11 was developed based on the precedence network in Fig. 6.10. It is seen that project A is partitioned into two phases, project B is partitioned into three phases, and project C is partitioned into two phases. Conventional CPM calculations show that only the first phases of projects A and B are on the critical path. Both phases of project C are on the critical path.

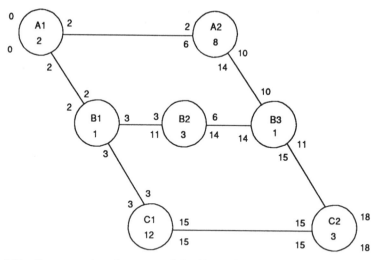

Fig. 6.11 Compressed project network for Example 2.

Figure 6.12 shows the corresponding Gantt chart for the expanded network. The Gantt chart is drawn on the basis of earliest start times. Examining the earliest start times, it is seen that project B is physically split at the boundary of B_2 and B_3 such that B_3 is separated from B_2 by four months. This implies that work on project B is temporarily stopped at time 6 after B_2 is finished and is not started again until time 10. Despite the four-month delay in starting B_3, the entire project is not delayed. This is

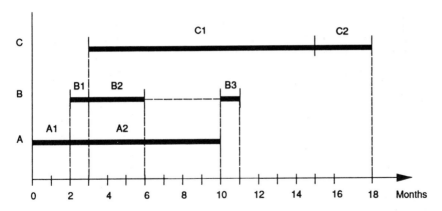

Fig. 6.12 Gantt chart based on earliest start times.

because B_3, the last phase of project B, is not on the critical path. In fact, B_3 has a total slack of four months. In a situation like this, the duration of project B can actually be increased from five months to nine months without any adverse effect on the overall project duration. This anomalous characteristic of PDM was discussed in detail by Wiest (1981). It should be recognized, however, that increasing the duration of a project may have negative implications on project cost and personnel productivity.

If physical splitting of projects is not permitted, then the best option available in Fig. 6.12 is to stretch the duration of B_2 so as to fill up the gap from time 6 to time 10. An alternative is to delay the starting time of B_1 until time 4 so as to use up the four-day slack right at the beginning of project B. Unfortunately, delaying the starting time of B_1 by four months will delay the overall project by four months since B_1 is on the critical path as shown in Fig. 6.11. Consequently, we need to evaluate appropriate trade-offs between splitting of projects, delaying projects, increasing project durations, and incurring higher project costs. If the Gantt chart for the network in Fig. 6.11 is drawn on the basis of latest start times, the result will be the schedule shown in Fig. 6.13.

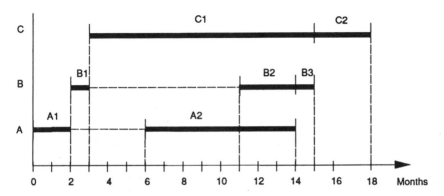

Fig. 6.13 Gantt chart based on latest start times.

Figure 6.13 indicates that it will be necessary to split both project A and project B even though the total project duration remains the same at eighteen months. If project splitting is to be avoided, then we can increase the duration of project A from ten to fourteen months and the duration of B from five to thirteen months without adversely affecting the entire project duration. The important benefit of precedence diagramming is that the ability to overlap projects facilitates some flexibilities in manipulating individual project times and compressing the project duration.

6.4.2 Resource-constrained scheduling

The preceding PDM example did not consider the limitations on resource availability in the scheduling process. Resource limitations can be incorporated into the compressed project network. For example, resource requirements can be specified for each of the seven nodes in Fig. 6.11. In this case, the compressed network nodes will be scheduled using existing resource-constrained project scheduling techniques. The topic of schedule compression has been extensively studied in the literature. There have been several mathematical techniques for resource-constrained project scheduling. Unfortunately, the mathematical formulations are not generally practical due to the complexity involved in implementing them for large projects. Many mathematical techniques have been studied and reported in the literature (Pritsker *et al.*, 1969; Elsayed, 1982).

A more practical approach to resource-constrained scheduling is the use of heuristics. If the circumstances of a problem satisfy the underlying assumptions, a good heuristic will yield schedules that are feasible enough for practical purposes. A major factor in heuristic scheduling is to select a heuristic whose assumptions fit the prevailing problem. Several resource-constrained scheduling heuristics exist for a wide variety of special cases. Thesen (1976), Elsayed (1982), Wiest (1967), and Badiru (1988b) present practical resource-constrained scheduling heuristics. The procedure for using heuristics to schedule projects involves the prioritizing of projects in the assignment of resources and time slots in the project schedule. As an example, the composite allocation factor (CAF) scheduling heuristic presented by Badiru (1988b) is briefly discussed in the following section.

6.5 COMPOSITE ALLOCATION FACTOR (CAF)

CAF is a resource allocation heuristic that takes into account both the resource requirements and the variabilities in task times. Tasks with higher values of CAF are given priority during the resource allocation process. For each task i, CAF is computed as a weighted and scaled sum of two priority measures: resource allocation factor (RAF) and stochastic activity duration factor (SAF). The heuristic has been successfully implemented in a scheduling program named STARC (stochastic time and resource constraints). Full details of CAF and STARC have been provided in the literature (Badiru 1991a,b). The computations for CAF are performed as presented below:

$$CAF_i = (w)RAF_i + (1 - w)SAF_i$$

where w is a weighting factor between 0 and 1. RAF is defined for each task i as:

$$RAF_i = \frac{1}{t_i} \sum_{j=1}^{R} \frac{x_{ij}}{y_j}$$

where x_{ij} = number of resource type j units required by task i, y_j = maximum units of resource type j required by any task in the project, t_i = the expected duration of task i, and R = the number of resource types involved. RAF is a measure of the expected resource consumption per unit time. A scaling procedure is used so that the discrepancies between the units of resource types and the units of time are eliminated. The set of RAF values is scaled from 0 to 100. Resource-intensive activities have larger magnitudes of RAF and, as such, require a greater attention in the scheduling process. To incorporate the uncertainty in task times, SAF is defined for each task i as:

$$SAF_i = t_i + \frac{s_i}{t_i}$$

where t_i = expected duration for task i, s_i = standard deviation of duration for task i, and s_i/t_i = coefficient of variation of the duration of task i. The SAF values are also scaled to eliminate the discrepancy in the units of the terms in the equation. It is on the basis of the magnitudes of CAF that a task is assigned a priority for resource allocation in the project schedule. A task that lasts longer, consumes more resources, and varies more in duration will have a larger magnitude of CAF. Such a task is given priority for resource allocation when activities compete for resources. The weighting factor, w, is used to vary the relative weights assigned to RAF and SAF. Consequently, the scheduling program, STARC, gives a project analyst the option of assigning more weight to the resource requirement aspects of a project and less to the probabilistic time aspects and vice versa. A simulation experiment may be conducted to find out the best value of w for a given project (Badiru, 1991c).

6.5.1 Computer simulation example

Suppose the compressed project network example in Fig. 6.11 is extended to consider resource requirements of each node in the network. Table 6.1 presents an example of resource-based project data. In order to achieve the expected durations used earlier in Fig. 6.11, the PERT time estimates (a, m, and b) are chosen as follows:

$m = t_e$ (expected duration as specified in Fig. 6.11)
$a = m - d$
$b = m + d$

In this example, d is set equal to 1 for convenience.

Table 6.1 Resource-based compressed project network data

Activity No.	*Name*	*Predecessors*	*PERT time estimates (a, m, b)*	*Units of resource type 1 required*	*Units of resource type 2 required*
1	A1	–	1, 2, 3	3	0
2	A2	A1	7, 8, 9	5	4
3	B1	A1	0, 1, 2	4	1
4	B2	B1	2, 3, 4	2	0
5	B3	A2, B2	0, 1, 2	4	3
6	C1	B1	11, 12, 13	2	7
7	C2	B2, C1	2, 3, 4	6	2

Units of resource type 1 available = 8
Units of resource type 2 available = 10

STARC computer program was used to simulate the schedule for the data in Table 6.1. The results are presented in the figures that follow. Figure 6.14 shows the unconstrained project schedule with a duration of eighteen

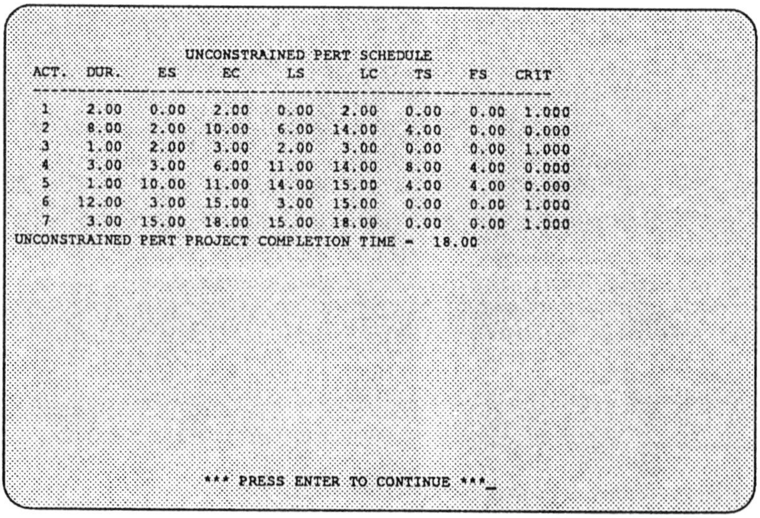

UNCONSTRAINED PERT SCHEDULE

ACT.	DUR.	ES	EC	LS	LC	TS	FS	CRIT
1	2.00	0.00	2.00	0.00	2.00	0.00	0.00	1.000
2	8.00	2.00	10.00	6.00	14.00	4.00	0.00	0.000
3	1.00	2.00	3.00	2.00	3.00	0.00	0.00	1.000
4	3.00	3.00	6.00	11.00	14.00	8.00	4.00	0.000
5	1.00	10.00	11.00	14.00	15.00	4.00	4.00	0.000
6	12.00	3.00	15.00	3.00	15.00	0.00	0.00	1.000
7	3.00	15.00	18.00	15.00	18.00	0.00	0.00	1.000

UNCONSTRAINED PERT PROJECT COMPLETION TIME - 18.00

*** PRESS ENTER TO CONTINUE ***_

Fig. 6.14 Unconstrained project duration.

months, which, as expected agrees with the duration in Fig. 6.11. Figure 6.15 shows that the shortest simulated resource-constrained schedule has a duration of 24.48 months. The Gantt chart associated with this shortest duration is shown in Fig. 6.16. The above example shows that the combination of precedence relaxation and heuristic scheduling can facilitate better management of concurrent projects. The partitioning of projects into phases can be done to any desired level of detail to suit specific management needs. Specific allocation of resources to individual phases facilitates better project planning and control.

```
                    SHORTEST SIMULATED SCHEDULED
ACT.   DUR.    ES       EC      LS       LC      TS      FS     CRIT
------------------------------------------------------------------------
 1     2.40   0.00     2.40    7.31     9.71    7.31    0.00   0.000
 2     7.31   2.40     9.71   14.07    21.37   11.67    3.99   0.000
 3     0.35   9.71    10.06    9.71    10.06   -0.00    0.00   1.000
 4     3.64  10.06    13.70   17.73    21.37    7.67    0.00   0.000
 5     0.93  13.70    14.63   21.37    22.30    7.67    7.67   0.000
 6    12.24  10.06    22.30   10.06    22.30    0.00    0.00   1.000
 7     2.18  22.30    24.48   22.30    24.48    0.00    0.00   1.000
SHORTEST SIMULATED PROJECT DURATION - 24.48



                    *** PRESS ENTER TO CONTINUE ***_
```

Fig. 6.15 Shortest simulated project duration.

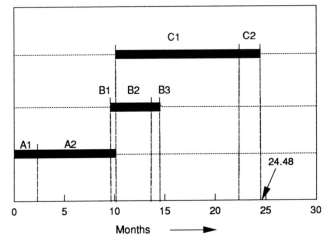

Fig. 6.16 Gantt chart for shortest simulated project schedule.

6.6 CONCLUSION

This chapter has presented an approach to using the existing technique of precedence diagramming to achieve precedence relaxation in concurrent manufacturing projects. A computational example of the precedence diagramming method (PDM) is presented. A scheduling heuristic based on resource requirements and task duration variabilities is presented. The heuristic, named CAF (composite allocation factor), is recommended for application to the compressed project network generated by the PDM procedure. A simulation analysis of a sample project network is presented.

6.7 REFERENCES

Apt, K.R., (ed.) (1985) *Logics and Models of Concurrent Systems*, Springer-Verlag, NY.

Badiru, A.B. (1988a) *Project Management in Manufacturing and High Technology Operations*, John Wiley & Sons, New York.

Badiru, A.B. (1988b) *IEEE Transactions on Engineeering Management*, **35**(2) 82–9.

Badiru, A.B. (1992) *Expert Systems Applications in Engineering and Manufacturing*, Prentice-Hall, New Jersey.

Badiru, A.B. (1991a) *Computers and Industrial Engineering*, **20**(3) 389–400.

Badiru, A.B. (1991b) *Simulation*, **57**(4) October, 245–55.

Badiru, A.B. (1991c) *Project Management Tools for Engineering and Management Professionals*, Industrial Engineering & Management Press, Norcross, GA.

Ben-Arieh, D. and Miron, I. (1991) *Computers and Industrial Engineering*, **20**(1) 45–58.

Cleland, D. (1991) *PM Network*, May 5–6.

Crandall, K.C. (1973) *Project Management Quarterly*, **4**,(3) September, 49–58.

Eitzinger, R. (1990) *Electrical Manufacturing*, **4**,(4) July 16–19.

Elsayed, E.A. (1982) *International Journal of Production Research*, **20**(1), 95–103.

Evans, W. (1988) *Mechanical Engineering*, **110**(2) February, 38–9.

Fondahl, J.W. (1962) 'A non-computer approach to the critical path method for the construction industry', Technical Report No. 9, Stanford University.

Foreman, J.W. (1989) *Proceedings of AUTOFACT '89*, Detroit, Michigan, October 30–November 2, Society of Manufacturing Engineers, 715–20.

Gill, S.B. and Whitman, N.M. (1991) *Project Management Journal*, **22**(3) 33–8.

Gopalakrishnan, B. (1989) *Proceedings of 1989 IIE Fall Conference*, Atlanta, GA, November 13–15, 78–84.

Gunasingh, K. Raja and Lashkari, R.S. (1991) *Computers and Industrial Engineering*, **20**(1) 111–17.

Harhalakis, G. (1990) *European Journal of Operational Research*, **49**, 50–9.

Hollingum, J. (1989) *Assembly Automation*, **9**(3) August, 128–31.

Lu, S., Subramanyam, S., Thompson, J.B. and Klein, M. (1989) *Proceedings of the 1989 ASME International Computers in Engineering Conference and Exposition*, Anheim, CA, July 30–August 3, 9–18.

Lu, S. and Subramanyam, S. (1988) *Proceedings of the 1988 ASME Conference on Advances in Manufacturing System Engineering*, Chicago, IL, November 27–December 2, 35–46.

Madsen, C. (1989) *Proceedings of AUTOFACT '89*, Detroit, Michigan, October 30–November 2, Society of Manufacturing Engineers, 742–7.

Moder, J.J., Phillips, C.R. and Davis, E.W. (1983) *Project Management with CPM, PERT and Precedence Diagramming*, (3rd edition) Van Nostrand Reinhold, New York.

Nevins, J.A. and Whitney, D.E. (eds) (1989) *Concurrent Design of Products and Processes*, McGraw-Hill.

Pritsker, A.B., Walters, L.J. and Wolfe, P.M. (1969) *Management Science*, **16**(1) September 93–108.

Shina, S.G. (1991) *Concurrent Engineering and Design for Manufacture of Electronics Products*, Van Nostrand Reinhold, New York.

St. Charles, D.P. (1989) *Proceedings of AUTOFACT '89*, Detroit, Michigan, October 30–November 2, Society of Manufacturing Engineers, 663–8.

Thesen, A. (1976) *Management Science*, **23**, December, 412–22.

Wiest, J.D. (1967) *Management Science*, **13**, February, B359–77.

Wiest, J.D. (1981) *Journal of Operations Management*, **1**(3) February, 213–22.

Wiest, J.D. and Levy, F.K. (1977) *A Management Guide to PERT/CPM With GERT/PDM/DCPM and Other Networks* (2nd ed.), Prentice-Hall, Englewood Cliffs, New Jersey.

Winner, R.I., Pennell, J.P., Bertrand, H.E. and Slusarczuk, M.M.G. (1988) *The Role of Concurrent Engineering in Weapons Systems Acquisition*, Report R-338, Institute for Defense Analysis, December.

PART TWO

Tools and Techniques of Concurrent Engineering

CHAPTER 7

Models of design processes

Ali Bahrami and Cihan H. Dagli

7.1 INTRODUCTION

Design is a process of developing plans or schemes of action; more particularly, a design may be the developed plan or scheme, whether kept in mind or set forth as a drawing or model ... Design in the fine arts is often considered to be the creative process per se, while in engineering, on the contrary, it may mean a concise record of embodiment of appropriate concepts and experiences. In architecture and product design the artistic and engineering aspects of design tend to merge; that is; an architect, craftsman, or graphic or industrial designer cannot design according to formulas alone, nor as freely as can a painter, poet, or musician.

(Britannica, 'Architecture, The Art of,' 1986)

How does a design come to be? How does a designer get in touch with his ideas and translate them from fuzzy mental images and abstract concepts to the crisp design?

Louis Kahn, the famous architect, viewed design as a process by which the transcendent forms of thinking and feeling produce the realization of form. To Kahn, form meant the essence created by a certain relationship of elements in a whole. The form of a chair, for instance, is a piece of furniture designed to accommodate one sitting person. It consists of a seat, a backrest and support system that elevates it from the floor. Despite whether the chair is made of plastic, wood or metal, it is recognizable as a chair as long as the seat, backrest, and legs remain in a certain relationship to one another (Tyng, 1984).

Design or problem solving is a natural human activity. We have been designing and acting as designers (sometimes unconsciously) throughout our lives. Design begins with the acknowledgment of needs and dissatisfaction with the current state of affairs and realization that some action must take place in order to correct the problem. When a small child moves a stool to an appropriate location so that she can use it to get to her toy, she has acted as a designer, a rudimentary design; by positioning the stool so that she can satisfy her need of playing with the toy.

Design is a purposeful activity; it involves a conscious effort to arrive at a state of affairs in which certain characteristics are evident (Coyne *et al.*, 1990). Suh defines design as four distinct aspects of engineering and scientific endeavor (see Fig. 7.1). These are:

1. Problem definition from fuzzy sets of facts and myths into a coherent statement of the question;
2. Creative process of devising a proposed physical embodiment of solutions;
3. Analytical process of devising a proposed physical embodiment of solutions; and
4. Ultimate check of the fidelity of the design product to the original perceived needs.

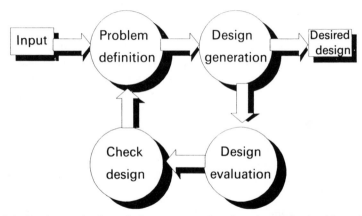

Fig. 7.1   Design as the four distinct aspects of engineering and scientific endeavor.

Every field of engineering involves and depends on the design or synthesis process, which allows us to fulfill needs through the creation of physical and/or informational structures, including machines, software, and organizations (Suh, 1990).

This chapter is organized into six sections. Section 7.2 describes the scientific views of design process. Section 7.3 reviews the design categories. Section 7.4 surveys the design models. Section 7.5 explains the theories and axioms of design process and Section 7.6 presents the summary and conclusion.

7.2 SCIENTIFIC VIEWS OF DESIGN ACTIVITY

Although design is a natural human activity and we all live in a man made environment and use the product of man's creativity, we cannot describe or understand the process that produces the outcome – design.

The main source of the confusion about design is that 'engineering design' lacks the sufficient scientific foundations. Dixon argued that without an adequate base of scientific principles, engineering design education and practice are guided too much by the specialized empiricism, intuition, and experience (Dixon, 1987).

Simon has argued that the science of design is possible and some day we will be able to talk in terms of well-established theories and practices (Simon, 1969). Kuhn (1970) concluded that design is at a prescience phase and it must go through several phases before it constitutes a mature science which is that state of a discipline in which there is a coherent tradition of scientific research and practice, embodying law, theory, application, and instrumentation. Coyne *et al.* (1990) in their book *Knowledge-Based Design Systems* argued that in order to achieve this kind of maturity, we must borrow the methodologies from other disciplines that have reached 'full-fledged' science. Basically there are two major approaches to increase our understanding in a particular discipline which lacks the scientific theories. These are case studies and models. The case studies approach was prevalent in such disciplines as early psychology, prior to the establishment of any appropriate experimental method. This technique is also the predominant approach in engineering design which lacks the solid foundation of scientific theories and relies mostly on the interpretation. The second approach is to use a model to define and understand the design process. Although models are less ambitious than theories and unlike the theories that attempt to explain observed phenomena and predict behaviors that are somehow connected, models are content with the explanation and prediction of phenomena. However, in most cases the mathematical relationships among the connected components are required in order to build a useful model.

Coyne *et al.* (1990) have written that: 'In modeling design we do not attempt to say what design is or how human designers do what they do, but rather provide models by which we can explain and perhaps even replicate certain aspect of design behavior.' A model does not constitute a theory; theory emerges when there is a testable explanation of why the model behaves as it does (Dixon, 1987).

Three types of models that can be used in the design process are *perspective, computational* and *cognitive* (Dixon, 1987). A perspective model stipulates how something should be done. A perspective model of design is concerned with improving the design by advocating how design should be done under certain circumstances. A cognitive model is representative of how people perform some mental task or activity and the inter-relationships of active intelligent human designers with computerized tools such as computer aided drafting systems. Finally a computational model delineates the methods by which a computer can perform a task by computing the variables which are presented in the model. This chapter addresses the first two, namely perspective and computational models.

By borrowing from other disciplines such as artificial intelligence and problem solving i.e., space search techniques, expert systems and neural networks, logic in general and fuzzy logic in particular, object oriented methodology, database and language theory, one can define, study and understand the design activity. Coyne *et al.* (1990) summarized elegantly the definition of science and design as:

> Science attempts to formulate knowledge by deriving relationships between observed phenomena. Design, on the other hand, begins with intentions and uses the available knowledge to arrive at an entity possessing attributes that will meet the original intentions. The role of the design is to produce form or more correctly, a description of form using knowledge to transform a formless description into a definite, specific description. Moreover, design is a pragmatic discipline, concerned with providing a solution within the capacity of the knowledge available to the designer. This design may not be 'correct' or 'ideal' and may represent a compromise, but it will meet the given intentions to some degree.

7.3 DESIGN CATEGORIES

Duvvuru *et al.* (1989) have classified the design process into four categories: *creative design, innovative design, redesign* and *routine design*. These classifications of design are process dependent and product independent. The description of these design classes that have been defined by them can be summarized as follows:

7.3.1 Creative design

Creative design is an abstract decomposition of the problem into a set of levels that represent choices for the component of the problem. An *a priori* plan for the solution of the problem does not exist. The key element in this design type is the transformation from the subconscious to conscious.

7.3.2 Innovative design

The decomposition of the problem is known, but the alternatives for each of its subparts do not exist and must be synthesized. Design might be an original or unique combination of existing components. They argue that a certain amount of creativity comes into play in the innovative design process.

7.3.3 Redesign

An existing design is modified to meet the required changes in the original functional requirements.

7.3.4 Routine design

An *a priori* plan of the solution exists. The subparts and alternatives are known in advance, perhaps as a result of either a creative or innovative design process. Routine design involves finding the appropriate alternatives for each subpart that satisfies the given constraints (Duvvuru *et al.*, 1989).

They explain that at the creative end of the spectrum, the design process might be fuzzy, spontaneous, chaotic, and imaginative. At the other end of the spectrum, namely the routine design, the design is precise, crisp, predetermined, systematic and mathematical (see Fig. 7.2).

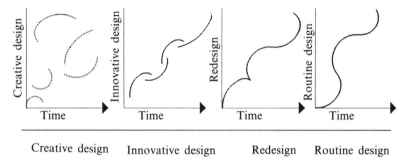

Creative design   Innovative design   Redesign   Routine design

Fig. 7.2   At the creative end of the spectrum, design is very fuzzy. As it moves to routine design it gets precise, crisp and determined.

7.4 SURVEY OF DESIGN MODELS

There is no single model that can provide a perfect or at least a satisfactory definition of the design process. One must look at the design from various angles in order to grasp a better understanding of the process. Many researchers have been trying to develop a scientific theory or at least a model of design. Their perspectives and views on the design are summarized in the following sections.

The models presented basically describe the same phenomena; therefore they may overlap in more ways than one. Hopefully by studying these models we may gain a better understanding of the design activity.

7.4.1 Perspective model of design

As explained earlier, the perspective model of design concerns with improving the design by recommending how design should be done. This section reviews two types of perspective models, namely the general and prototype models of design process.

(a) General model of design process

A general model of design can be visualized as a feedback loop of synthesis and analysis. The general model of design process is inherently iterative; the designer repeatedly goes back to refine and improve his design until the design satisfies the requirements (Serbanati, 1987).

Design is initiated by specifying and analyzing the given problem. The first step in most cases consists of decomposing the problem into subproblems. The designer must then search for the practical solutions. In some cases, this requires ingenuity which may result in a new and original design. In other cases the designer may recall the design solutions that have been employed in the previous problems of a similar nature and apply one of those solutions (Anderson, 1977).

Three basic phases of design described by Coyne *et al.* (1990), Asimow (1962) and Luckman (1967) are analysis, synthesis and evaluation. Analysis is concerned with defining and understanding the 'WHATs' that must be translated by the designer to an explicit statement of functional requirements (goals). Synthesis is involved with finding the feasible solutions among the alternatives, and finally the evaluation phase is concerned with assessing the validity of the solutions relative to the original functional requirements (Coyne *et al.*, 1990). Analysis and synthesis are on the forward path of the design loop, and on the backward path, the designer verifies his assumption through the evaluation process (Serbanati, 1987). A cycle is iterated so that the solution is revised and improved by reexamining the analysis. It has been argued that these three phases of the design formed the basis of the framework for planning, organizing, and evolving design activity.

Figure 7.3 depicts a more comprehensive version of the general model of design process. This version of the model consists of the following tasks:

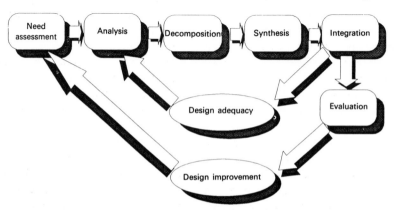

Fig. 7.3 Comprehensive version of general model of design.

need assessment, analysis, decomposition, synthesis, integration and evaluation. The first task is concerned with the assessment of the desired need and requirements which are usually fuzzy in nature. Analysis involves the specification, identification and preparation of the problem and producing an explicit statement of goals. Decomposition is concerned with breaking the problem into parts and defining the boundaries of a space in which a fruitful search for the solution can take place. Synthesis is concerned with discovering the consequences of putting the new arrangement into practice (Coyne *et al.*, 1990). Task evaluation is judging the validity of the solutions relative to the goals and selecting among alternatives. The inner cycle implies that the outcome out of the integration phase is revised and improved by reexamining the analysis. The outer cycle demonstrates that the solution is evaluated that might revise the perceived needs.

(b) Prototype model of design

A prototype typifies, or exemplifies, a class of designs, and thus serves as a generic design. A description of a class of designs also may be prototyped or knowledge or rules may even constitute a prototype (Coyne *et al.*, 1990). Prototype model of design is very similar in nature to the object oriented paradigm. The fundamental contribution of the prototype model is the abstract data type approach. Rather than describing the individual designs, one must concentrate on the patterns that are common to a whole class of design. A particular design can then be instantiated or exemplified from the class (prototype) of designs.

Lansdown (1987) argues that innovation arises from incremental modification of existing 'tried and true' ideas rather than entirely new approaches. Lansdown believes that the transformation from initial to final description is continuous and design is more like fine-tuning a set of already working ideas rather than inventing something new, although the results might not resemble anything previously imagined. The prototype model of design expiates the common form of design and it can be argued that it is more useful than searching for highly innovative design (Lansdown, 1987).

Gardiner describes the prototype model of design as: 'one that brings together several new divergent lines of development to form a new "composite" design, which is internally adjusted to form a new consolidated design which is then further developed as a variety of "stretched" design' (Gardiner, 1986).

A prototype model consists of three activities: *modification, adaptation* and *creation*. Prototype creation, which is known as conceptual or original design, is where the new prototypes materialize as in design of the first airplane (Coyne, 1990). Prototype modification consists of working within the constraints of a particular class of designs. Prototype adaptation pertains to extending the boundaries of a particular class of design.

7.4.2 Computational models of design

A computational model views design as a problem-solving process of searching through a state space where the states represent the design solution. The main task of a designer is to make decisions based on the functional requirements (goals) and design constraints. This decision-making process in the computational models of design is the heart of the matter. Optimization and simulation can assist designers in the decision-making process of design.

(a) Optimization

Mathematical programming techniques can be used to identify the potential design configuration by optimizing it based on the functional requirements and/or goals.

In general, in these methods the solution of the problem is developed by solving the mathematical model consisting of an objective function that is to be optimized and a set of constraints representing the limitations on the resources. Any model can be represented in a standard form as follows:

$$
\begin{aligned}
\text{Maximized} \quad & \mathbf{Z} = f(\mathbf{X}) \\
\text{Subject to} \quad & \mathbf{g}(\mathbf{X}) = 0 \\
\text{Where} \quad & \mathbf{X} \geqslant 0 \\
& \mathbf{X} = [x_1, x_2, \ldots, x_n]
\end{aligned}
$$

Optimization can provide a vital solution in cases where a design problem can be formulated based on the objective and functional requirements. The main problem with optimization techniques is that they do not address the question of how to arrive at the objective functions. Designers do not think in terms of the objective functions or optimization and they do not easily relate to these techniques.

(b) Simulation

Simulation is based on a problem-solving method that has been in use for many years, sometimes referred to as the model-building method or more commonly the scientific method.

Design can utilize simulation in the same fashion as it exploits optimization techniques to compute decision variables. These variables are then evaluated against the functional requirements to determine whether the design satisfies the input requirements. If the design is not found to be acceptable, then the decision variables may be modified and the process is repeated. However, the main shortcoming of simulation technique is that there is no way to tell how good the final design is in comparison with the other possible designs.

7.5 DESIGN THEORIES AND AXIOMS

It is important to keep in mind that there is a significant difference between the theoretical statements and the real world system that theory tries to explain. The theoretical systems never match the real systems perfectly, but they can come close in describing the real world phenomena (Dixon, 1987).

Dixon believes that the major components of theories are:

1. Data or observations from the real system;
2. Generalization of data or observations; and
3. Explanations of why the generalizations are what they are.

Keep in mind that generalizations are not theories. Theories emerge from describing and figuring out why generalization of phenomena is the way it is (Dixon, 1987).

In the following section an attempt is made to concentrate on the development of theories of design.

7.5.1 Toward design axiom and scientific theory of design

Hongo (1985) in his paper 'On the significance of the theory of design' provides the definitions for the design science or scientific study of design activities. Scientific study of design activity is a collection of the logically connected knowledge such as design methodology and design technique. Design technique consists of three sections: the applied knowledge from the natural and human science, the theory of machine systems, and, finally, the theory of design processes.

The theory of design can infer from the following observation:

1. The theories of design must be derived from the facts;
2. During design process only the law of nature is absolute and unchangeable;
3. Design methodology or method that can be used by a designer to attain his desired goal must be the main guide of the designer;
4. The design methodologies are useful only in the conscious mode of design activities; and
5. Intuitive design activity should supersede all other activities including the design methodology (Hongo, 1985).

7.5.2 Design axioms

Suh defines design as the culmination of synthesized solutions in the form of product, software, processes or system by the appropriate selection of DPs (design parameters) that satisfy perceived needs through the mapping from FRs (functional requirements) in the functional domain to DPs in the structure domain. This mapping process is not unique and more than one design may result from the generation of DPs that satisfy the FRs. Therefore, there can be an infinite number of feasible design solutions and mapping techniques. See Fig. 7.4. Suh's design axioms provide the principles that the mapping technique must satisfy the input requirements in order to produce a good design, and provide a framework for comparing and selecting design (Suh, 1990). Suh's design axioms are *independence axiom* and *information axiom* (Suh *et al.*, 1978; Suh, 1984). These axioms are defined as follows:

> *Independence Axiom:* In an acceptable design, the DPs and the FRs are related in such a way that specified DP can be adjusted to satisfy its corresponding FR without affecting other functional requirements.
> *The Information Axiom:* The best design is a functionally uncoupled design that has minimum information content.

Design mapping paradigm

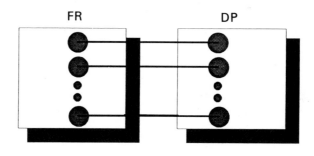

Fig. 7.4 Design mapping paradigm.

Suh's information axiom coincides with an important principle in design known as 'simplicity'. Simplicity is a word that has been used by designer and aestheticians for many centuries and remains a principle of primary concern. What does the designer mean when he applies the word 'simple' to the work of art? He is not speaking of mental deficiency. Simple in the design means being able to reduce it to the fewest possible lines, shapes and subparts, without reducing its functional requirements or violating its specifications. Simple design will reduce the assembly time and product cost, and increase reliability by many orders of magnitude. A good design is one to which no more can be added, and at the same time, one from which nothing can be subtracted without causing an incomplete design. In other words, good design must not be restrained with the nonessential details.

The independence axiom does not imply that a part has to be broken into two or more separate physical parts, or that a new element has to be added to the existing design. Functional coupling may be achieved without physical separation, although in some cases such physical decomposition may be the best way of solving the problem.

Information axiom implies that the misconception that my design is better than yours because it does more than what it was intended is misguided (Suh, 1990). A design should only fulfill the precise needs defined by the FRs and nothing more and nothing less.

7.5.3 General theory of design

Yoshikawa and Warman (1981) have developed the general design theory and its major achievements are the mathematical formulation of the design process and the justification of the knowledge representation techniques in a certain situation. General design theory according to Yoshikawa and Warman is a descriptive model that tries to explain how design is conceptually performed in terms of knowledge manipulation. In general design theory, a design process is regarded as a mapping from the functional space onto the attribute space both of which are defined on an entity concept set. Yoshikawa and Warman argue that from this formalization based on an axiomatic set theory one can mathematically derive interesting theory that can well explain a design process.

Yoshikawa and Warman have developed the design theory under two circumstances. Design under the *ideal knowledge* and design under the *real knowledge*. They describe that in the ideal knowledge we know all of the elements of the entity set and each element can be described crisply by the abstract concept without ambiguity. Another important property of the ideal knowledge is that design can be viewed as a mapping process from the functional space onto attribute space which immediately terminates when the specifications are described. Since everything is known in the ideal knowledge, one can completely describe the specifications in terms of function and the solution is obtained in terms of attributes. Figure 7.5 depicts the Yoshikawa and Warman view of design as a mapping process. The following are the design axioms in the ideal knowledge.

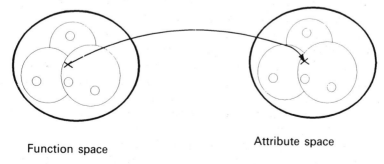

Function space   Attribute space

Fig. 7.5 Design as mapping in the ideal knowledge.

Axiom 1 (Axiom of recognition): Any entity can be recognized or described by attributes and/or other abstract concepts.

Axiom 2 (Axiom of correspondence): The entity set S' and the set of entity concept (ideal) S have one-to-one correspondence.

Axiom 3 (Axiom of operation): The set of abstract concepts is topology of the set of entity concepts.

The real knowledge unlike the ideal knowledge is fuzzy and one must take into consideration the following characteristics:

1. Design is not a simple mapping process but rather a stepwise refinement where the designer seeks the solution that can satisfy constraints;
2. Use of the behavior instead of function; and
3. The ideal knowledge does not take the physical constraints into consideration.

In the ideal knowledge, design is a direct mapping process from the functional space to the attribute space, while in the real knowledge, design is a stepwise, evolutionary transformation process (see Fig. 7.6).

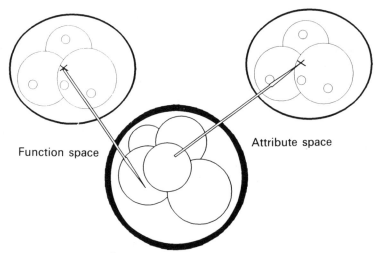

Fig. 7.6 Design as a stepwise refinement process in the real knowledge.

Yoshikawa and Warman admit that the general design theory does not completely support the cognitive model of design, and the cognitive model of design process can be derived from the results of design experiments. They argue that design process can be decomposed into small design cycles. Each cycle has the following subprocess:

1. *Awareness:* Problem identification by comparing the object under consideration and specification.
2. *Suggestion:* Suggesting the key concept needed to solve the problem.

3. *Development:* Developing candidates from the key concepts by using design knowledge.
4. *Evaluation:* Evaluating the alternatives in various ways such as structural computation, simulation of behavior, etc. If a problem is found as the result of evaluation, it also becomes a new problem to be solved in another design cycle.
5. *Conclusion:* Selecting a candidate for adaptation and modification.

7.6 SUMMARY AND CONCLUSION

To conclude this chapter it is useful to reiterate that design contains a wide range of concepts. Design science is a collection of many different logically connected knowledges and disciplines. Design begins with the acknowledgment of needs and dissatisfaction with the current state of affairs and realization that some action must take place in order to solve the problem.

Design is a very difficult subject to understand and research. The main source of the confusion about design is that engineering design lacks sufficient scientific foundations. Design is at a prescience phase and it must go through several phases before it constitutes a mature science.

Design has been classified into four categories. These are creative design, innovative design, redesign and routine design. At the creative end of the spectrum, design is fuzzy, imaginative, spontaneous and chaotic. At the other end of the spectrum, namely the routine design, design is precise, predetermined and systematic.

One must look at the design from various perspectives in order to apprehend a better understanding of the process. Although there is no single model that can furnish a perfect definition of the design process, these models provide us with the powerful tools to explain and understand the design process.

7.7 REFERENCES

Anderson, J.R. (1977) *Cognitive Science,* **1**(2), 125–57.
Asimow, W. (1962) in *Emerging Methods in Environment Design and Planning,* (ed. Moore G.T.), MIT Press, Cambridge, Massachusetts, 285–307.
Coyne, R.D., Rosenman, M.A., Radford, A.D., Balachandran, M. and Gero, J.S. (1990) *Knowledge-Based Design Systems,* Addison-Wesley, Reading, Massachusetts.
Dixon, J.R. (1987) *AI EDAM,* **1**(3), 145–57.
Duvvuru, S., Stephanopouls, G., Logcher, R. *et al.* (1989) *AI Magazine,* **10**(3), 79–96.
Gardiner, J.P. (1986) in *Design, Innovation and Long Cycles in Economic Development,* (ed. Freeman, C.), Frances Pinter, London.
Hongo, K. (1985) in *Design and Synthesis,* (ed. Yoshikawa H.), Elsevier Science Publishers B.V., North-Holland.

Kuhn, T.S. (1970) *The Structure of Scientific Revolutions*, University of Chicago Press, Chicago.

Lansdown, J. (1987) *Design Studies*, **8**(2), 76–81.

Luckman, J. (1967) *Operational Research Quarterly*, **18**(4), 345–58.

Serbanati, L.D. (1987) *IEEE 9th International Conference of Software Engineering*, 190–7.

Simon, H.A. (1969) *The Science of the Artificial*, MIT Press, Cambridge, MA.

Suh, N.P., Bell, A.C. and Gossard, D.C. (1978) *Journal of Engineering for Industry, Transactions of ASME*, **100**(2), 127–30.

Suh, N.P. (1984) *Robotics and Computer Integrated Manufacturing*, **1**(3/4), 399–455.

Suh, N.P. (1990) *The Principles of Design*, Oxford University Press, New York, Oxford.

Tyng, A. (1984) *Beginnings*, John Wiley & Son, New York.

Yoshikawa, H. and Warman, E.A. (eds) (1987) *Proceedings of the IFIP W.G. 5.2 Working Conference 1985 (Tokyo)*, North-Holland, Amsterdam.

CHAPTER 8

A decision-based approach to concurrent design

Farrokh Mistree, Warren Smith and Bert Bras

Modern, computer-based concurrent design requires a holistic approach that integrates the representation, management and processing of information. Integration is possible through the 'standardization' of information management within a design process. We approach standardization from the perspective of decision-based design (DBD), namely, that 'the principal role of an engineer, in the design of an artifact, is to take decisions'. Given that decisions are foundational, we enable concurrent design processes through the simultaneous analysis, synthesis and resolution of multiple decisions.

In this chapter, we introduce the fundamental paradigms of DBD and describe a decision-based design methodology called the decision support problem technique (DSPT). Specifically, we start by providing some background and by stating the axioms needed to characterize 'decisions' as decision support problems (DSPs). Introduced next is the formal syntax and semantics of DSPs. This generic protocol ensures the applicability of the DSPT across varying domains of application by providing uniform and structured mappings between the designers' view of the world and the particular syntax needed to facilitate solution. Finally, we present some examples from marine design to explicate our approach.

8.1 SHIP DESIGN CASE STUDIES – NOMENCLATURE

Variables:

L	or LBP	length between perpendiculars in meters
B	or BEAM	ship design beam in meters
T	or DRAFT	ship design draft in meters
D	or DEPTH	ship design depth in meters
C_b	or CB	block coefficient (ship's hull) $= \dfrac{\text{Displaced volume}}{L \times B \times T}$

C_p	or CP	prismatic coefficient (ship's hull)	$= \dfrac{\text{Displaced volume}}{\text{Midship section area} \times \text{L}}$
C_w	or CW	waterplane coefficient (ship's hull)	$= \dfrac{\text{Waterplane area}}{\text{L} \times \text{B}}$
LCB		longitudinal center of buoyancy in meters forward of midships	
LCF		longitudinal center of flotation in meters forward of midships	
SDKHT		standard height between decks in meters	
VK		maximum sustained speed in knots	
VKCR		endurance speed in knots	

Other parameters:

CODOG	combination of diesel engine (for VKCR) or gas turbine (for VK) – propulsion plant
CODAD	combination of diesel engine and gas turbine (for VKCR and VK) – propulsion plant
COGAG	combination of gas turbine and gas turbine (for VKCR and VK) – propulsion plant
GM	vertical distance between the center of gravity and the transverse metacenter
ROI	economic return on investment

8.2 OUR FRAME OF REFERENCE

It is our contention that to increase both the efficiency[1] and effectiveness[2] of the process of design a contemporary paradigm for design is needed. We offer such a paradigm from the perspective of decision-based design. The paradigm which encompasses systems thinking and embodies the ideas of concurrent engineering design for the life cycle is based on the foundational premise that 'the principal role of an engineer, in the design of an artifact, is to make decisions'. We demonstrate concurrency by the simultaneous solution of multiple decisions and through the integration and holistic treatment of design analysis and synthesis. By choice, we focus on the early stages of project initiation. This is not to say that the tools we develop and employ are limited to applications within the early stages. Our motivation to work in the early stages is that it offers the greatest potential to affect the design process and the artifact, since this phase dramatically shapes what is to follow.

As a design process progresses and decisions accumulate the freedom to make changes is reduced. At the same time, the knowledge about the object of design increases. This increase in knowledge is characterized by a transformation of 'soft' information into 'hard' information. By soft information, we refer to the heuristic and qualitative information that stems from a designer's judgment and experience whereas hard information tends to be based on scientific principles and to be more quantitative in nature. Given this nature of design information, what a concurrent design approach facilitates is 'to know' more about the design early on, that is, increase in qualitative ratio of hard to soft information. This relative improvement in the quality of information is expected to lead to equivalent or better designs that are completed in less time and at less cost than those designed using a traditional sequential process.

Compared to traditional engineering design in which synthesis of the product plays the central role, the dominant feature in concurrent engineering is the synthesis of the process (which includes design, manufacture and support aspects). With the synthesis of the process, it is expected that the synthesis of the product will follow naturally. Certainly some aspects of the design are by necessity pursued sequentially. For example, the preliminary design event will generally follow the conceptual design event. What we mean specifically by the synthesis of the process and concurrent design is that the decisions that can be simultaneously resolved are simultaneously resolved. Indeed, what we seek is a holistic integrated model that yields a solution to all of the relevant decisions simultaneously.

Given this argument and as evidenced by the host of design research initiatives being undertaken worldwide, design science is an emerging discipline and the attitudes toward design are changing. The fundamental reasons for this can be attributed to two singular events; a new emphasis on systems thinking and the pervasive presence of computers. However, independent of the approaches or methods used to plan, establish goals and model systems; designers are and will continue to be involved in two primary activities, namely, processing symbols and making decisions. So, what characterizes a decision?

The characteristics of decisions are governed by the characteristics of design of real-life engineering systems. These characteristics may, in part, be summarized by the following descriptive sentences:

- Decisions involve information that comes from different sources and disciplines;
- Decisions are governed by multiple measures of merit and performance;
- All the information required to make a decision may not be available;
- Some of the information used in making a decision may be hard and some information may be soft; and
- The problem for which a decision is being made is invariably loosely defined and open.

Virtually none of the decisions are characterized by a singular, unique solution. The decision solutions are less than optimal and are called satisficing solutions.

From a decision-based design perspective, decisions help bridge the gap between an idea and reality. They serve as markers and units of communication to identify the progression of a design from initiation, through implementation to termination and they exhibit both domain-dependent and domain-independent features. Focusing upon decisions leads to a description of the design processes written in a common 'language' for teams from the various disciplines – a language useful in the process of designing. Our formal definition of the term designing is as follows (Kamal, *et al.*, 1987; Mistree *et al.*, 1989):

> Designing is a process of converting information that characterizes the needs and requirements for a product into knowledge about a product.

In this definition, we use the term product in its most general sense. We believe that perhaps the most significant design products are the design processes themselves.

8.3 THE DECISION SUPPORT PROBLEM TECHNIQUE

The implementation of decision-based design can take different forms. We call our approach the decision support problem technique (DSPT). It is being developed and implemented, at the University of Houston, to provide support for human judgment in designing systems. The DSPT consists of three principal components: a design philosophy rooted in systems thinking, an approach for identifying and formulating decision support problems (DSPs), and the supporting software, DSIDES. The DSPT comprises two phases, namely, a meta-design phase and a computer-based design phase.

> Meta-design is a metalevel process of designing systems that includes partitioning the system for function, partitioning the design process into a set of decisions and planning the sequence in which these decisions will be made.

For meta-design to represent dynamic partitioning and planning the connotation we place on meta is derived from the work of Klir (Klir, 1985). He states that meta can have three meanings:

- *after* – meta X occurs after X; thus X is a prerequisite of meta X;
- *change* – meta X indicates that X changes and is a general name of that change; and
- *above* – meta X is above (superior to) X in the sense that it is more highly organized, of a higher logical type or viewed from an enlarged perspective (transcending).

We have adopted this third meaning. This notion of higher has also been adopted by the computer scientists, for example, in terms like meta-knowledge, meta-domain, etc.

During Phase I (meta-design), the detailed product specific decisions are not made or even pursued. Rather, what is designed is the process to be implemented in Phase II. In Phase II (design), major decisions are modeled as DSPs and solutions to these DSPs are sought. Phase I of the DSPT is based on the primary axioms of DBD. These axioms map the particular design tasks to characteristic decisions and provide a domain-independent framework for the representation and processing of domain relevant design information (Kamal, 1990).

Axiom–1   Existence of decisions in the DSPT
The application of the DSPT results in the identification of decisions associated with the system (and subsystems that may be relevant).

Axiom–2   Type of decisions in the DSPT
All decisions identified in the DSPT are categorized as Selection, Compromise, or a combination of these.

Selection and compromise are referred to as primary decisions. All other decisions which are represented as a combination of these are identified as derived decisions. The selection decision, in the context of the DSPT, is defined as follows.

Definition–1   The selection decision
The selection decision is the process of making a choice between a number of possibilities taking into account a number of measures of merit or attributes.

The emphasis in selection is on the acceptance of certain alternatives through the rejection of others. The goal of selection in design is to reduce the alternatives to a realistic and manageable number based on different measures of merit. These measures, called attributes, represent the functional requirements and may not all be of equal importance. Some of the attributes may be quantified using hard information and others may be quantified using soft information. Similarly, the compromise decision, in the context of the DSPT, is defined as follows.

Definition–2   The compromise decision
The compromise decision requires that the 'right' values (or combination) of design variables be found to describe the best satisficing system design with respect to constraints and multiple goals.

The emphasis in compromise is on modification and change (e.g., dimensional synthesis) by making appropriate tradeoffs based on criteria relevant to the feasibility and performance of the system.

The second axiom is explained using set notation. The set of all primary decisions in the DSPT is given by, $Decision := \{S, C\}^3$ where S denotes Selection and C denotes Compromise. All derived decisions result from operations on this set. Some derived decisions are illustrated in Fig. 8.1. The coupled selection-compromise decision (Fig. 8.1a) is represented by the

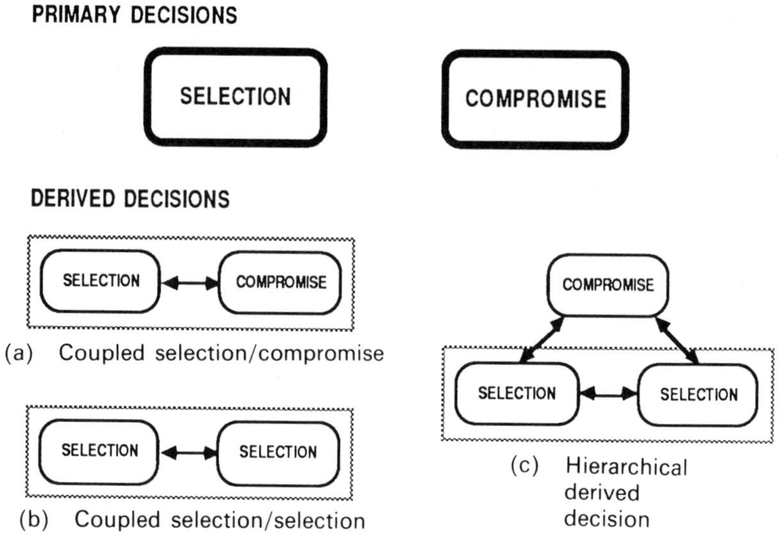

PRIMARY DECISIONS

SELECTION   COMPROMISE

DERIVED DECISIONS

(a)   Coupled selection/compromise

(b)   Coupled selection/selection

(c)   Hierarchical
      derived
      decision

Fig. 8.1   Examples of derived decisions.

operator SC (S, C)⁴ where S and C are contained in the set *Decision*. Similarly a coupled selection–selection decision (Fig. 8.1b) is represented by the operator SS(S, S). A hierarchical decision (Fig. 8.1c) is represented by CSS (C, SS(S, S)) where SS is as defined above. The efficacy of using coupled DSPs is discussed in (Karandikar, *et al.*, 1991; Karandikar and Mistree, 1992).

A corollary to Axiom–1 and Axiom–2 links the decisions to DSPs.

Corollary to Axiom–1 and Axiom–2
Decision support problems are utilized to provide decision support for the decisions identified (within the DSPT).

That is, decisions (primary and derived) are resolved using specialized constructs known as decision support problems. For instance, the coupled selection-compromise DSP is used to provide decision support for the decision shown in Fig. 8.1a. Within the DSPT the nature of decision support problems is qualified through the following two axioms (Kamal, 1990).

Axiom–3   Domain independence of DSP descriptors and keywords
The descriptors and keywords used to model DSPs need to be domain-independent with respect to processes (e.g., design, manufacture, maintenance) and disciplines (e.g., mechanics, engineering management).

Axiom–4   Domain independence of the means to resolve DSPs
The techniques used to resolve DSPs (to actually provide decision support) need to be domain-independent with respect to processes (e.g., design, manufacture, maintenance) and disciplines (e.g., mechanics, engineering management).

Summarized in Table 8.1 are the keywords and descriptors associated with the selection and compromise DSPs. The keywords are the 'verbs' that classify domain relevant information, and identify the relationships between that information. In the DSPs listed in Table 8.1, the keyword 'Given' is a heading under which the background or known information is grouped. Keywords embody in themselves the domain independent 'procedural knowledge'[5] for DSPs. Descriptors are objects organized under the relevant keywords within the DSP formulation. Again, they also help transform the problem from its discipline specific description to a discipline independent representation. For example, to select a material (using the selection DSP) based on strength, color and cost, the material choices are listed as 'alternatives' and the selection criteria as 'attributes'. Descriptors represent 'declarative knowledge'[6] (Rich, 1983).

Table 8.1   DSP keywords and descriptors

DSP	*Keywords*	*Descriptors*
Selection	Given	Candidate alternatives
	Identify	Attributes
		Relative importance
	Rate	Alternatives w.r.t. attributes
	Rank	Order of preference
Compromise	Given	Information
	Find	System variables
		Deviation variables
	Satisfy	System constraints
		System goals
		Bounds
	Minimize	Deviation function

Axiom–4 may seem self-evident as many solution techniques (e.g., linear programming, nonlinear optimization and expert systems) are applicable to problems from different domains. However, this condition supplements Axiom–3 by stating that decision support models using domain-independent constructs should be solved in a domain-independent manner.

To facilitate the design of engineering systems, our approach is to make available tools (analogous to the palette of a painter) that a human designer can use in various events of the design timeline. The decision support problem technique palette was first published in (Mistree *et al.*, 1990). Some refinement and expansion of the concept occurred and the current palette is fully described in (Bras and Mistree, 1991). The palette contains three different classes of entities, namely, potential support problem entities, base entities and transmission entities. The icons representing these entities are shown in Fig. 8.2. A model or network of a process is created by connecting entities in a systematic fashion. An extensive

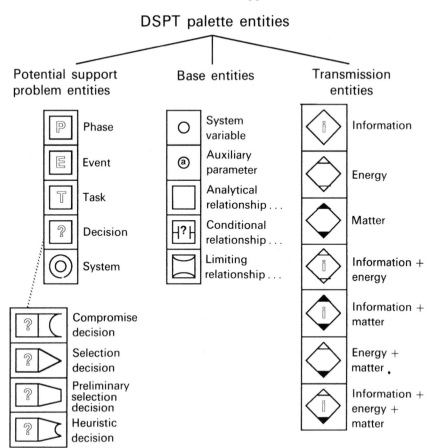

Fig. 8.2 The DSPT palette for modeling processes (Bras and Mistree, 1991).

example using the palette in the design of a frigate is given in (Mistree, et al., 1990).

A designer working within the DSPT has the freedom to use submodels or subnetworks of a design process created and stored by others (prescriptive models) and/or to create original models of the intended plan of action (descriptive models). The icons can easily accommodate different traditions and cultures and should not be considered limited to the examples presented herein. Further issues associated with designing models of design processes are discussed in (Bras *et al.*, 1990).

In Phase II of the DSPT, the focus is on structuring and solving the DSPs that correspond to the decisions identified in Phase I. The organization of information with keywords and descriptors results in a general knowledge representation scheme with an associated sense of data abstraction. Using the keywords and descriptors from Table 8.1, the selection DSP is reiterated as follows.

Given
A set of candidate alternatives.
Identify
The principal attributes influencing selection.
The relative importance of attributes.
Rate
The alternatives with respect to their attributes.
Rank
The alternatives in order of preference based on the computed merit function values.

Similarly, the compromise DSP is stated in words as follows:

Given
An alternative to be improved through modification.
Assumptions used to model the domain of interest.
The system parameters (fixed variables).
The constraints and goals for the design.
Find
The independent system variables values (they describe the artifact's physical attributes).
The deviation variables values (they indicate the extent to which the goals are achieved).
Satisfy
The system constraints that must be satisfied for the solution to be feasible.
The system goals that must achieve, to the extent possible, a specified target value.
The lower and upper bounds on the system variables and bounds on the deviation variables.
Minimize
The deviation function that is a measure of the deviation of the system performance from that implied by the set of goals and their associated priority levels or relative weights.

Since the selection DSP can be reformulated as a compromise DSP, the compromise DSP is considered the principal mathematical DSPT formulation (Bascaran *et al.*, 1989). This transformation of selection to compromise makes it possible to formulate and solve coupled selection–selection DSPs and coupled selection–compromise DSPs (Smith, *et al.*, 1987; Karandikar, 1989; Bascaran, 1990). Indeed, an augmented compromise DSP is used to solve any derived decision. Therefore, let us examine the underpinnings for the mathematical formulation of the compromise DSP.

The compromise DSP formulation is a multiobjective programming model which we consider to be a hybrid formulation (Mistree *et al.*, 1992). It incorporates concepts from both traditional mathematical programming and goal programming. It also makes use of some new features. The term 'Goal programming' was used, by its developers (Ignizio, 1982, 1983, 1985), to indicate the search for an 'optimal' program (i.e., a set of policies to be implemented), for a mathematical model that is composed solely of goals. This does not represent a limitation as any mathematical programming model (e.g., linear programming), may find an alternate representation via GP. Not only does GP provide an alternative representation, it often provides a representation that is more effective in capturing the nature of real world problems.

The compromise DSP is similar to GP in that the multiple objectives are formulated as system goals (involving both system and deviation variables) and the deviation function is solely a function of the goal deviation variables. This is in contrast to traditional mathematical programming where multiple objectives are modeled as a weighted function of the system variables only. The concept of system constraints is retained from the traditional constrained optimization formulation. However, the compromise DSP places special emphasis on the bounds of the system variables unlike traditional mathematical programming and GP. The compromise DSP constraints and bounds are handled separately from the system goals, contrary to the GP formulation in which everything is converted into goals. In the compromise formulation, the set of system constraints and bounds define the feasible design space and the set of system goals define the aspiration space (see Fig. 8.3). For feasibility the system constraints and bounds must be satisfied. A satisficing solution then is that feasible point which achieves the system goals as far as possible. The solution to this problem represents a tradeoff between that which is desired (as modeled by

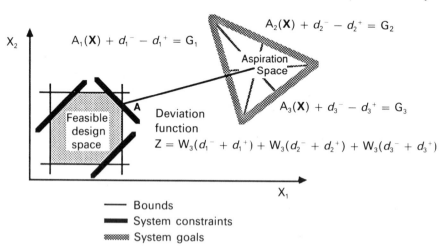

In this case, it is assumed that $W_1 = W_2 = W_3$

Fig. 8.3 Graphical representation of a two-dimensional DSP, Archimedean formulation.

the aspiration space) and that which can be achieved (as modeled by the design space).

Compromise DSPs are written, in general, in terms of n system variables. The vector of variables, $\mathbf{X}$ may include continuous variables and boolean (1 if TRUE, 0 if FALSE) variables. System variables are independent of the other descriptors and can be changed to alter the state of the system. System variables that define the physical attributes of an artifact must be positive.

A system constraint models a limit that is placed on the design. The set of system constraints must be satisfied for the feasibility of the design. Mathematically, system constraints are functions of system variables only. They are rigid and no violations are allowed. They relate the demand placed on the system, $D(\mathbf{X})$, to the capability of the system, $C(\mathbf{X})$. The set of system constraints may be a mix of linear and nonlinear functions. In engineering problems the system constraints are invariably inequalities. However, occasions requiring equality system constraints may arise. The region of feasibility defined by the system constraints is called the feasible design space.

A set of system goals is used to model the aspiration a designer has for the design. It relates the goal, G_i, of the designer to the actual performance, $A_i(\mathbf{X})$, of the system with respect to the goal. The deviation variable is introduced as a measure of achievement since we would like the value of $A_i(\mathbf{X})$ to equal G_i. Constraining the deviation variables to be non-negative, the system goal becomes:

$$A_i(\mathbf{X}) + d_i^- - d_i^+ = G_i; \ i = 1, \ldots, m$$

where
$$d_i^- \cdot d_i^+ = 0 \text{ and } d_i^-, d_i^+ \geq 0.$$

The product constraint $(d_i^- \cdot d_i^+ = 0)$ ensures that at least one of the deviation variables for a particular goal will always be zero. If the problem is solved using a vertex solution scheme as a matter of course then this condition is automatically satisfied.

Bounds are specific limits placed on the magnitude of each of the system and deviation variables. Each variable has associated with it a lower and an upper bound. Bounds are important for modeling real-world problems because they provide a means to include the experience-based judgment of a designer in the mathematical formulation.

In the compromise DSP formulation the aim is to minimize the difference between that which is desired and that which can be achieved. This is done by minimizing the deviation function, $Z(d^-, d^+)$. This function is always written in terms of the deviation variables. All goals may not be equally important to a designer and the formulations are classified as Archimedean or Preemptive – based on the manner in which importance is assigned to satisficing the goals. The most general form of the deviation function for m goals in the Archimedean formulation is

$$Z(d^-, d^+) = \sum_{i=1}^{m} (W_i^- d_i^- + W_i^+ d_i^+)$$

where the weights $W_1^-, W_1^+, W_2^-, W_2^+, \ldots, W_m^-, W_m^+$ reflect the level of desire to achieve each of the goals. Generally, these weights would be chosen to sum to unity. However, it may be difficult to identify truly credible weights. A systematic approach for determining reasonable weights is to use the schemes presented in (Kuppuraju *et al.*, 1985; Bascaran *et al.*, 1989).

The most general approach to assigning priority is a preemptive one where the goals are rank ordered. Multiple goals can be assigned the same rank or level, in which case, Archimedean styled weights may be used within a level. This assignment of priority is probably easier in an industrial environment or in the earlier stages of design. The measure of achievement is then obtained in terms of the lexicographic minimization of an ordered set of goal deviations. Ranked lexicographically, an attempt is made to achieve a more important goal (or set of goals) before other goals are considered. The mathematical definition of lexicographic minimum follows (Ignizio, 1982, 1983).

Lexicographic minimum
Given an ordered array f of non-negative elements f_k's, the solution given by $f^{(1)}$ is preferred to $f^{(2)}$ if $f_k^{(1)} < f_k^{(2)}$ and all higher-order elements (i.e., $f_1, \ldots, f_{k-1}$) are equal. If no other solution is preferred to f, then f is the lexicographic minimum.

If we consider $f^{(r)}$ and $f^{(s)}$, where $f^{(r)} = (0, 10, 400, 56)$ and $f^{(s)} = (0, 11, 12, 20)$, then $f^{(r)}$ is preferred to $f^{(s)}$. Hence, the deviation function for the preemptive formulation is written as

$$Z = \{f_1(d_i^-, d_i^+), \ldots, f_k(d_i^-, d_i^+)\}.$$

For a four goal problem, the deviation function may look like

$$Z(d^-, d^+) = \{(d_1^- + d_2^-), (d_3^-), (d_4^+)\}.$$

The numerical solution of a preemptive formulation requires the use of a special optimization algorithm developed to solve this type of problem. One such algorithm, Multiplex, has been developed by Ignizio (1985) and has been incorporated into DSIDES.

The Mathematical Formulation of the Compromise DSP is as follows:

Given

n	number of system variables
$p+q$	number of system constraints
p	equality constraints
q	inequality constraints
m	number of system goals
$g_i(\mathbf{X})$	system constraint function $g_i(\mathbf{X}) = C_i(\mathbf{X}) - D_i(\mathbf{X})$
$f_k(d_i)$	function of deviation variables to be minimized at priority level k for preemptive case
W_i	weight for Archimedean case

Find

System variables

$$X_i \qquad i = 1, \ldots, n$$

Deviation variables

$$d_i^-, d_i^+ \qquad i = 1, \ldots, m$$

Satisfy

System constraints (linear, nonlinear)

$$g_i(\mathbf{X}) = 0; \qquad i = 1, \ldots, p$$
$$g_i(\mathbf{X}) \geqslant 0 \qquad i = p+1, \ldots, p+q$$

System goals (linear, nonlinear)

$$A_i(\mathbf{X}) + d_i^- - d_i^+ = G_i; \qquad i = 1, \ldots, m$$

Bounds

$$X_i^{\min} \leqslant X_i \leqslant X_i^{\max}; \qquad i = 1, \ldots, n$$
$$d_i^-, d_i^+ \geqslant 0; \qquad i = 1, \ldots, m$$
$$(d_i^-, d_i^+ = 0; \qquad i = 1, \ldots, m)$$

Minimize

Case a: Preemptive (lexicographic minimum)

$$Z = \{f_1(d_i^-, d_i^+), \ldots, f_k(d_i^-, d_i^+)\}$$

Case b: Archimedean

$$Z = \sum_{i=1}^{m} (W_i^- d_i^- + W_i^+ d_i^+)$$

As identified above, the selection DSP can be reformulated as a compromise DSP as follows:

Given

M   candidate alternatives
N   attributes
I_j relative importance of the *j*th attribute
R_{ij} the normalized rating of the *i*th alternative with respect to the *j*th attribute

$$\sum_{j=1}^{N} I_j R_{ij} = MF_i = \text{merit function of alternative } i$$

Find

Design variables

$$X_i; \quad i = 1, \ldots, M$$

Deviation variables

$$e^-, e^+$$

Satisfy

Selection system constraint

$$\sum_{i=1}^{M} X_i = 1 \tag{1}$$

Selection system goal

$$\sum_{i=1}^{M} MF_i X_i + \theta^- - \theta^+ = 1 \tag{2}$$

Bounds

$$0 \leqslant X_i \leqslant 1; \quad i = 1, \ldots, M$$

Minimize

$$Z = e^- + e^+$$

The solution to the 'reformulated' selection DSP which is a linear, $0-1$ variable optimization problem can be found without necessitating the use of specialized integer programming codes. We use the ALP algorithm incorporated in DSIDES (Mistree *et al.*, 1981; Mistree *et al.*, 1992). In respect to the guaranteed boolean behavior of X for a single selection DSP, Bascaran (Bascaran *et al.*, 1989) argued the case in the following way. Considering the constraint that the product of e^- and e^+ is equal to zero, there are three equality equations in $M+2$ unknowns (X_{i-}, for $i = 1, \ldots m$ M; e^- and e^+). By assuming normalization of the merit function values, that X_k is not perfect ($MF_k \neq 1$), and that the maximum merit function value is unique and nonzero ($MF_k > MF_i$ for all $i \neq k$); the equation for the product of the deviation variables leads to e^+ being zero, equation (2) leads to e^- being nonzero, and equation (1) to only one X_i being nonzero. The value of the nonzero X_i will then be unity as dictated by equation (1). A rigorous proof can also be provided using monotonicity analysis.

With care, this argument to guarantee boolean behaviour of continuous variables and uniqueness can be extended to coupled multiple-selection problems. For each selection problem, a particular uniqueness condition and a corresponding goal will exist in the formulation. However, a heuristic adjustment of the goals is recommended to ensure that e^- remains nonzero. Depending on the selection goal priorities and the right hand side (RHS) values of the goals, it is otherwise possible to satisfy the uniqueness constraints with a fractional, nonboolean X vector. The suggested adjustment to equation (2) is in a new RHS such as

$$\sum_{i=1}^{M} MF_i X_i + e^- - e^+ = M + \text{delta}.$$

When dealing computationally with complex coupled problems, additional system constraints and techniques may be helpful and/or necessary depending on the user's problem formulation. Indeed, an optional branch-and-bound zero-one algorithm is available within DSIDES for use with specially defined 'selection' variables. Alternatively, if a designer wishes to use continuous variables exclusively and to impose boolean behavior, constraints such as

$$\sum_{i=1}^{M} X_i(1 - X_i) = 0$$

are strongly recommended. This modification has proved useful in practice.

Additional system constraints may be required to model exclusionary behavior. For example, certain combinations of alternatives in a coupled selecting–selection DSP may be infeasible. Consider the case where if material A is selected ($X_1 = 1$), manufacturing process C cannot be used ($X_7 \neq 1$). This condition can be modeled as

$$X_1 + X_7 \leqslant 1.$$

The use and formulation of such exclusionary constraints is demonstrated in (Karandikar, 1989; Bascaran, 1990).

As a final word here on selection, while an engineering solution can be sought in this way, the designer must examine the sensitivity of the information used. The variances of the rating (or grading) systems used may in reality be too great to truly discriminate one solution from another. (Remember that the goal of the DSPT is to provide decision support and not to automate the decision process). For further details regarding this issue, (see Kuppuraju *et al.*, 1985; Bascaran *et al.*, 1989).

As stated earlier, in our case, concurrent design is achieved within a decision or subsystem through the integration and holistic treatment of design analysis and synthesis. Further, concurrency is modeled by the simultaneous consideration of subsystems, or in DBD terms, the simultaneous resolution of derived decisions. Consider a derived decision involving two compromise DSPs and one selection DSP. Some characteristics of concurrent design are embodied in the system synthesis, a mathematical formulation of which is provided in Table 8.2. Each subsystem is shown to impact the others through the subsystem constraints and goals being functions of all of the subsystem variables. Concurrency is also emphasized in the minimization of a composite deviation function.

8.4 APPLICATIONS OF THE DECISION SUPPORT PROBLEM TECHNIQUE

Applications of DSPs include the design of ships, damage tolerant structural and mechanical systems, the design of aircraft, mechanism, thermal

Table 8.2 Modeling concurrency through system synthesis–succinct math formulation

Compromise DSP 1	*Compromise DSP 2*	*Selection DSP*
Find		
$\mathbf{X}, d^-, d^+$	$\mathbf{Y}, e^-, e^+$	$\mathbf{S}, h^-, h^+$
Satisfy		
$g_i(\mathbf{X}, \mathbf{Y}, \mathbf{S}) \geqslant 0$	$g_i(\mathbf{X}, \mathbf{Y}, \mathbf{S}) \geqslant 0$	$\sum S_i = 1$
$A_i(\mathbf{X}, \mathbf{Y}, \mathbf{S}) + d_i^- - d_i^+ = G_i$	$A_i(\mathbf{X}, \mathbf{Y}, \mathbf{S}) + e_i^- - e_i^+ = G_i$	$\sum MF_i(\mathbf{X}, \mathbf{Y})S_i + h^- - h^+ = 1$
$X_i^{min} \leqslant X_i \leqslant X_i^{max}$	$Y_i^{min} \leqslant Y_i \leqslant Y_i^{max}$	$0 \leqslant S_i \leqslant 1$
$d^-, d^+ \geqslant 0$	$e^-, e^+ \geqslant 0$	$h^-, h^+ \geqslant 0$

$$\textit{Minimize (lexicographically)}$$
$$Z = \{h^-, f_1(d^-, d^+, e^-, e^+), \ldots, f_k(d^-, d^+, e^-, e^+)\}$$

Notes: X, Y – system variables
S – selection variables
$d^-, d^+, e^-, e^+, h^-, h^+$ – deviation variables
g_i–ith constraint function for subsystem
A_i – ith goal function for subsystem
G_i – ith goal target value for subsystem
MF_i – ith merit function for alternative s_i
Z – deviation function (Preemptive form, k priority levels)

energy systems, design using composite materials and data compression. A detailed set of references to these applications is presented in (Mistree *et al.*, 1990). However, in the following sections, a collection of representative examples from the marine field are presented. They are based on a general analysis of current practice, but no attempt is made to constrain all design life cycles to fit these models. For the examples, a conscious decision has been made not to document one design problem from beginning to end. Rather, to emphasize the versatility of the approach, a number of problems are partially addressed. These examples highlight aspects such as meta-design, the representation of ship subsystems as DSPs, design synthesis, possible results and the parametric use of results. Collectively, the examples are used to illustrate the efficacy of the DSPT in supporting a human designer. To aid understanding, the examples are provided in a pseudo-chronological form in keeping with the two phases of the DSPT.

Experience indicates practical success in using DSPs to solve ship design problems. In addition to the studies discussed herein, five case studies involving comparisons of preliminary ship designs developed using RAPID (a DSIDES template[7]) with designs produced commercially are detailed in (Lyon and Mistree, 1985). The first two case studies involve actual proven designs that RAPID matched in all technical aspects. The remaining three case studies represent comparisons with other computer-assisted ship design methods. In these cases, RAPID designs were considered superior overall, both in approach and results.

8.4.1 A meta-design example: modeling the process of design

As a representation of meta-design at the highest level, a portion of a hypothetical timeline for designing a frigate is shown in Fig. 8.4. This is referred to as the phase-event-information diagram or the P-E-I diagram for short. From left to right, the qualitative relationship between hard and soft information increases. The design phases, events and product specific information are shown in different sections within the figure. The storybook graphics at the top represent the strategic need, the various concepts, the selected basic concept, the preliminary design, the contract negotiations, the manufacturing, the finished ship, the ship after the half-life refit and the decommissioned ship.

The timeline is partitioned into four major design phases for this example. Typically, the end of each phase is not abrupt and it is often difficult to see when a new phase starts. Therefore, the phases in Fig. 8.4 overlap each other. Within these phases we identify a number of events. Events are not restricted to one phase. For instance, the preliminary design event is found in designing for concept, designing for manufacture and designing for maintenance. The horizontal bars provide an indication of the duration, in physical time, of phases and events. Input to the design process is a strategic need or foreign policy and during the life of the frigate more and more hard information becomes available (e.g., drawings and documentation). Thus, the ratio of hard to soft information is seen to increase as the timeline is traversed from left to right.

The output of each event augments the product specific information. This is shown in the fourth section of Fig. 8.4. The first event identified is the development of the Naval Staff requirements. This event results in a significant document referred to as the 'Statement of requirements'. This document plus general design information then forms the primary input for the conceptual design event. The conceptual design event feeds forward a basic concept while initiating a feedback loop to the development of the naval staff requirement event. The basic concept and the general design knowledge provide the necessary information for the preliminary and contract design events. Note that again an overlap between these two events occurs. The preliminary design event provides the top level specification and the ship characteristics, whereas the contract design event provides the general specification and the guidance drawings.

As stated, the output of each event augments the product specific information. This is useful in planning a project. A designer of the design process could ask the following question: If I had some particular information, will I be able to use it to further my design? If the answer is yes it may be entered in the product specification line. Of course there are other questions that this designer will pose and answer before arriving at the correct statement of the product specification section. Some examples: Can I obtain this information in some easier manner? What are the consequences of my forgoing this information? In summary, the P-E-I diagram provides a good foundation for developing design process schedules and barcharts.

While a concurrent approach in respect to model synthesis is ideal in all design phases and there is a similarity in the decision templates across the timeline of a design, one 'super' template cannot model all phases effectively. What is sought is the development of phase and event based 'master derived decision templates' that account for, at an appropriate level, as many aspects of the design life-cycle as is possible and/or reasonable. For a closer look at the preliminary design event of Fig. 8.4, imagine 'double clicking' in the event bar chart on the preliminary design bar. 'Opened' as the subordinate layer of process documentation would be the model of this event as shown in Fig. 8.5. The master decision template is the 'preliminary ship synthesis' entity at the center, shown under the magnifying glass. All other activity identified in the network of Fig. 8.5 is essentially associated with gathering, structuring or disseminating input or output information. Concurrently, each primary aspect to be considered in preliminary ship synthesis is mapped into relevant selection and compromise DSPs. As an aside, the general principles for developing or finding the 'best' network is a matter of current research. Some related work in progress is reported in (Bras and

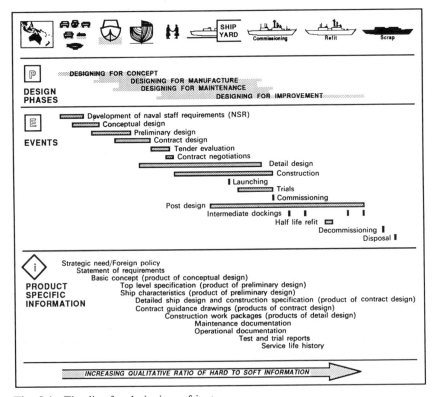

Fig. 8.4 Timeline for designing a frigate.

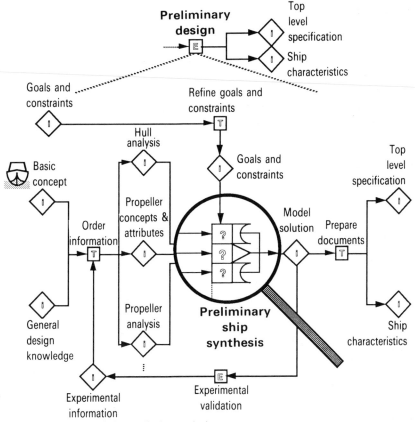

Fig. 8.5 A model of the preliminary design event.

Mistree, 1991). Upon implementation on a computer we envisage that there could and should be similar models underlying all entries shown in the P-E-I diagram.

8.4.2 Preliminary ship synthesis – concurrent design of the hull–propeller–machinery system

From a concurrent design perspective, what we seek is a holistic integrated model that yields a solution to all of the relevant decisions simultaneously. The model of the preliminary design event depicted in Fig. 8.5 demonstrates this. As shown, it remains open ended as the individual decisions to be addressed in the preliminary ship synthesis could be many. Restricting our discussion to the gross design elements, in traditional preliminary design the hull is designed first. The propeller is then designed and minor sequential and iterative modifications are made to integrate the hull and the propeller. Finally, the machinery is selected to match the system as well as it can. In Fig. 8.5 we illustrate a coupled DSP for preliminary ship synthesis that involves these elements of hull design, machinery selection and propeller design. The key notion here is that the coupled DSP enables concurrent solution by taking into account the interactions between the subsystems. The solution represents the holistic design of a hull–propeller–machinery system. A representative word formulation for the hull subsystem compromise DSP follows.

Given

Naval operational requirements
- Ship type   – frigate
- Speed   – maximum sustained in knots
               – endurance in knots
- Range   – at endurance speed in nautical miles
- Endurance – days at sea
- Seakeeping – maximum seastate for normal ship and helicopter operations
- Payload   – number and type of weapons, aircraft, command, control and surveillance equipment

Desirable characteristics
- Machinery type (CODOG or CODAD or COGAG or steam turbine)
- Location of machinery space (aft or midships)
- Number of propeller shafts
- Hull and superstructure material (steel and/or aluminum)

Design Assumptions
- General arrangement 'geometry'

and a starting design:

$$\mathbf{X}^0 = \{X_i^0 : i = 1, \ldots, M\}$$

Find

The values of system variables

$\mathbf{X}^* = \{X_i^* : i = 1, \ldots, M\}$ as defined by the principal ship dimensions:

L	length between perpendiculars in meters
B	ship design beam in meters
T	ship design draft in meters
D	ship design depth in meters

The form coefficients:

C_b	block coefficient
C_p	prismatic coefficient
C_w	waterplane coefficient
C_m	midship section coefficient

and parameters:

LCB   longitudinal center of buoyancy in meters forward of midships
LCF   longitudinal center of flotation in meters forward of midships
SDKHT   standard height between decks in meters

and the deviation variables:

$$d^+ \text{ and } d^-$$

Satisfy

The following system constraints:

Space
- Displacement is equal to or greater than the estimated weight
- Internal volume is equal to or greater than the required internal volume
- Total deck area is equal to or greater than the required deck area
- Length must exceed aerial, weapon and ship system separation requirements
- The double bottom height is between 0 and 1.5 meters

Stability
- The intact stability in the minimum operating condition exceeds the minimum requirement determined in accordance with the stability criteria.

Seakeeping
- The seakeeping rank is greater than that required for the specified sea state for normal ship operations and for helicopter operations
- The natural period of roll is greater than the period of encounter
- The natural period of heave is greater than 120% of the period of encounter relevant to heave, i.e., ship operates in supercritical region
- The natural period of pitch is greater than 120% of the period of encounter relevant to pitch, i.e., ship operates in supercritical region
- Freeboard is greater than the required freeboard at midships

Form
- The prismatic coefficient is within limits defined by the speed–length ratio
- The LCF is at least a given percentage aft of the LCB
- The form coefficient relationship, $C_b = C_p^* C_m$
- Minimum and maximum values for:

  L/B, L/D, L/T, B/D, B/T, T/D, C_b/C_w, C_b/C_p and C_w/C_p

The following system bounds ($X^{min} \leqslant X \leqslant X^{max}$):
- All variables, except LCB and LCF, must be positive

The following system goals:
- The capital cost is equal to or smaller than the target value

- The seakeeping quality is equal to or greater than the target value
- The endurance speed powering is equal to or smaller than the target value
- The maximum sustained speed powering is equal to or smaller than the target value
- The displacement is equal to or smaller than the target value
- The height between decks is equal to or greater than the target value

Minimize

The deviation function

$$Z = \{f_1(d^-, d^+), \ldots, f_k(d^-, d^+)\}$$

Intuitively, a library of such DSP templates can be created and used widely on relevant problems. For example, an appropriate template for propeller design could be used in conjunction with a multitude of hull form templates.

And now to model the subsystems as a coupled DSP as depicted mathematically in Table 8.2. Generally, a ship's subsystems (e.g., hull design, propeller design and machinery selection), are highly dependent in respect to the total ship system. For example, the attribute ratings for the machinery selection are influenced by the geometric system variables of the ship's length, beam, depth, etc. Similarly, the values of the hull goals are dependent upon the machinery selected and the performance of the propeller. Essentially, this interaction is modeled by coupling the decision subsystems and/or establishing a decision hierarchy. This form of interdependency and hierarchy of decisions is typical of real-world design problems and has been addressed in (Smith *et al.*, 1987). The mathematical justification for this coupling is presented in (Karandikar *et al.*, 1991).

With little imagination, it is easy to see the interplay between subsystems considering the goals of each. The goals for the hull subsystem are identified above as:

- Achieve the desired capital cost;
- Achieve the desired seakeeping quality;
- Achieve the desired endurance speed powering;
- Achieve the desired maximum sustained speed powering;
- Achieve the desired displacement; and
- Achieve the desired height between decks.

Similarly, for the propeller subsystem the goals may be:

- Achieve the desired efficiency; and
- Achieve the desired matching of the propeller and ship thrust coefficients.

Finally, the propulsion machinery subsystem goals could be:

- Achieve the desired capital cost;
- Achieve the desired machinery weight;

- Achieve the desired volume of machinery spaces;
- Achieve the desired range; and
- Achieve the desired reliability.

The same goal (and constraints) may occur in multiple subsystems and the assignment of specific goals to specific subsystems is a question open to debate. However, when subsystems are integrated, the goals are concatenated together in a single set and duplicates are discarded. This leads to a composite deviation function.

Moving to mathematics, the succinct statement of the coupled hull/propeller/machinery system math formulation would be similar to that provided in Table 8.2. The bounds on the system variables (e.g., $100\,m \leqslant X_1 \leqslant 200\,m$: $X_1 = L$) and the linear constraints representing hull geometry 'design lanes' (e.g., $1.220 \leqslant 1.0942*X_5 + 1.0*X_7 \leqslant 1.260 : X_5 = C_b$ and $X_7 = C_w$) are easily represented algebraically. However, the nonlinear constraints and goals cannot be so easily represented. Each is typically encoded in a subroutine or set of subroutines.

In the foregoing, we have addressed an example of meta-design, detailed in words a decision template and discussed the modeling of subsystems that result in concurrent design. Let us now turn our attention to some numerical examples.

8.4.3 A container ship example

The problem statement for this example is as follows. Assume that a market exists for transporting containers between two ports 6,000 nautical miles apart. A forecast of the operating costs and market revenues has been made and a 15% minimum rate of return on investment (ROI) is desired. A 60% average load factor is assumed for the vessel since the market is seasonal and a larger cargo capacity will be necessary at certain times of the year. Due to economic considerations, the owner has specified a ship that has a cargo capacity of 650 TEUs (twenty foot equivalent units), where the average weight of a unit is assumed to be 20 tons. Determine an initial estimate of the principal dimensions and associated ship characteristics for a container ship in such service. Assume that the technical goals stated for the design are:

- Achieve the desired deadweight to displacement ratio;
- Achieve the target cargo capacity of 650 TEUs;
- Achieve the target GM of 2.0 meters;
- Achieve only the classification freeboard;
- Achieve the desired resistance; and
- Achieve the target speed.

The economic goals for the design are:

- Achieve at least the target ROI of 15%; and
- Achieve the desired vessel cost.

A DSP template was created for this decision such that an estimate of through-life economic characteristics could be made, and that the hull lowering information could be calculated with or without consideration of the propeller design (by using different algorithms). The highest priority was placed on achieving at least the target ROI and the target speed. The second level sought to achieve the required cargo capacity of 650 TEUs. Some characteristic data from a parametric study using this DSP is tabulated in Table 8.3. Cases were run for various target speeds over a range from 10 to 24 knots assuming one of the following:

- Concurrent design of hull and propeller with economic considerations ('H/P/E'); or
- Design of hull only with economic considerations ('H/E'); or
- Design of hull only without economic considerations ('H only').

The units of the data presented in Table 8.3 are as follows:
- L, B, T, D – meters;
- C_b – block coefficient (dimensionless);
- SHP – metric horsepower in thousands;
- VK – vessel speed knots;
- ROI – return on investment as a percentage; and
- CAP – cargo capacity in TEUs.

The subsequent plots of this data are, in the main, given against speed-length ratio (V/LO^5), with speed in knots and length in feet. The examples cover a range of speed-length ratio from 0.72 to 0.92 that corresponds to Froude numbers between 0.21 and 0.28.

As a function of the container ship DSP template used, the powering algorithm invoked for 'H/E' and 'H only' scenarios is based on the Series 60 work of Shaher Sabit (1972). In contrast, 'H/P/E' calls on a composite algorithm that utilizes the power prediction method proposed by Holtrop and Mennen (Holtrop and Mennen, 1982; Holtrop, 1984), coupled with the Wageningen B-screw series data presented by Oosterveld and van Oossanen (1975). The economic model, when invoked, for 'H/E' and 'H/P/E,' is identical. All other analytical calculations are identical across scenarios.

We include the containership example not to illustrate the design of a perfect ship but rather to demonstrate the process of designing under the DSPT umbrella. Therefore, we suggest that the data be looked at qualitatively rather than quantitatively. Now, given the template and the set of parametric data, what are some of the characteristic observations that can be made about the model and the decision support provided by using the model?

Various hull form design lanes were specified as constraints in this model (e.g., $(L/B)^{min} \leqslant L/B \leqslant (L/B)^{max}$). However, philosophically, a design with specifications outside these design 'trend' ranges is not necessarily infeasible or unacceptable. These constraints only represent statistical data compiled for 'similar' ships and remaining within such accepted and proven ranges

Table 8.3 Parametric data for container ship

	Ship speed (knots)									
	10	12	14	15	16	17	18	20	22	24
Hull and propeller with economic considerations ('H/P/E')										
L	123.3	117.3	123.8	123.1	122.0	123.7	124.8	125.1	124.7	123.8
B	22.02	20.94	22.10	21.97	21.79	22.09	22.28	22.34	22.17	22.11
T	9.390	10.47	9.360	9.400	10.78	11.04	11.14	11.17	10.89	10.94
D	11.23	14.24	11.22	11.24	12.51	12.81	12.93	12.96	12.68	12.72
CB	0.737	0.729	0.734	0.742	0.673	0.647	0.630	0.625	0.650	0.651
SHP	7.720	6.520	6.970	7.620	9.040	10.57	13.91	12.28	11.92	12.54
VK	14.90	14.45	14.87	15.00	16.00	17.00	18.00	17.84	17.50	17.55
ROI	17.50	18.00	18.50	18.50	25.00	26.50	22.50	25.00	25.00	24.00
CAP	646.0	652.0	650.0	648.0	663.0	658.0	624.0	640.0	644.0	638.0
Hull only with economic considerations ('H/E')										
L	123.4	124.7	116.8	125.7	122.3	124.3	124.9	126.4	125.6	125.6
B	22.04	22.27	20.85	22.44	21.84	22.20	22.31	22.53	22.43	22.43
T	9.260	9.380	10.41	9.310	10.50	10.71	11.15	11.26	11.21	11.21
D	11.10	10.71	13.77	11.16	12.26	12.49	12.92	13.09	13.02	13.02
CB	0.735	0.704	0.720	0.707	0.657	0.631	0.692	0.610	0.618	0.618
SHP	6.120	5.160	5.330	6.620	6.880	8.340	10.44	11.50	11.66	11.66
VK	15.00	14.57	14.55	15.55	16.01	17.00	18.04	18.55	18.55	18.55
ROI	20.50	18.00	21.00	20.50	25.50	26.50	26.50	29.50	29.50	29.50
CAP	649.0	648.0	649.0	642.0	646.0	639.0	649.0	649.0	650.0	650.0
Hull only without economic considerations ('H only')										
L				126.8	123.8	127.0	127.4			
B				22.21	22.04	22.21	22.62			
T				9.380	10.43	10.83	11.30			
D				11.30	12.29	12.69	13.15			
CB				0.698	0.648	0.619	0.593			
SHP				5.810	6.880	8.340	10.09			
VK				15.10	16.06	17.01	17.99			

only assures the designer that the generated design will not be extraordinary. Indeed, the experience and knowledge gained from past designs is important for directing the creation of a new design, but a new design should not be limited by past trends. A close analysis of the solutions generated must be made so as not to limit the design to the ordinary when the extraordinary is called for and is possible. For instance, if a particular design trend constraint is active in the final design for several, slightly varied models, then a close examination of the reasons for this limiting activity should be made. If appropriate the limiting constraint should be relaxed to allow the design to move to another, perhaps more meaningful, limiting constraint.

A design trend constraint that was consistently active in the parametric study was $(L/B)^{min}$; it was set at 5.6. The L/B ratio plotted against speed–length ratio for each case is shown in Fig. 8.6. It is observed that nearly all of the designs generated which include economic goal constraints ('H/E' and 'H/P/E') are limited by $(L/B)^{min}$. One might correctly reason that the economic considerations are driving the design to a minimum length. However, further investigation is probably warranted in order to establish the true lower bound on (L/B). It is suggested under such circumstances that the active design trend constraint be treated like a secondary goal. From a designer's perspective, trends should not be adhered to as strictly as system constraints, but they should be allowed to focus the problem. As a final comment here, the use of trend data is extremely useful and is not unique to naval architecture. The same type of experience-based design is used in almost all major engineering systems.

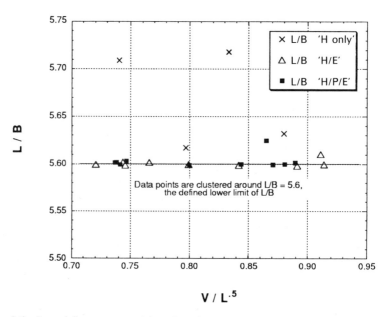

Fig. 8.6 Length/beam v. speed length ratio.

Remembering that the parametric study covered a target speed range of 10 to 24 knots, it is noted that results for 'H only' are only provided for target speeds of 15 through 18 knots. At either end of the spectrum, satisfactory convergence could not be achieved based on technical efficiency alone. While results for the runs involving the explicit economic goal produced results for cases with target speeds outside this range, the achieved speeds were still effectively held to this range. This phenomenon is depicted clearly in Fig. 8.7. In effect, at low speeds, the economics of the scenario said, 'it's all right to increase speed'. Similarly, at the other end of the spectrum, the best tradeoff between economic and technical efficiencies limited speed to about 18 knots.

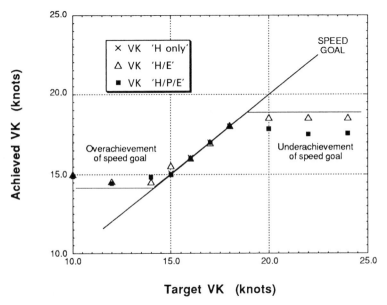

Fig. 8.7 Achieved speed v. target speed.

Some other observations are that at low speed ratios, the final designs as plotted seem to indicate two possible solutions, at least with respect to dimensional ratios. By exploring the convergence history for each solution (though not done here), we could perhaps gain some further insight. This would be particularly so if we were to identify converging subsequences generated by the optimization algorithm. Given that these subsequences were identified, what does this imply to the designer? Simply, it identifies that the 'alternate' solutions are equivalent in terms of goodness as measured by the current deviation function; while at the same time being dimensionally and technically different. However, this dichotomy could be resolved by modifying the current set of goals better to reflect the designer's aspirations. In contrast, at the higher speed ratios, there appears to be convergence across all scenarios to a single solution. This is clearly demonstrated by the plot of B/D against speed-length ratio as shown in Fig. 8.8. At low speed ratios, a B/D of either 1.5 or 2.0 is identified while at higher speeds, a value of 1.75, plus or minus 'delta,' is strongly suggested. Further, at these higher speed ratios there is still an observable trend to decrease B/D as speed–length increases. This is in keeping with conventional wisdom where depth increases at a faster rate than beam or length. It is well-known that length is a very expensive parameter to increase. Indeed, many estimates of cost are based on length alone. Increases in length and/or beam generally increase resistance. The effect of increasing depth on the other hand tends to affect construction costs only.

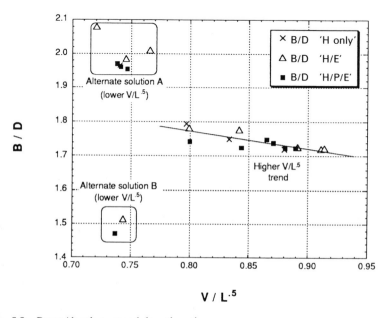

Fig. 8.8   Beam/depth v. speed–length ratio.

Finally, to lend further evidence to the claim that the model is reasonable for preliminary design studies, let us examine the trend in the displacement characteristics. Since the owner requires a fixed capacity of 650 TEUs, it is intuitively obvious that if this is achieved, displacement will increase with ship speed due to higher resistance, larger engines and greater fuel bunkering requirements. This trend is indeed evident in the model as shown in Fig. 8.9.

In summary, some anticipated trends have been highlighted lending weight to the validity of the template. Second, some insight regarding the interpretation of results has been provided, and third, some discussion regarding the use of the template to support the design process and its decision making has been entered into.

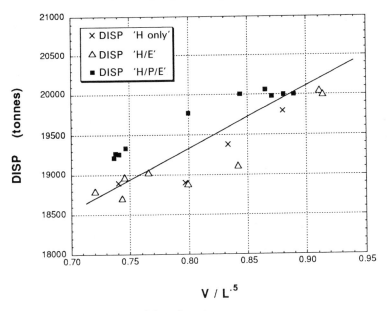

Fig. 8.9 Displacement v. speed–length ratio.

8.5 SOME COMMENTS ON IMPLEMENTATION

Presented is a decision-based approach to concurrent design called the decision support problem technique. Concurrency of design is achieved in one respect through the simultaneous resolution of 'derived' or multiple decisions (see Fig. 8.1). In another respect, all decision models, primary and derived, inherently integrate the required analysis and synthesis components of design and thereby support concurrent design activity (see Table 8.2). General experience with this approach across a range of disciplines has been very encouraging. This is evidenced by the example highlighted herein and those case studies cited through the reference list.

The development and validation of significant decision models is a non-trivial task and as with any modeling process, the law of 'garbage in – garbage out' applies. For meaningful results, a commitment to the development of the appropriate tool-box for a domain specific field of application is mandatory. This is particularly true during the initial DSPT implementation phase within an organization while the domain specific analysis tool library is developed into a modular and DSPT usable form. Further, education programs that encourage people to learn how to cope with and to be agents of change need to be developed. Designers need to learn how to make adjustments in their personal and/or organization's design paradigms and best practices. We are all creatures of habit. But, we must be prepared to adopt and merge new ideas in order to survive. We believe that once the tools and thought processes are in place, fruit from the concurrent act of designing complex artifacts should be quickly forthcoming.

8.6 GLOSSARY

Decision-based design (DBD): a term to emphasize a perspective for design (Shupe, 1988; Mistree *et al.*, 1989).

Decision support problem technique (DSPT): an implementation of decision-based design (Muster and Mistree, 1988; Shupe *et al.*, 1988).

Compromise – a primary DSP – the determination of the 'right' values (or combination) of design variables to describe the best satisficing system design with respect to constraints and multiple goals.

Derived DSPs – a combination of primary DSPs to model a complex decision, e.g., selection/selection, compromise/compromise and selection/compromise decisions (Smith, 1985; Smith *et al.*, 1987; Bascaran *et al.*, 1989; Karandikar, 1989; Bascaran, 1990).

Meta-design: a metalevel process of designing systems that includes partitioning the system for function, partitioning the design process into a set of decisions and planning the sequence in which these decisions will be made.

Designing: a process of converting information that characterizes the needs and requirements for a product into knowledge about a product.

8.7 ACKNOWLEDGMENTS

The parametric container ship data presented was derived from a report by Randy Emmons titled 'Preliminary ship design: technical and economic considerations'. This report was completed to satisfy the 'capstone' design requirement for his undergraduate degree.

Warren Smith as a naval architect of the Australian DoD (Navy) is currently attached to the Systems Design Laboratory. Bert Bras is similarly attached and is supported by MARIN: Maritime Research Institute Netherlands. The inherent contribution of each author's parent organization is duly recognized and appreciated.

The financial contribution of our corporate sponsor, The BF Goodrich Company, to develop further the Decision Support Problem Technique is gratefully acknowledged. An NSF Equipment Grant 8806811 is also gratefully acknowledged.

8.8 NOTES

1. We consider efficiency to be a measure of the swiftness with which information required by a designer is generated.
2. We consider effectiveness to be a measure of the quality of a decision (correctness, completeness, comprehensiveness) made by a designer.

3. {..., ...} indicates a set.
4. (..., ...) indicates arguments for the operator preceding the left parenthesis. The arguments separated by a ',' must be present for the operator to be valid.
5. Procedural knowledge is the knowledge about the process, i.e., knowledge about how to represent (and process) domain information (for design synthesis).
6. Declarative knowledge is the set of facts represented (usually) according to the protocol defined by the procedural knowledge. It is the knowledge about the product, i.e., the representation of problem relevant information, facts and background knowledge about the domain.
7. Decision support problems provide a means for modeling decisions encountered in design and the domain specific mathematical models so built and implemented on a computer are called templates.

8.9 REFERENCES

Bascaran, E. (1990) *A Conceptual Model for the Design of Thermal Systems: concurrent decisions in designing for concept*, Department of Mechanical Engineering, University of Houston, Houston, Texas.

Bascaran, E., Bannerot, R. B. and Mistree, F. (1989) *Engineering Optimization*, **14**, 207–38.

Bras, B., Smith, W.F. and Mistree, F. (1990) in *CFD and CAD in Ship Design*, (ed G. v. Oortmerssen) Elsevier Science Publishers B.V., Wageningen, The Netherlands, 221–31.

Bras, B.A. and Mistree, F. (1991) Designing design processes in decision-based concurrent engineering, *Proceedings SAE Aerotech '91*, SAE Publication SP-886, Paper No. 912209, Long Beach, California, SAE International, 15–36.

Holtrop, J. (1984) *International Shipbuilding Progress*, **31(363)**.

Holtrop, J. and Mennen, G. G. J. (1982) *International Shipbuilding Progress*, **29(385)**.

Ignizio, J.P. (1982) *Linear Programming in Single and Multi-Objective Systems*, Prentice-Hall, Englewood Cliffs, New Jersey.

Ignizio, J.P. (1983) *Computers and Operations Research*, **5(3)**, 179–97.

Ignizio, J.P. (1985) *Introduction to Linear Goal Programming*, Sage University Papers, Beverly Hills, California.

Ignizio, J.P. (1985) *European Journal of Operational Research*, **22**, 338–46.

Kamal, S.Z. (1990) *The Development of Heuristic Decision Support Problems for Adaptive Design*, Department of Mechanical Engineering, University of Houston, Houston, Texas.

Kamal, S.Z., Karandikar, H. M., Mistree, F. and Muster, D. (1987) in *Expert Systems in Computer-Aided Design*, (ed. J. Gero) Elsevier Science Publishers B.V., Amsterdam, 289–321.

Karandikar, H.M. (1989) *Hierarchical Decision Making for the Integration of Information from Design and Manufacturing Processes in Concurrent Engineering*, Department of Mechanical Engineering, University of Houston, Houston, Texas.

Karandikar, H.M. and Mistree, F. (1992) in *Structural Optimization: Status and Promise*, (ed. M.P. Kamat) AIAA, Washington, D.C.

Karandikar, H.M., Rao, J.R. and Mistree, F. (1991) in *Advances in Design Automation*, (ed. G.A. Gabrielle) ASME, New York, 361–9.

Klir, G.J. (1985) *Architecture of Systems Problem Solving*, Plenum Press, New York.

Kuppuraju, N., Ittimakin, P. and Mistree, F. (1985) *Design Studies*, **6(2)**, 91–106.

Lyton, T.D. and Mistree, F. (1984) *Journal of Ship Research*, **29(4)**, 251–69.

Mistree, F., Hughes, O.F. and Bras, B.A. (1992) in *Structural Optimization: Status and Promise*, (ed. M.P. Kamat) AIAA, Washington, D.C.

Mistree, F., Hughes, O.F. and Phuoc, H.B. (1981) *Engineering Optimization*, **5(3)**, 141–4.

Mistree, F., Muster, D., Shupe, J.A. and Allen, J.K. (1989) *A Decision-Based Perspective for the Design of Methods for Systems Design*, Recent Experiences in Multidisciplinary Analysis and Optimization, Hampton, Virginia, NASA.

Mistree, F., Muster, D., Srinivasan, S. and Mudali, S. (1990) *Mechanism and Machine Theory*, **25(3)**, 273–86.

Mistree, F., Smith, W.F., Bras, B., Allen, J.K. and Muster, D. (1990) in *Transactions, Society of Naval Architects and Marine Engineers*, Jersey City, New Jersey, 565–97.

Muster, D. and Mistree, F. (1988) *The International Journal of Applied Engineering Education*, **4(1)**, 23–33.

Oosterveld, M.W.C. and Oossanen, P. v. (1975) *International Shipbuilding Progress*, **22(251)**, 251–62.

Rich, E. (1983) *Artificial Intelligence*, McGraw Hill.

Shaher Sabit, A. (1972) *International Shipbuilding Progress*, **19**.

Shupe, J.A. (1988). *Decision-Based Design: taxonomy and implementation*. Department of Mechanical Engineering, University of Houston, Houston, Texas.

Shupe, J.A., Muster, D., Allen, J.K. and Mistree, F. (1988) in *Expert Systems, Strategies and Solutions in Manufacturing Design and Planning*, (ed. A. Kusiak) Society of Maufacturing Engineers, Dearborn, Michigan, Chapter 1, 3–37.

Smith, W.F. (1985) *The Development of AUSEVAL: An automated ship evaluation system*, Department of Mechanical Engineering, University of Houston. Houston, Texas.

Smith, W.F., Kamal, S.Z. and Mistree, F. (1987) *Marine Technology*, **24(2)**, 131–42.

CHAPTER 9

Concurrent optimization of product design and manufacture

Masataka Yoshimura

9.1 INTRODUCTION

Recently, the circumstances in product design and manufacturing of machine products have greatly changed. The times in which computer-aided systems such as computer-aided design (CAD), computer-aided manufacturing (CAM), computer-aided engineering (CAE) and computer-aided process planning (CAPP) were independently developed are changing to one in which these fields are integrated and product design and manufacturing are rationally and efficiently conducted using computer systems. So, CIM (computer-integrated manufacturing), concurrent engineering (Brazier and Leonard, 1990; Haug, 1990) and simultaneous engineering (Foreman, 1989) have attracted special interest recently. The major goals of these technologies are to realize higher product performance, lower manufacturing cost, shorter lead time and automation of a variety of low-volume production systems, etc. (Hitomi, 1979).

This chapter describes fundamental methodologies for concurrently optimizing decision making items concerning product design and manufacturing. Concurrent optimization is a key for realizing a product having higher product performance and lower product manufacturing cost from a global viewpoint.

9.2 FLOW AND RELATION OF RESEARCH AND DEVELOPMENT, PRODUCT DESIGN, MANUFACTURING AND MARKETING DIVISIONS

Figure 9.1 shows a conventional product manufacturing flow of research and development, product design, manufacturing and marketing and the

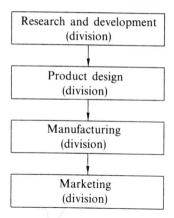

Fig. 9.1 Relation of research and development, product design, manufacturing and marketing divisions.

relations of these divisions. This is a sequential flow from an upper stream to a lower stream which does not allow mutual communication between divisions. Within each division, particular decision making is conducted according to the information from the upper divisions. Requirements and matters decided in the upper divisions are the constraints for the decision making in the lower division also. Conflicting requirements may exist among divisions but they cannot be solved by the sequential flow such as shown in Fig. 9.1.

9.3 DIRECTION FOR PRODUCT DESIGN

Figure 9.2 shows the direction for product design. Usually, designers are required to design products having higher product performance and lower product manufacturing cost. The arrow in Fig. 9.2 shows direction toward which designers seek to go.

A goal of product design is to design a product having a higher product performance and a lower product manufacturing cost. For realizing this goal, the sequential flow and relation of divisions such as shown in Fig. 9.1 is not sufficient. When an attempt to reduce manufacturing costs is made after the details of product designs have been decided, the amount of reduction is very small because the product design almost fixes the manufacturing cost and not much room for reducing manufacturing costs exists at the process design stage in which the manufacturing methods are determined.

Figure 9.3 shows the relation between the product performance and the product manufacturing cost. The shaded part corresponds to the region feasible using the present technologies, knowledge and/or theories. Designers are always seeking a higher product performance and a lower product manufacturing cost as shown in Fig. 9.2. That is, they are searching for

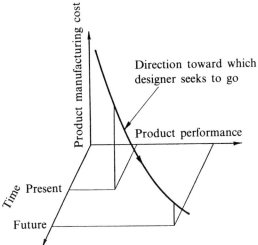

Fig. 9.2   Direction for product design.

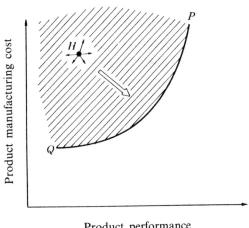

Product performance

Fig. 9.3   Relationship between the product performance and the product manufacturing cost.

designs along the direction of the big arrow shown in Fig. 9.3. The heavy solid line PQ corresponds to the Pareto optimum design solutions (the feasible design solutions in each of which there exists no other feasible design solution that will yield an improvement in one objective without causing a degradation in at least one other objective) (Cohon, 1978; Zeleny, 1982; Yoshimura *et al.*, 1984a; Stadler, 1988; Eschenauer *et al.*, 1990) of a multiobjective optimization problem having two objectives: maximization of the product performance and minimization of the product manufacturing cost. The solutions are a set of design points where both further improvement of the product performance and further reduction of the product manufacturing cost cannot be realized. The designers ultimately search for a design solution on the heavy line. All divisions shown in Fig. 9.1 are fundamentally related with arriving at a design solution on the Pareto optimum solution.

Now, let point H in the shaded region correspond to a present design point or an initial design point. Usual design improvements and design changes mean shifting of the design solution in radial directions from point H. In the state, before arriving at any specific level, improvement of the product performance and reduction of the product manufacturing cost can cooperate with each other, but after arriving at a specific level, that is, at a solution level on the Pareto optimum set shown by the heavy line in Fig. 9.3, these may conflict. If design decision-making is conducted without formulating systematically and integratedly the design problem as a concurrent optimization problem, it is often impossible to move a design solution to that on the Pareto optimum solution set of the PQ line.

9.4 FUNDAMENTAL CONCEPTS OF CONCURRENT OPTIMIZATION

In order to have good communications among divisions, representatives from individual divisions should gather for a meeting as shown in Fig. 9.4. All representatives can discuss design alternatives and design proposals together. Communications between divisions are actively made. Conflicting requirements can be solved by mutual consent.

The fundamental philosophy of concurrent engineering is to conduct mutual communications and concurrent decision making using computer

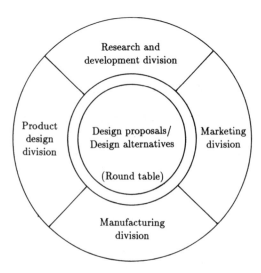

Fig. 9.4 Meeting of representatives from individual divisions for discussing design proposals laid out on a table.

systems. Design alternatives laid out on a table are replaced by ones on the computer displays and computer database. Communications are conducted by computer networks. In this case, representatives do not need to gather together in one room, but the same purposes are more efficiently and effectively conducted using computer systems.

The following means are necessary for conducting concurrent engineering:

1. Communication among divisions using network systems;
2. Sharing of common database; and
3. Simultaneous evaluation and decision making.

It is most important to focus on the techniques and methodologies of factor 3 in order to obtain a solution on the Pareto optimum solution set in Fig. 9.3, that is, to obtain an optimum solution from both viewpoints of product performance and product manufacturing cost. Therefore, this chapter's contents are devoted mainly to the study of these techniques and methodologies.

9.5 DECISION-MAKING ITEMS IN PRODUCT DESIGN AND PROCESS DESIGN

In the product design, configurations or shapes of products are determined under the evaluative criterion of the product performance. On the other hand, in the process design for manufacturing, methods for realizing the designed shapes are determined under the criterion of the manufacturing cost. Table 9.1 tabulates an example of decision-making items to be determined during the design process (Yashimura *et al.*, 1989). D_j are decision-making items in the usual product design and P_j are those in the usual process design.

9.6 SIMULTANEOUS EVALUATION OF PRODUCT PERFORMANCE AND PRODUCT MANUFACTURING COST

Simultaneous evaluation of decision-making items which include ones from different fields generally corresponds to larger-scale decision-making problems, but it also means looking at things from a systematic or integrated viewpoint. This may be effective for discovering a regularity from confused phenomena and may make determining optimum solutions easier.

In usual design optimization, consideration is only given to the improvement of the product performance which is indicated on the horizontal axis in Fig. 9.3. The product manufacturing cost and factors concerning manufacturing processes are scarcely considered. However, factors concerning manufacturing of the product parts processed according to the results of the

Table 9.1 Decision-making items in the product design and the process design

	Decision-making items in product design
D_1	Mechanism
D_2	Parts constitution of product
D_3	Connected relations among parts
D_4	Purchased parts or manufactured parts
D_5	Shape of part
D_6	Shape accuracy
D_7	Surface roughness
D_8	Dimension
D_9	Tolerance
D_{10}	Material
	Decision-making items in process design
P_1	Preparation process of raw material
P_2	Machining methods
P_3	Machine tools
P_4	Tools and jigs
P_5	Machining sequence
P_6	Quality of finished surface
P_7	Cutting conditions
P_8	Raw material shape
P_9	Heat treatment process

product design have a great influence on the product performance while also the product manufacturing cost can be said to be mostly dependent on the product design. Therefore, the product performance and the product manufacturing cost have close relationships.

Minimization of the structural weight corresponds directly to the reduction of the building cost of the structure in civil or architectural structures because the material cost of the structures represents a great portion of the total cost. However, in manufacturing machine products, the ratio of material cost to manufacturing cost is usually not as high. Especially in machines requiring high precision and efficiency, the cost of the manufacturing processes, such as machining and welding, fills a large portion of manufacturing costs. Therefore, in the design optimization of machine structures, evaluation of manufacturing costs is also important.

In the following, examples in which evaluation of product performance and evaluation of manufacturing cost have close relations for decision making of product design and manufacturing are explained.

9.6.1 Design decision for partitions and ribbings in structural members

In order to realize structural members having higher rigidity under a constraint of the constant weight, partitions and/or ribbings are added inside the structural members. But addition of partitions and ribbings results in increase of manufacturing cost such as welding cost. Figure 9.5 shows examples of design patterns of longitudinal partitions in cross-sections of structural members, where N is the total number of welded parts. Design of partitions and ribbings of structural members should be determined by both evaluating the rigidity of the structural members (product performance) and the welding cost (manufacturing cost) (Yoshimura *et al.*, 1984b).

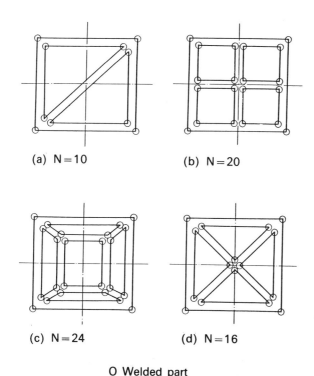

(a) N=10   (b) N=20

(c) N=24   (d) N=16

O Welded part

Fig. 9.5 Design patterns of longitudinal partitions in cross-sections of structural members (*N*: number of welded parts).

For example, in the minimization of the welding cost, detailed shapes of structural members are determined so that the welding cost ψ_0 is minimized under the constraint that the maximum value Δ_{max} of the local deformation must not exceed the permitted value Δ^U as follows:

$$\text{minimize } \psi_0 = \sum_{i=1}^{N} Q_i L_i$$

$$\text{subject to } \psi_1 = \Delta_{max} - \Delta^U \leqslant 0$$

where Q_i and L_i are the welding cost per unit length and the length of welded part i, respectively.

9.6.2 Deciding contact surface shape and machining method
for the contact surface

Machining accuracies of machine parts are related with both the product characteristics (performance) and the manufacturing cost. Detailed shapes of joints and machining methods of the contact surfaces are determined by evaluating the machining cost of the contact surfaces and the product characteristics (Yoshimura *et al.*, 1984b). The maximum surface roughness is related to the machining cost per unit contact area and the equivalent spring stiffness of joints, that is, the rigidity of joints.

Three kinds of methods for machining contact surfaces: super finishing, grinding and milling, are considered here. Figure 9.6 is an example of the relationship between maximum surface roughness (maximum height of irregularities) and the relative machining cost coefficient. The relative cost coefficient is the ratio of the machining cost of various machining methods to the machining cost of a milled surface having a roughness of H_{max} of 1.2×10^{-5} [m]. Along the heavy portion of the curves in Fig. 9.6, the minimum machining cost is realized for the same maximum surface roughness.

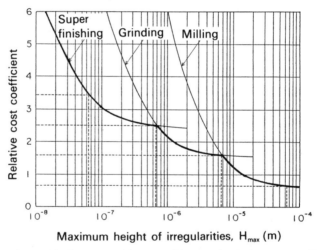

Fig. 9.6   Relationship between surface roughness and relative cost coefficient.

The formulation for this design optimization is, for example, to minimize the machining cost required for machining contact surfaces of structural members under the constraints of the rigidity of joints. The decision variables are the surface roughness (which corresponds to machining methods) and the shapes of the contact surfaces.

9.6.3 Design optimization using the concept of group technology

When manufacturing costs are not included in the product design decision making, all design variables have different values. That results in high product manufacturing cost. Around such optimum solutions, the product performance and the product manufacturing cost have conflicting relations.

Group technology (GT) plays a great role in decreasing the manufacturing cost of products. GT is a technique and philosophy which aims at increasing production efficiency by grouping a variety of parts having similarities in shape, dimension, and/or process routing. Group technology has been primarily used for improving efficiency in manufacturing (Ham *et al.*, 1985). In order to further promote standardization of machine products and parts, it is necessary to apply the concept of GT early at the product design stage. The application of GT at the design stage will allow for a positive use of this technique in the total production process from the design stage through the manufacturing of products (Yoshimura and Hitomi, 1986).

The use of GT from the product design stage yields the following advantages in the production processes:

1. In design: the number of parts newly designed is decreased and design automation and NC (numerical control) programming for automatic machining are easily conducted.
2. In production planning and management: process planning and scheduling are simplified.
3. In manufacturing: the type of production is changed from jobbing production to lot or mass production. Because several jobs are grouped and processed in sequence, the same jigs and tools may be employed, thus reducing setup time and setup cost for each job.

The foregoing factors lead to an increase in productivity and a reduction in manufacturing costs.

The use of identical or similar machine parts as often as possible in a product or series of products corresponds to the minimization of manufacturing costs. However, the use of GT brings about the decrease in the performance or characteristics of machines. In the design decision process, the conflicting relationships between the advantages and the disadvantages must be evaluated.

When a requirement for performance or characteristics of a product is selected as an objective function, the grouping having the minimum objective function will generally correspond to a product in which all machine parts have different design variables. Then, the incremental quantity of the objective function for each grouping from the minimum value of the objective function is calculated. Of the groupings in which the value of the incremental quantity of the objective function is less than the maximum allowable value, the one having the minimum manufacturing cost should be adopted as the optimum design solution.

Table 9.2 shows an example of groupings for the case in which the number of design variables is 5. Each of G1, G2, G3, G4 and G5 expresses the identical or a similar machine part. For example, in the case of group No. 3, the total number of groupings is 3 (G1, G2, G3) as parts 2, 3, and 4

Table 9.2   Examples of grouping of machine parts in a product

Group No.	Part No.				
	1	*2*	*3*	*4*	*5*
1	G 1	G 2	G 3	G 4	G 5
2	G 1	G 2	G 2	G 3	G 4
3	G 1	G 2	G 2	G 2	G 3
4	G 1	G 2	G 2	G 3	G 3
5	G 1	G 2	G 2	G 2	G 2
6	G 1	G 1	G 1	G 1	G 1

are identical or similar. The product consists of three kinds of machine parts. Figure 9.7 shows the schematic figure of a practical manipulator. Figure 9.8 shows the simulation model for structural analysis, where numbers in the figure denote machine part numbers. In this example, the total mass of the machine parts is required to be minimum under the constraint of the static deflection at the holding point H of a workpiece. This constraint guarantees handling performance, positioning accuracy, etc.

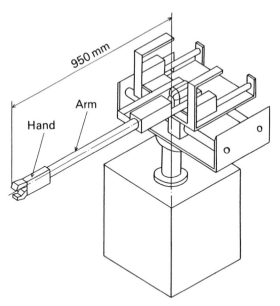

Fig. 9.7   Schematic figure of a manipulator.

Figure 9.9 shows the relationship between the levels of grouped parts listed in Table 9.2 and the incremental quantity of objective function (the total mass of the complete structure). A larger group number corresponds to the design in which the number of different cross-sectional shapes of parts is small. The incremental quantity of the objective function up to group 3 is

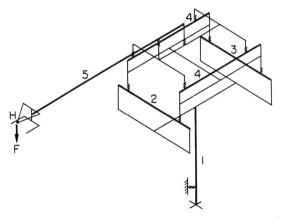

Fig. 9.8 Simulation model of the manipulator for structural analysis (numbers in the figure denote machine part numbers).

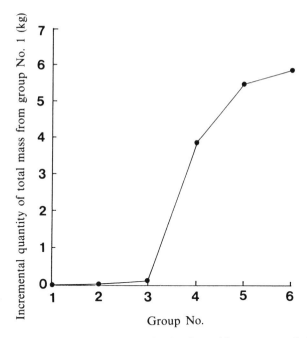

Fig. 9.9 Relations between grouping level of machine parts and incremental quantity of total mass from group No. 1.

very small but from group 4 the value of the objective function increases greatly. The design of group 3 may be selected for decreasing the manufacturing cost without great degradation of the product performance. Simultaneous evaluation of reduction of manufacturing cost and degradation of product performance generated by grouping the parts designs is effective for design decision making.

9.7 PROBLEMS AND FEATURES OF CONCURRENT AND INTEGRATED DECISION MAKING

It can be understood from the foregoing that simultaneous decision-making is effective. Generally, the decision-making problem is stated by adopting an evaluative factor to be minimized or maximized as the objective function and by setting requirements concerning the other evaluative factors and the lower and upper bounds of design variables as constraints. The following patterns of formulation exist for integrated optimization of product design and manufacturing.

Formulation 1: product performance→maximize
under the upper bound constraint of
the product manufacturing cost
Formulation 2: product manufacturing cost→minimize
under the lower bound constraint of
the product performance
Formulation 3: multiobjective problem
[product performance→maximize, product manufacturing cost→minimize]

The most suitable formulation is selected depending on the environmental conditions of design practice. Basically, Formulation 3 is most natural.

The concurrent and integrated optimization of product design and manufacturing has generally some or all of the following problems and features:

9.7.1 Large-scale nonlinear-programming design optimization

In the integrated design optimization, both the number of decision-making items and the number of decision variables increase greatly. Hence, the optimization problem is far larger than the usual design optimization problems. In the formulation of design optimization, the objective function and constraints are usually nonlinear functions of design variables. In such cases, a lot of local optimum solutions exist in the feasible design space. That is, it is very difficult to obtain the global optimal solution (a solution on the line PQ in Fig. 9.3).

9.7.2 Multiobjective design optimization

The integrated design optimization problem is in itself a multiobjective one as indicated in Formulation 3.

9.7.3 Discrete variable optimization

Discrete decision variables, for example, materials and manufacturing costs, are often included in the decision-making items concerning product design and manufacturing. That fact makes decision making difficult.

9.8 FUNDAMENTAL METHODOLOGIES FOR CONCURRENT DECISION-MAKING

For conducting effectively concurrent (simultaneous) decision making of many evaluative factors, multiobjective optimization procedures and multiphase decision making procedures can be used. These are fundamental techniques for solving the difficulties described in the foregoing section.

9.8.1 Application of multiobjective optimization methods

In the integrated decision making, many characteristic and evaluative factors are considered. Those often interact mutually and in a complicated fashion. In some cases, conflicting relationships exist among them. The relationships make evaluation and analyses of characteristics for design decision-making difficult. Specifically, designers are required to design a product having a higher performance and a lower manufacturing cost as shown by the big arrow in Fig. 9.3. However, the two objectives for improving the product performance and lowering the product manufacturing cost generally have a conflicting relationship. Conflicting relations often exist also among evaluative characteristics of product performance (Yoshimura, 1987). When the trade-off relations among the conflicting evaluative factors affect the design decision-making, analyses are effective which are based on a multiobjective design optimization problem in which the multiple evaluative factors are selected as objective functions (Yoshimura *et al.*, 1989).

9.8.2 Application of multiphase optimization procedures

Multiphase decision-making procedures are most effective when they are based on careful objective consideration of the subjects.

In the decision making of product designs, many characteristics and many kinds of costs are considered as indicated in Table 9.3 (Yoshimura and Takeuchi, 1991). Among them, characteristics and costs which can be obtained using simpler models should be evaluated at the upper level of the multiphase procedures.

The weight of structural members can be evaluated even if the modeling of machine designs is simple. Static and dynamic displacements can be evaluated using simplified models of structures while detailed shapes of structures are necessary for evaluating stress/strain distributions.

The material cost is obtained by multiplying the weight of members by the material cost per unit volume. Hence the material cost can be evaluated even if the modeling of machines is simple. However, for evaluation of machining costs and welding costs, information of detailed shapes of parts is necessary as already explained.

In a sequential flow shown in Fig. 9.1, after decision-making items corresponding to a division are determined, decision-making items

Table 9.3 Relations of simplicity levels of shape modelings with evaluable structural characteristics and manufacturing costs

Simplicity levels of shape modeling	Structural characteristics to be evaluated	Manufacturing costs to be evaluated
Modeling of general shapes	• Rough estimate of structural weight for candidate materials	• Rough estimate of manufacturing costs for candidatate manufacturing methods
Modeling of principal shapes	• Static and dynamic displacement • Natural frequency • Thermal deformations • Weight of structural members	• Manufacturing cost for making principal shapes and material cost
Modeling of detailed shapes	• Static and dynamic stress strain distributions	• Machining costs and welding costs for making parts having detailed shapes

corresponding to the succeeding division are determined. On the other hand, in the multiphase optimization procedures, more detailed decision variables are determined from fundamental ones by step according to the multiphase procedures. Feasible regions of decisions variables are decreased by step according to the multiphase procedures.

9.9 INTEGRATED DECISION-MAKING PROCEDURES OF PRODUCT DESIGN AND MANUFACTURING

Practical procedures for concurrently optimizing decision-making items concerning product performance and manufacturing cost are described.

9.9.1 Large-scale multiobjective optimization

Integration of product design and process design requires fundamentally simultaneous decision-making of the product performance and the product manufacturing cost. Decision-making problems for this integration correspond to Formulation 3 in section 9.7.

In the product design stage, configurations or shapes of products are determined under the evaluative criterion of the product performance. On the other hand, in the process design stage, methods for realizing the designed shapes are determined under the criterion of the manufacturing cost. Requirements for the product performance at the product design stage and those for the product manufacturing cost at the process design stage often have conflicting relationships. Therefore, usual sequential decision making at both stages may not bring about the global optimum design solution even if individual optimization is conducted at each stage. That is, optimization of both the product design and the process design should be integratedly conducted.

Each of the decision-making items has a relation with another. Therefore, they are not determined independently. Figure 9.10 shows an example of precedence relations among the decision-making items listed in Table 9.1. Determining a decision-making item not only depends on the items

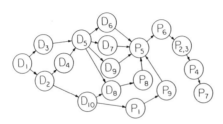

Fig. 9.10 An example of precedence relations among the decision-making items given in Table 9.1.

previously determined, but also it must satisfy the necessary conditions required by the precedence relations.

Decision-making procedures in the integrated design process are constructed as follows (Yoshimura *et al.*, 1989):

Step 1   A product configuration is defined from the product specification. The product has n_0 constituent parts, among which n_1 parts are manufactured in the work shop.

Step 2   Evaluative characteristics are extracted based on information of the product specification and parts configuration and are categorized into two kinds: (a) objective functions and (b) constraints.

Step 3   The conditions required for each decision-making item are determined based on the precedence relationship among decision-making items.

Step 4   The decision variables which satisfy specific conditions or those to be determined by decision rules are identified.

Step 5   An integrated optimization problem is constructed; the optimal solutions are obtained for decision variables which have not yet been determined.

The optimal solutions are obtained for decision variables. The integrated design optimization problem has the following features:

1.   The problem is a kind of multiobjective optimization.
2.   Optimal solutions can be obtained according to a specified order of priorities of multiple criteria.
3.   The shaded regions on the objective-function space as shown in Fig. 9.11 correspond to feasible design regions for various combinations of part materials. As shown in Fig. 9.11, each combination produces a Pareto optimum solution. The number of Pareto optimum solution sets increase with the number of candidate materials. Hence, an optimum solution must be effectively decided from among those Pareto optimum solution sets.
4.   In case of machine-product design, a specific satisficing level (satisfactory level) may be settled on each evaluative characteristic. Then, the design solutions are determined such that all the objective functions be within the required levels.

The outline of the procedures for obtaining the optimum solution is as follows:

1.   Obtain a Pareto optimum solution set which has a high possibility of containing the integrated optimum solution;
2.   Obtain a tentative solution within the satisficing level by considering the degree of relative importance among objective functions; and
3.   Check whether or not more preferable solutions exist near the above solution.

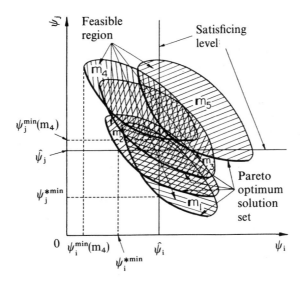

Fig. 9.11 Feasible regions and Pareto optimum solution sets for various materials on the objective-function space.

The integrated design optimization procedures were applied to the design of a cylindrical-coordinate robot shown in Fig. 9.12. Characteristics selected as objective functions in this example are: the static compliance at the installation point (H) of the hand on the arm, $f_S(=\psi_1)$, the

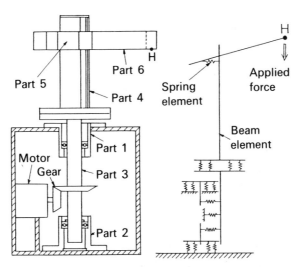

(a) Schematic construction

(b) Simulation model for structural analysis

Fig. 9.12 A cylindrical-coordinate robot to be designed: (a) schematic construction; (b) simulation model for structural analysis.

total weight of the structure, $W_T(=\psi_2)$, and the product manufacturing cost, $C(=\psi_3)$.

The integrated optimum design of the cylindrical-coordinate robot is formulated as a three-objective optimization problem as follows:

Minimize $\psi = [f_S, W_T, C]$
subject to the constraints concerning surface roughness,
tolerance and dimensions.

Main parts, 1, 2, and 3 of the robot shown in Fig. 9.13 are objects of optimization design, since those parts take an important role in the product performance. Candidate materials of those parts are cast iron (FC) and low carbon steel (SC). The Pareto optimum solution sets for $f_{S}v$. W_T and $f_{S}v$. C are shown in Fig. 9.14 (a) and (b). Point Q is a tentative solution on the Pareto optimum solution which was first searched for and point A corresponds to the final optimum design solution.

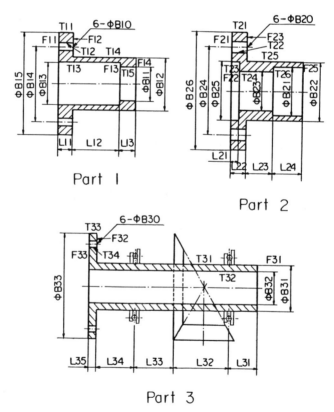

Part 1

Part 2

Part 3

Fig. 9.13 Symbols of machined surfaces and dimensions for parts to be designed (B, F, L and T are symbols added to surfaces and dimensions).

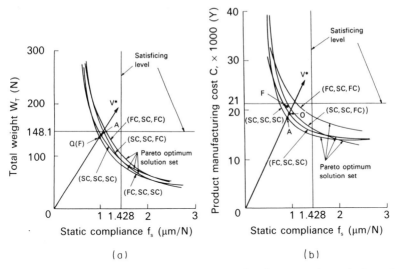

Fig. 9.14 The Pareto optimum solution sets, the first tentative solutions (Q), and the final optimum solutions (A) (the three symbols in parentheses indicate materials used for parts 1, 2 and 3 in that order): (a) $f_s - W_T$ relation; (b) $f_s - C$ relation.

9.9.2 Multiphase design optimization strategy using simplified structural models

The decision making process of design is generally divided into two phases: fundamental design stage and detailed design stage. Usually, design optimization is scarcely applied to the fundamental design stage, but applied to the detailed design stage. However, for a design in which a practical design configuration is given, room for design change is small, and great improvement of the product performance (and/or great reduction of the product manufacturing cost) cannot be expected as a matter of course. Even if the global optimal solution can be obtained in a design optimization problem using a detailed design model, it may only be a local optimal solution when the optimization is considered from the stage of fundamental design. In order to obtain a great improvement of the product performance, design optimization or design decision making starting from a simplified or an idealized model is necessary.

(a) Minimization of manufacturing cost

In order to obtain optimum designs of practical machine structures, a multiphase design optimization method using simplified structural models has been presented (Yoshimura *et al.*, 1983). As shown in Fig. 9.15, the process of the design optimization procedures is divided into three

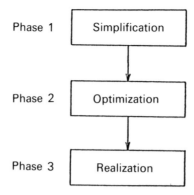

Fig. 9.15 The flow of a multiphase design optimization method using simplified models.

phases, 'Simplification', 'Optimization' and 'Realization'. Design optimization using simplified structural models has the following advantages:

1. Since a machine structure is simplified into a structural model having a smaller number of design variables in the first phase, mathematical programming methods for optimization can easily be applied.
2. In most design optimization problems, the objective function and the constraints are nonlinear functions of design variables. In such cases, it is usually difficult to obtain a global optimum. Use of suitable simplified models may reduce the possibility of poor convergence into a local optimum.
3. Design optimization which uses the structural model faithfully according to its detailed shape as an initial model leaves little room for design choices and design changes, so it is difficult to expect great improvements in the objective function. In the design optimization using simplified structures, great improvements are expected because large design changes are made through a simplified model, allowing much room for design choices and design changes.

In the first phase, 'simplification', mathematical models or simulation models which have structural properties equivalent to the practical structures are constructed. The left-hand side of Fig. 9.16 shows examples of cross-sectional shapes for column members of practical machine structures. These are transformed into simplified box-type members having equivalent cross-sectional characteristic values. On the other hand, practical joints having complicated contact surface shapes are modeled by structural elements having equivalent rigidities and energy dissipation ability.

In the second phase 'optimization', design optimization is conducted for the simplified model, using a mathematical programming method. Since the number of design variables is greatly reduced in this model compared to that of the practical structure, application of mathematical programming

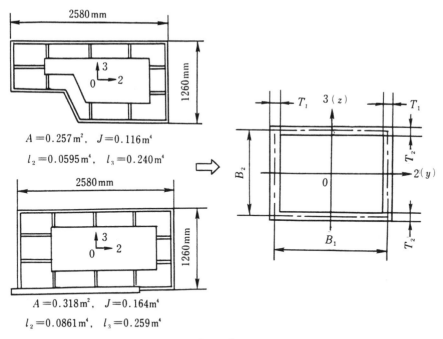

$A = 0.257 \, \mathrm{m^2}, \quad J = 0.116 \, \mathrm{m^4}$

$l_2 = 0.0595 \, \mathrm{m^4}, \quad l_3 = 0.240 \, \mathrm{m^4}$

$A = 0.318 \, \mathrm{m^2}, \quad J = 0.164 \, \mathrm{m^4}$

$l_2 = 0.0861 \, \mathrm{m^4}, \quad l_3 = 0.259 \, \mathrm{m^4}$

Fig. 9.16 Simplification of structural members.

methods is easy and the possibility of a poor convergence into some local optimum solution can be reduced.

Finally, in the third phase 'realization', concrete shapes, configurations and dimensions of a practical machine structure are determined so that all requirements defined in the optimization phase are satisfied (if it is difficult to satisfy all the requirements because of having too many requirements, permissible ranges are defined for some of them by evaluating sensitivity of the objective function). In this realization phase, requirements for additional designs and functions are considered. For example, ribs and partition plates are attached to practical structural members in order to avoid local deformations of the members, and various elements are added from the standpoint of functional requirements. Figure 9.5 shows an example of practical cross-sectional shape patterns. After a cross-sectional pattern is selected, practical design variables are determined so that the optimized cross-sectional characteristic values are realized. In the realization stage of a joint design, a suitable joint configuration pattern is selected from among various joint shape patterns, considering functional requirements such as sliding and fixing. Those 'realized' structural members and joints are synthesized into a complete machine structure having detailed shapes.

The multiphase design optimization procedure shown in Fig. 9.15 can be applied to design problems minimizing the product manufacturing cost (material cost, machining cost and welding cost) of machine-tool structures under the constraints concerning machining accuracy and machining productivity. The outline of the procedures is shown in Table 9.4 (Yoshimura *et al.*, 1984b).

Table 9.4 Design optimization procedures for minimizing the manufacturing cost

Phase 1	Simplification	Simplified machine structural models are constructed
Phase 2	Optimization	The material cost of the complete structure is minimized subject to the constraints of machining accuracy and machining productivity
Phase 3	Realization	1 Design of partitions and ribbings of structural members is determined so that the welding cost is minimized
		2 A machining method for contact surfaces is determined so that the machining cost is minimized

The material cost is minimized by evaluating rigidity and vibrational characteristics for the simplified model of a complete machine structure. The welding cost and the machining cost are minimized at the same time that the practical detailed shapes are determined.

(b) Application of group technology

A multiphase design optimization strategy using the simplified machine-structural models shown in Fig. 9.15 is applied to the design optimization using the concept of group technology (Yoshimura and Hitomi, 1986). In usual design optimization, a machine structure is modeled according to the detailed shape of the final design. However, since in such modeling the detailed practical shape has been determined, it is difficult for identical or similar machine parts to be used as often as possible after the design optimization. In a design optimization using a simplified machine structure, the use of the identical or similar machine parts is easy, because the detailed design is determined in the realization phase.

The design decision process consisting of the three phases; simplification, optimization, and realization, corresponds to the simplified parts classification and coding system according to group technology, as shown in Fig. 9.17.

In the simplification phase, machine parts are divided into two basic groups; box-type parts and cylindrical or bar-type parts. One of these basic

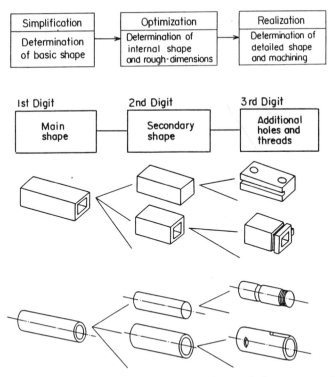

Fig. 9.17 Simplified parts classification and coding, and the corresponding geometric part examples.

shapes is selected as a simplified model part. In the optimization phase, internal shapes and rough dimensions of the parts are determined. According to the degree of difference in the shapes and dimensions, the machine parts are divided into more detailed groups. Finally, in the realization phase, detailed shapes of the parts are determined by considering additional design and functional requirements not considered in the optimization phase. The difference in detailed designs corresponds to the differences in internal shapes, plane surface machining, auxiliary holes, threadings and so on. According to the degree of difference in the manufacturing processes, the machine parts grouped in the optimization phase may be divided into groups which allows for more specific details.

9.9.3 Integrated optimization procedures of product design and manufacturing using multiphase modeling

In order to conduct more effectively integrated optimization of product design and manufacturing, multiphase procedures in which decision making is conducted from the fundamental design through detailed design in steps were constructed (Yoshimura and Takeuchi, 1991). Here, modeling machine parts corresponding to each phase acts an important role.

Table 9.3 shows relations of simplicity levels for shape modelings with evaluable structural characteristics and manufacturing costs. These are examples which can be applied to machine products such as machine tools and industrial robots requiring higher accuracy and operating efficiency. When general shapes of machine structures and manufacturing methods such as welding and casting are given, rough estimates of the weight of the structures and the manufacturing cost are possible. At the lower level of the table, more detailed design shape information is defined.

The evaluative factors categorized in Table 9.3 can be determined corresponding to the multiphase decision-making procedures. The multiphase decision making procedures in integrated product design and manufacturing are as follows:

1st phase: General shapes and alternative manufacturing methods for machine products are determined.

2nd phase: Main dimensions and manufacturing methods of machine parts for complete machine products are determined. Machining costs for principal contact surfaces of parts are roughly calculated for evaluating the product manufacturing cost. Main dimensions and manufacturing methods are determined by evaluating the principal product performance and the principal manufacturing cost (machining costs for principal contact surfaces) for complete machine products.

3rd phase: Detailed shapes and machining procedures for machine parts are determined. The detailed shapes and the machining procedures considered here do not have a direct effect upon the principal product performance nor the principal product manufacturing cost.

Each phase procedure corresponds to smaller design problems: (Phase 1) a design problem for simple shape models in which the principal product performance and the principal manufacturing cost can be roughly evaluated; (Phase 2) a design problem for fundamental design models in which the principal dimensions of machine parts and the principal manufacturing methods can be determined; and (Phase 3) a design problem for detailed models in which secondary product performance and secondary product manufacturing costs can be evaluated for each machine part.

In the process of the multiphase decision-making, the optimum solution can be effectively attained by a gradual reduction in steps of the feasible region of the design variables. The shape description method giving main consideration to contact surfaces facilitates the simultaneous evaluation of the product performance and the product manufacturing cost.

9.10 CONCLUDING REMARKS

In this chapter, methodologies for simultaneously determining both decision-making items of product design and manufacturing were described. The methodologies are principal techniques in concurrent engineering and bring about optimum designs from both sides of product performance and product manufacturing cost. Further integrated decision-making including other divisions such as the marketing division shown in Fig. 9.4 is considered to be effective for the advancement of concurrent optimization of product design and manufacturing.

9.11 REFERENCES

Brazier, D. and Leonard, M. (1990) *Mechanical Engineering*, January, 52–3.

Cohon, L.L. (1978) *Multiobjective Programming and Planning*, Academic Press.

Eschenauer, H., Koski, J. and Osyczka, A. (Eds.) (1990) *Multicriteria Design Optimization*, Springer-Verlag.

Foreman, J.W. (1989) Gaining competitive advantage by using simultaneous engineering to integrate your engineering, design, and manufacturing resources, *Proceedings of Autofact '89*, October, 14.7–14.20.

Ham, I., Hitomi, K. and Yoshida, T. (1985) *Group Technology – Applications to Production Management*, Kluwer-Nijhoff.

Haug, E.J. (ed.) (1990) Concurrent engineering of mechanical systems, proceedings of the 1990 ASME Design Technical Conferences – 16th Design Automation Conference, September.

Hitomi, K. (1979) *Manufacturing Systems Engineering*, Taylor & Francis.

Stadler, W. (ed.) (1988) *Multicriteria Optimization in Engineering and in the Sciences*, Plenum Press.

Yoshimura, M., (1987) *ASME Journal of Mechanisms, Transmissions, and Automation in Design*, **109**(1), 143–50.

Yoshimura, M., Hamada, T., Yura, K. and Hitomi, K. (1983) *ASME Journal of Mechanisms, Transmissions, and Automation in Design*, **105**(1), 88–96.

Yoshimura, M., Hamada, T., Yura, K. and Hitomi, K. (1984a) *ASME Journal of Mechanisms, Transmissions, and Automation in Design*, **106**(1), 46–53.

Yoshimura, M., Takeuchi, Y. and Hitomi, K. (1984b) *ASME Journal of Mechanisms, Transmissions, and Automation in Design*, **106**(4), 531–7.

Yoshimura, M. and Hitomi, K. (1986) *ASME Journal of Mechanisms, Transmissions, and Automation in Design*, **108**(1), 3–9.

Yoshimura, M., Itani, K. and Hitomi, K. (1989) *International Journal of Production Research*, **27**(8), 1241–56.

Yoshimura, M. and Takeuchi, A. (1991) Integrated optimization in computer-aided design and manufacturing of machine products based on shape descriptions of contact surfaces, Proceedings of the 1991 ASME Design Automation Conference.

Zeleny, M. (1982) *Multiple Criteria Decision Making*, McGraw-Hill, New York.

CHAPTER 10

Computer-based concurrent engineering systems

Michael J. O'Flynn and M. Munir Ahmad

10.1 INTRODUCTION

Realisation of the concurrent engineering concept requires intimate cooperation and integration across the upstream product development and
downstream functional groups. Such integration can present numerous
difficulties since organizations have evolved into very specialized disciplines
often geographically distributed. Exchange of design information in a
diversity of formats without appropriate mechanisms of communication,
cooperation and coordination can lead to immense bottlenecks, suboptimal
designs and avoidable flaws. Such complex integration and cooperation, in
today's engineering design environment, can only be achieved by means of
computer-based systems. They provide a means for the efficient accumulation and distribution of information across geographic boundaries. By
means of specialized computer applications and intelligence-based tools,
these data can be structured, analyzed and molded into the design, enabling
the designer to achieve the optimal configuration. Computer-based tools
help simplify the effort and shorten the time required to implement concurrent engineering (CE) and design for manufacture (DFM). However, there is
a lack of extensive development in this field today (Lu *et al.*, 1988).

Noncomputerized concurrent engineering methodologies can be rather
cumbersome and time consuming to use. There is a large dependency on the
designer's understanding of the benefits of employing such methods and a
willingness to expend off-line effort. If a designer has to search large libraries
of data to acquire the information, it is likely the efficiency of the design will
be reduced. Ideally, such techniques should be integrated with the design
system. Automatically advising the designer on the direction in which to
proceed with the product design requires comprehensive knowledge on the
functionality of each part and an adequate model of the downstream
processes. The rapid development in recent years of artificial intelligence

techniques and knowledge-based systems provided a powerful method for managing such tasks. The purpose of this chapter is to explore computerized systems that support the concurrent engineering product development environment. It is divided into two main sections:

1. In Section 10.2, a DFM cost-based system aimed specifically at printed wiring board (PWB) assembly and test, is presented (this system, which is in the prototype stage, has been developed at the Digital facility in Galway, Ireland); and
2. A number of other computerized systems currently in use or under development in industry, that support the concurrent engineering environment are briefly reviewed in Section 10.3.

10.2 A DFM COST-BASED SYSTEM FOR PWB ASSEMBLY

This section describes a cost-based DFM software system, developed at the Digital facility in Galway, Ireland. It is aimed specifically at PWB assembly and test although its logic is applicable to other electronic design and manufacture processes such as PWB fabrication, semiconductor packaging, computer systems design and assembly. This system consists of a number of integrated modules (Fig. 10.1). A data extraction module (DEM) accesses the design files and extracts the information necessary to analyze the PWB. A placement assignment module (PAM) assigns each component to a PWB assembly placement step and determines the optimum soldering method(s). A process flow generator (PFG) then adds the ancillary process steps and determines the overall sequence of operations.

Sufficient information is generated by these modules on the product and its processing requirements to analyze the manufacturing cost impact. This analysis is carried out by the process cost module (PCM) using the activity based costing (ABC) approach. The output cost data can then be analyzed, particularly high cost contributors, which may result in alternative design concepts being proposed. The DFM system can be re-run to quantify the benefits of the proposed changes in terms of manufacturing costs. This system is currently in the prototype stage.

10.2.1 A cost-based DFM strategy

The general purpose of a DFM system is to analyze a product design, assess its impact on manufacturing, and provide effective and meaningful feedback to the design team. Data provided by the DFM system enables product/ process tradeoff decisions to be taken. The analysis and evaluations should be jointly carried out by the product designers and manufacturing.

A number of DFM methods have been devised that are qualitative in nature. One common example are those that employ DFM rules or

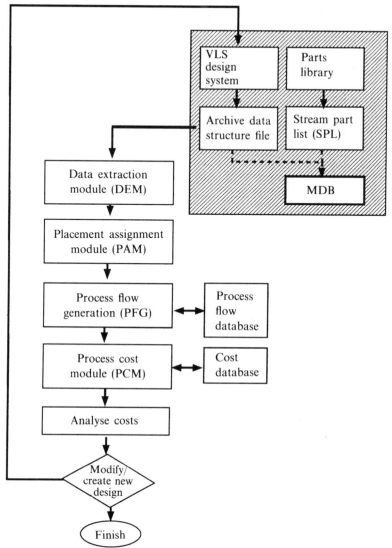

Fig. 10.1 Cost-based DFM system.

guidelines. These provide the designer with statements of good design practice that have been derived from past experience. They generally do not state in quantitative terms the impact of not adhering to the guidelines. It is therefore very difficult for the designer to determine the benefits of adopting the guidelines. The Pugh method of concept selection is another example (Pugh, 1981). Based on a set of criteria, this system indicates which are the better concepts. No quantitative information is provided on the superiority of a chosen design.

DFM systems that carry out analysis and provide quantitative information on the proposed design are significantly more effective. They provide real data on the goodness of the design, allowing for easier comparisons of design proposals. A number of these systems are based on minimizing the number of parts, part variation, or the number of manufacturing operations (Boothroyd, 1988). Others analyze specific design and manufacturing features of the concepts and generate scores which can be used for comparative purposes (Miyakawa and Ohashi, 1986; Miles, 1989).

All DFM systems, whatever the approach, share a common goal, that is to minimize the total product life-cycle costs. For example, systems which minimize part quantity and variation lower the material cost of the product. Providing quantitative cost data to the designer is a very effective means of impacting positively the product development process. The manufacturing cost can be reduced as less operations may be required. Conradson (1988) identifies cost-based DFM tools as an effective means of determining the impact of a design on manufacturing and providing information in an appropriate level of detail. Mazzullo (1989) stresses the need to maximize total product performance by establishing cost estimates during concept formulation. It is at this stage that the greatest cost flexibility exists.

The purpose of the PWB DFM system described here is to provide cost-based information to the development team on the manufacturing impact of the proposed design. The system can be used during the conceptual phase of the development cycle.

10.2.2 Sourcing design data

DFM systems require data on the product design as it evolves. Acquiring this data in the right format during the early design phases is not always easy, particularly if the design and manufacturing facilities are globally distributed. Ideally the data should be extracted electronically. The Vax Layout System (VLS) is the Digital CAD system for creating and modifying PWB designs. The system is VAX based with a complete set of applications that are typical of most PWB design systems. It can be linked to other systems to carry out performance and behavioral modeling, timing analysis, and simulation. VLS offers flexibility to the designer by allowing the use of dummy components if qualified ones are not available.

Digital uses a manufacturing database (MDB) to provide manufacturing with data on a PWB design. It contains a number of detailed files that describe the product. However the MDB is not available during the earlier design stages and for this reason it was not used as input to this DFM system. VLS generates an archive data specification (ADS) file containing detailed data on the proposed design. This file can be created during the initial design stages even before placement and routing are complete. It may contain dummy parts if qualified ones are not available. The DFM system proposed here uses this file as the sole source of design data. This allows the system to be executed before any hardware commitment is made. Its effectiveness is therefore increased as it is during the earliest design stage that manufacturing inputs have the greatest impact on reducing product costs.

10.2.3 Data extraction module (DEM)

The ADS file, which is in ASCII notation, describes the physical and electrical properties of the components (parts) and networks (signals) and the basic features of the proposed design. It is divided into a number of sections which in turn are further split into records, subrecords, and fields each containing design data. The precise information in each field is specified in the ADS specification document (Digital Equipment Corporation, 1990). Much of the data are in coded format and can be deciphered by referencing this document.

The DEM generates two output files. The first provides data on the printed wiring board (PWB). The vertices that describe the PWB outline are included in here. The second supplies data on each component assembled onto the board. Two types of component data are provided:

1. Generic data on each part number used on the PWB, e.g. lead quantity, part size; and
2. Data on the PWB-component relationship, e.g. position on PWB, rotation.

The component output file contains one record of data for each part on the PWB. The PWB and component files provide the DFM system the information necessary to evaluate the design and assess its cost impact on manufacturing.

The DEM provides the user with two options. Option one extracts data from the ADS file and stores it in an output file structured specifically for input to the other modules of the DFM system. The second option permits the use of the DEM as a stand alone module. This option generates customized data at the request of the user. Data of this type enables the manufacturing engineer to carry out activities such as set-up analysis, preliminary machine programming, capacity analysis and equipment justification.

10.2.4 Placement assignment module (PAM)

The purpose of the placement assignment module (PAM) is to assign each component on the PWB to an assembly placement step (e.g. auto DIP insert, SMT pick and place) and determine the optimum soldering method(s) (e.g. reflow, wave solder, standing wave). The two files generated by the DEM are the primary inputs. Analysis is carried out on the component package type, technology mix and board layout. Component placement assignment is carried out in three phases.

In phase 1 each component is individually checked and assigned to a placement step based on the package type. The relationship with other components, the technology mix and overall layout are not considered during this phase. These can have a significant bearing on the final placement assignment and the soldering method. Analysis of this type is carried out during phase 3. Components are first assigned to a placement step by referencing the component insertion code (IC). In general, all *qualified* components on the Digital master parts file will have a specified IC. Each IC has a defined set of package characteristics (Digital Equipment Corporation, 1991). Some parts may be assigned a default IC. In this instance it can only be established if the part is PTH or SMT. Other parts may not have any or have invalid codes. In this case additional component data extracted by the DEM are analyzed, and using a set of guidelines, placement assignments are carried out.

Phase 2 provides an option for manually assigning components that are not assigned during phase 1. This may be required if little or no data on a component is available in the ADS file. Experiments on a sample of ten test case PWB designs have shown this to be extremely rare. It was felt, however, that the option should be available as it may be required on occasion.

During phase 3 the PWB as a whole is analyzed. The relationship between components, their relative position on the PWB and the technology mix (PTH, 50-mil, 25-mil) are analyzed. This can alter the assignments made during phases 1 and 2. For example, if there are PTH and SMT components on the same PWB, some of the side 2 SMT components may need to be assigned to the Epoxy operation. The second primary purpose of phase 3 is to determine the soldering method(s). The default soldering operation for SMT is reflow and wave solder for PTH. However, some or all the PTH components may be soldered manually or, depending on the technology mix and board layout, the standing wave may be required. It is quite common also for side 2 SMT components (discrete only) to be wave soldered. Both the soldering methods and reassignments are established using a set of defined rules.

The PAM generates two output files, one each for the process flow generator and process cost module.

10.2.5 Process flow generator (PFG)

The PFG generates the process flow for the design under evaluation. It determines the additional non-placement steps (e.g. screen print, epoxy cure, aqueous clean) and the correct sequence for the complete set of operations, both placement and nonplacement. The manufacturing technical baseline for the PWB has then been established.

A database of all permutations of placement steps is generated. The PFG adds the appropriate ancillary steps (PTH and SMT) using a set of defined rules. The rules are based on the placement steps and the soldering operations used. An output file outlining the manufacturing operations required to assemble and test the PWB is generated for use by the process cost module.

The user is provided with two additional options. The first allows for the addition of nonstandard process steps (e.g. post reflow heat sink attach). As nonstandard operations are generally more costly, designs that require them should be discouraged. This should be done during the conceptual design phase when the cost of change is at its lowest. The second option provides the facility of adding etch cut and wire attach operations. This information is available in the ADS file but as these steps may be eliminated in future design revisions it was decided to allow the user the option of including or excluding them.

10.2.6 Process cost module (PCM)

Sufficient information has now been made available on the product and its processing requirements to analyze the cost impact on manufacturing. Product and process data generated by the PAM and PFG is used as input. The PCM utilizes the activity based costing (ABC) approach to determine the product costs.

Activity based costing (ABC) provides meaningful product cost and profitability information (Jeans and Morrow, 1989). It is a method that identifies the activities performed, traces cost to these activities, and uses cost drivers to trace the cost of the activities to the product. An activity is defined as a process or procedure that causes work and therefore consumes resources. A cost driver is a measure of the frequency and intensity of the demands placed on the activities by the product. Some systems use multiple cost drivers and allocate pools of cost to each cost driver. Each cost pool contains the resources consumed by several activities.

Traditional PWB assembly costing methods do not reflect the actual costs borne by a product, particularly overhead costs (Druary, 1989). They are generally based on cost per placement and do not account for variances in technology (e.g. plated through hole (PTH) surface mount technology (SMT)), volume, and batch sizes. The ABC system more accurately reflects the actual manufacturing cost burden of a product. Cost drivers at each process step are analyzed at a subactivity level to determine set-up, run-time and overhead costs.

Costs are determined under a number of categories. The primary categories are direct labour, quality (inspection, repair, retest costs), consumable, depreciation, occupancy, and indirect labour. The user has the option of excluding specific cost categories. For example, manufacturing may deem it appropriate to exclude depreciation costs if it is known that the capital base will not be impacted by the introduction of the proposed product.

The process cost model requires detailed data relating to the manufacturing process. These are provided via the process parameter editor (PPE) in three separate files (Fig. 10.2):

- The first contains the set up and run times for each process step – there can be several categories of these two parameters for each step;
- The second outlines component part numbers that are part of the fixed machine set-up; and
- Finally, the third file contains general process and material data. Examples include, batch size, consumable cost, depreciation and occupancy rates, and component defect data.

A set of files containing default data is available. Alternatively the process parameter editor is used to modify parameters and create a new set of files tailored specifically for the product and manufacturing process under analyses.

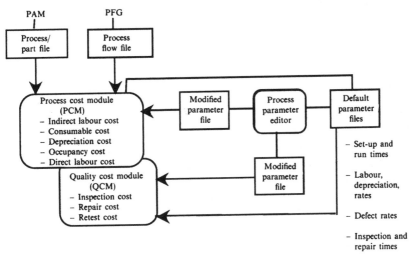

Fig. 10.2 Process cost module (PCM).

10.2.7 Quality cost module (QCM)

The PCM has an integrated quality cost module (QCM) (Fig. 10.2). This generates inspection, repair and retest costs. These vary significantly by PWB technology and package type. For example, 25-mil components yield poorer quality and are more costly to inspect and repair than 50-mil devices.

The defect rates and inspection and repair times at post-placement and post-reflow for the various component categories are determined based on historical data. A quality parameter file is generated and input to the quality cost module (QCM). As with other process data, this can be modified using the parameter editor. The QCM analyzes the components and, based on the quality parameters and process steps used, calculates the quality cost impact.

10.2.8 Material cost module (MCM)

It is planned to develop a material cost module and integrate it with the DFM system. Material cost represents a significant portion (up to 90%) of the overall product cost. It can therefore have a major impact on design and technology selection decisions. The purpose of the MCM will be to calculate the material costs. It will access the Digital corporate parts database which provides cost data on qualified components. An option will be provided for manually entering cost data on proposed new parts. Cost information on these parts may not be available on the corporate part lists. A cost premium will be levied if new part numbers are required on a design. The cost of introducing a new part number is significant. Each has to be sourced and qualified for reliability and process compatibility. Also, additional tooling such as placement machine feeders may be required.

10.3 COMPUTER-BASED CONCURRENT ENGINEERING SYSTEMS

The following sections briefly describe a number of other computer-based systems that support the concurrent engineering environment. A knowledge-based design for assembly (DFA) system is first presented. The features-based approach, including an application for designing casting products, is explored. Other techniques, including computer aided design with integrated suggestive systems, and variation simulation analysis (VSA) systems are briefly reviewed. Finally, a printed wiring board (PWB) DFM system, developed by Texas Instruments, is outlined.

10.3.1 A knowledge-based solution to DFA

A computer-based solution for concurrent engineering must have an inherent facility for incorporating into the design problem and solution, inputs from all functions both upstream and downstream. Qualitative techniques such as design rules and guidelines, and quantitative techniques, for example, the Boothroyd and Dewhurst (1987) DFA evaluation method, do not offer the designer a set of alternatives or suggest any improved concepts. Kroll *et al.* (1988) have developed a knowledge-based solution to the DFA problem, which automatically advises the development team on how to optimise the product design (Fig. 10.3). A brief overview of the system is outlined below.

The purpose, inputs and outputs, rather than a detailed working description of each module are presented. In the following sections of the chapter, a

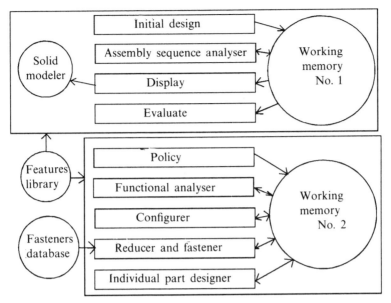

Fig. 10.3 Framework of DFA knowledge-based system.

more detailed functional analysis of intelligence-based techniques, that could typically be included as modules in this system, is described. The system is written in Turbo-PROLOG.

(a) Initial design module

An existing or new design concept is entered utilizing this module. Each part is represented by the following attributes; part number, name, quantity, type, material, geometry and topology. A database of existing parts is stored in libraries and retrieved upon request. For each feature or part, the user is prompted for the required parameters. Default values are shown where feasible. The topology describes the mating conditions of the features. Once completely entered, the assembly is stored in the main memory.

(b) Assembly sequence analyzer

The primary purpose of this module is to identify the sequence in which the parts of the product are assembled. The product is disassembled or exploded and based on a series of rules and mating conditions, the parts are sorted into the assembly sequence. This module also identifies the assembly direction of each component and subassembly which do not satisfy predefined constraints. All data is again stored in the main memory.

(c) Display

The product design or any aspect of the design can be presented or displayed on a graphics terminal. It is translated from the qualitative description to a constructive solid geometry (CSG) representation.

At this point the design for assembly session is activated. The system has five expert modules arranged in a hierarchical 'blackboard' architecture. Each of these modules treats the product at a lower abstraction level than its predecessor. When a module cannot satisfy the goals dictated to it from further up the hierarchy, the failure is returned to the higher level module for reconsideration of the decision.

(d) Policy module

The designer is asked to input the goals of the design process. This generates a set of specifications that need to be satisfied at a later stage in the design cycle.

(e) Functional analyzer

This module applies extensive knowledge about different machine elements and their function, and identifies parts which support primary functions and those that support secondary ones.

(f) Configure module

This is one of the more complex modules of the system and it plays an important role. It attempts to restructure the various sections of the product design by providing recommendations to the designer based on the functional module and as directed by the policy level goals, specifications and other information already stored in the main memory. It identifies conflicts, presents them to the user, and makes suggestions. It is driven by rule-based expert knowledge derived from the various functional disciplines.

(g) Reducer and fastener

This module has two functions. First, it attempts to reduce the number of parts in the assembly and, second, it determines the most suitable fastening method. Some parts are eliminated by combining them with other parts. This module uses a knowledge base plus information gathered from earlier stages to make the recommendations. The specifications will determine the fastening method. The program consults a library file of fastening methods, identifies potential ones, and ranks them. The designer makes the final choice from the list of suggestions.

(h) Individual part designer

This module integrates all previously accepted recommendations. The program is highly interactive, consulting the user at each step. General improvements in the handling, orientation, and assembly characteristics can be input.

(i) Evaluate module

This is the final process step. The original product design is compared and evaluated against the new improved one. At this stage a decision is made on whether or not to proceed with the modified design. The designer may decide to incorporate a portion of the newer features, but adhere for the most part with the original design.

10.3.2 Features-based design methodology

A feature can be defined as a geometric or non geometric attribute of a discrete part whose presence or dimensions are relevant to the product's function, manufacture, engineering analysis, use, etc., or whose availability as a primitive or operation facilitates the design process (Fazio, 1990). The features closely match the way the process planners view the portions of the object. Data is represented by functional and manufacturing descriptions that have unique meaning. The designer works in manufacturing a (machining, casting, assembly) mode applying operations that transform incoming materials into parts with the desired characteristics. A machined part might be designed by starting with a blank, and specifying a high level sequence of operations such as make-hole, make-slot, make-pocket. The features a designer needs are dictated by the particular domain. In casting for example, typical features include slabs, bosses, and holes. Whatever the domain, it is important that the features are logically symbolized in a manner the designer understands.

Features-based design methodology provides an effective means for bridging the gap between design and manufacturing. It is based on the premise that the best way to assure design for manufacturability is to develop products simultaneously and in conjunction with manufacturing plans. Features facilitate process planning by providing more information to manufacturing. Cutkosky *et al.* (1989) argues that process design after product design is both difficult and sub-optimal. It is more effective to do it interactively as the design process proceeds.

(a) A features-based design system for casting

Luby *et al.* (1986) have developed a features database system, called Casper, for concurrent design of casting products. Production problems in casting are closely related to certain geometric features such as slabs, bosses, and holes. Knowledge of these features and problems associated with them can be helpful when designing a casting. Casper is a system that builds up a design using two categories of features, macrofeatures and cofeatures. Macrofeatures are geometric building blocks such as boxes, L-brackets and cylinders. Cofeatures are details, such as holes, bosses and ribs, which can be added to the macrofeatures.

The system flow chart is outlined in Fig. 10.4. Using the process selection module, a casting method such as sand, die or investment is chosen. This module suggests a method depending on the quantity, size, shape, minimum section thickness, etc. The designer can accept the recommendation or decide on an alternative. The remaining operations will be described by illustrating the design and evaluation of a simple product (Fig. 10.5).

The add/create menu is invoked. This provides a list of macro- and cofeatures which can be used during design creation. Macrofeatures must first be selected. An L-bracket shape is chosen and this automatically invokes a second menu which requests data pertinent to this feature. Design rules are integrated which warn the designer if manufacturing guidelines are violated. For example, if the thickness is too narrow, the molten metal may solidify too quickly and prevent it from filling all sections of the part. Using similar methods, default dimensions are derived and supplied, which can be used at the discretion of the designer. When the data sheet is completed by the user, the program returns to the main menu.

In the same manner, cofeatures are added. The location of these on the macrofeatures are indicated using a graphics cursor. A coordinate grid can be placed on the face for more precise positioning. Again design rules are

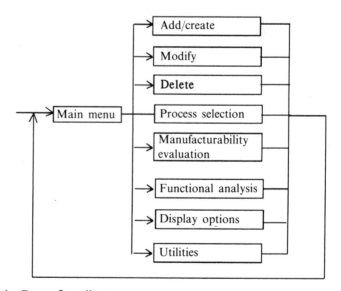

Fig. 10.4 Casper flow diagram.

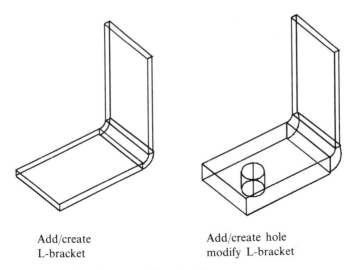

Add/create   Add/create hole
L-bracket   modify L-bracket

Fig. 10.5 Design of casting assembly using Casper.

used. For example, default dimensions for holes are presented in drill size increments. This greatly simplifies the manufacturing operation. If the shape of the desired part varies from the standard features provided, the modify module is used. This is used if the L-bracket requires the two slabs to be of different thickness.

Manufacturing evaluation is made possible by expressing features in terms of manufacturing knowledge. The evaluation module assesses the product for ease of production. It identifies problems such as thicknesses that are less than the minimum allowable for that type of casting method, potential sites of hot spots and areas of feeding difficulty. For example, feeding analysis focuses on the flow of metal between the different parts of the casting during solidification where the metal shrinks by up to 10%. The casting is divided into simple elements and the modulus of solidification is calculated to determine the freezing time. This analysis is carried out to ensure the reservoir of liquid metal is sufficient to fill the void left by the contraction.

10.3.3 Computer aided design with integrated suggestion systems

Current CAD systems are generally employed to capture design representations on specific product concepts. Eventually one concept is chosen and detailed design is performed. These systems do little to aid the designer to improve the design or ensure it is optimal in terms of manufacturability. The main role is recording designer intent.

The concurrent engineering approach is based on parallel activity and the incorporation of downstream inputs at the early design phases. Many concurrent engineering methods are aimed at gathering and providing information to the development team. It is important that this information be applied and effective means are available for this to happen. If a designer has to search large libraries of data to acquire the information, it is likely the efficiency of the design will be reduced. The more integrated this data is with the design system the more likely it will be applied. The approach presented here outlines the use of algorithms that provide suggestions to the designer as the design process proceeds (Jakiela *et al.*, 1984). The idea is that, as the designer is prompted to explore the possibility of implementing a particular suggestion, he is encouraged to think of different ways it can be done.

Early on in the development cycle, designers tend to pursue a single conceptual design idea. This idea, although altered and adjusted through-out, remains as the central theme and no matter how bad it proves to be. Integrated suggestion making systems improve the quality of the initial design and promote the abandonment of bad ones. As an example (Fig. 10.6), consider the design of a steel block on which a groove is required on the top and bottom surfaces in the Y direction. The designer inputs the groove on the underside. The top groove is now added. The systems suggestive algorithm recommends the groove be moved (shaded feature) which makes the design entirely symmetric about the Z axis. This will simplify the machining operation and make for easier material handling. The designer now checks the functional specifications to see if the suggested alteration can be accommodated. The final figure shows the recommended change having been accepted.

It is very important that the system be easy to use. If it places a large burden on the designer it loses its effectiveness. The designer should spend less time thinking about the system and expend most of the effort on the design

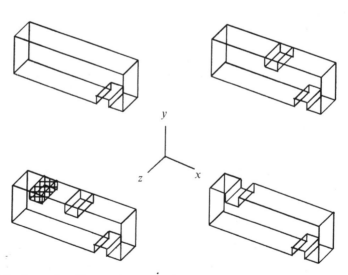

Fig. 10.6   Example of interactive suggestive process.

problem. The initial systems developed were text-based. These proved disappointing as the suggestions were often too brief, uninformative and sometimes difficult to interpret. Research was carried out into the effectiveness of incorporating the suggestions during versus after design activity (Jakiela and Papalambros, 1984). There was strong evidence that designers who interactively received suggestions during the design process created the more optimal configurations.

(a) Application of Boothroyd assembly charts

The Boothroyd and Dewhurst method (1987) uses charts and equations to evaluate proposed designs. It calculates a design efficiency measurement and generates the assembly costs. Jakiela *et al.* (1984) developed the Boothroyd and Dewhurst method, using the charts and equations as a base, into suggestion programs which guide the designer toward the optimal configuration.

10.3.4 Variation simulation analysis

Variation simulation analysis (VSA) provides the design engineer with the ability to evaluate the geometric effect of variation in an assembly due to component tolerances, processing relationships, assembly methods, and assembly sequence. The VSA technique, when fully integrated, provides an effective tool to structure communication between design and manufacturing. VSA can be defined as a simulation process that models the assembly process of a three-dimensional component. VSA positions the designed part in its place on the assembly line, and then models the process as it attaches the various components to the new part. The objective is to create the design that can 'absorb' the largest amount of variation, without affecting the quality and function of the product, thereby reducing design and assembly costs. Critical tolerances are controlled or decreased, tolerances on noncritical components are increased, and variation is redistributed by altering the product design or manufacturing process (Coffman, 1987).

In a standard manufacturing process, the dimensions will vary from product to product. The cost of decreasing tolerances rises exponentially as it approaches zero. To ensure a product meets its functional specification, design engineers traditionally assign tolerances to key dimensions on components. These specify the allowable variation from a nominal dimension that is acceptable. A 'tolerance stack' is calculated by summing all the dimensions using maximum and minimum tolerances. This determines if an assembly problem is expected to occur under extreme tolerance conditions.

There are some weaknesses with this method. Assembly sequence is not taken into account. The approach does not comprehend the three dimensional geometry variation (skewness) effects associated with the assembly methods. There is difficulty in quickly evaluating multiple design proposals and identifying which variables are significant contributors to nonconformance. Above all, it is a very conservative, costly approach to design, as the product is designed assuming the worst case scenario. Tolerancing standards have been developed for standard processes and attachment schemes. Design engineers frequently use these without understanding their application and effects on the manufacturing process.

The impact of dimensional variation on an assembly can be derived through analysis or review of previous, similar, designs, or through prototype build. The latter of these two normally occurs too late in the design cycle to have a significant impact. The cost of changing a design at this stage is normally too expensive. Simulation techniques provide a viable alternative. They allow a design to be evaluated in a simulation mode before the product and manufacturing process is finalized and significant portions of cost committed.

VSA is a simulation technique and software tool, based on the 'Monte Carlo Simulation Analysis' method, which predicts the dimensional variation that occurs in a product due to component and manufacturing process (method and sequence) variation (Craig, 1989). It uses a random number generator, combined with statistical probability curves to determine the probability of occurrence for any range of values. The following are the steps involved:

1. A VSA mathematical model is interactively created (Fig. 10.7) using a 3-D graphics preprocessor called a VSA-builder. The component geometry is represented by a set of defined points, located on a surface. Each

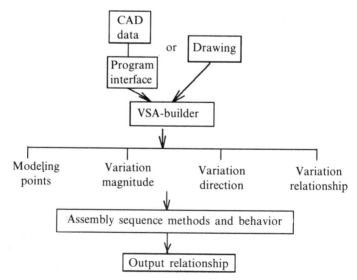

Fig. 10.7 Variation simulation analysis (VSA) model generation.

point is represented by its nominal x, y, and z-coordinate location, a statistical distribution representing the variation of the point, and a mathematical statement defining the relationship between the point with respect to other points. The mathematical model also represents manufacturing methods and sequence.

2. Simulation is then performed (Fig. 10.8). A random number generator randomly selects values for each input variable, that is, it considers the probability of occurrence of each input variable-based on the statistical distribution assigned. As an example, consider a design with a high number of components. These are stored in large lots along the assembly line. The VSA model statistically determines the size variation of each component in each lot, based on the design tolerance and manufacturing variations defined in the previous step. It then simulates the production of a predefined number of assemblies, incorporating the component and manufacturing variations.

3. The results of the simulation model are analyzed. They are presented in the form of statistical information which predicts the location or relationship of critical points. The percentage out of specification is determined for each output variable. The results can also provide percentage contribution analysis which predicts what certain input variables contribute to a given output variable. This provides the designer with important information about the key components and areas of the design. Emphasis can be placed on changing the largest contributing factors or increasing the smallest contributing tolerances. The results of the simulation can be used to determine the necessary component tolerance and manufacturing specifications.

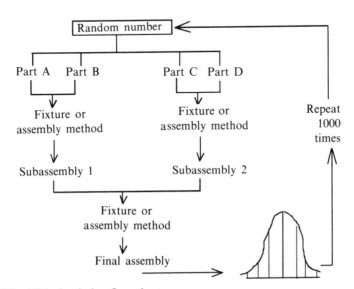

Fig. 10.8 VSA simulation flow chart.

The model can be employed during the conceptual and detailed design stage. As the information available during the conceptual stage is less defined, only major subassemblies and processes at a macrolevel (robotics versus manual, batch versus in-line) can be modeled. Each subassembly is treated as if a single component.

Numerous design and assembly proposals can be evaluated very quickly. The procedure is normally an iterative one. Several models are run and the results analyzed. The proposals may be modified and the process repeated. The use of VSA aids the designer in the selection of the optimal design configuration, assembly method and sequence. It should be remembered that the accuracy of VSA is dependent on the integrity of the input information.

10.3.5 A printed wiring board (PWB) DFM system

Concurrent engineering places many demands on the design process. The volume of information being dealt with is potentially enormous. The product is designed with consideration of all relevant downstream information. To satisfy these information management requirements, much emphasis must be placed on efficient management of data representation distribution and storage. In addition, information control is required to prevent analysis on an out-of-date revision of the design.

The following are the objectives an effective concurrent engineering system for electronic design and manufacture should strive to achieve (Amundsen and Hutchison, 1990):

- An interface that facilitates all levels of engineering expertise;
- The capture of information from the speciality disciplines into a knowledge base which can then be used as a design advisor to guide the design process toward the optimal configuration;
- An intelligent design environment with an integrated tool suite (this increases the productivity and efficiency of the design engineer);
- A single design representation which can be shared across all functions and tools;
- Data integrity and consistency across the system; and
- The system should accommodate changing circuit technology and complexity without redesign or reprogramming.

Texas Instruments has developed a design environment which consists of a number of the traditional CAD/CAE tools, a representation that supports PWB design from concept to manufacture and an inference engine (Amundsen and Hutchison, 1990). One of the unique features of the system is that each application shares a single common representation (Fig. 10.9). This implies that all parameters can be accessed by any of the tools. Any change to a parameter is immediately reflected in all applications. An object-oriented data representation approach is used that supports hierarchical

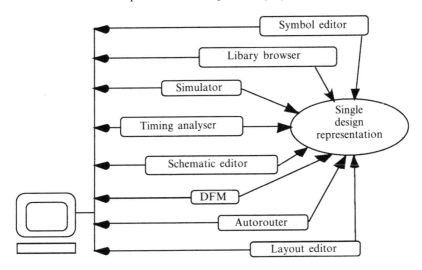

Fig. 10.9   Texas Instruments CE environment.

configuration, and multiple levels of data abstraction. The knowledge base contains modules that provide knowledge on downstream as well as upstream functions. The inference engine applies this knowledge to solving the design problems. A good deal of effort was expended on the development of the user interface. As a result, it is efficient and does not require a great deal of system expertise to use.

The system allows the designer to evaluate the impact of design decisions against life-cycle specifications, early in the development cycle. As the designer modifies the design representation the inference engine is invoked. This automatically assesses the design against a set of life-cycle constraints. These facilities represent a significant extension of the traditional CAD/CAE tools. A brief outline, rather than a detailed functional analysis, of some of the more important features of the Texas Instruments system is presented.

(a) Data representation

In general most CAD/CAE applications operate independently, each having their own optimized data representation structure. Data is captured relevant to the specific applications needs with little reference to other tools. It is quite common for the output files from one application to be required as part of the input to another. It is here that the unique representation structure can lead to communication difficulties, as it must be translated into the receiving tool's data format. Layout editors, for example, must transfer their output data to an automatic router. Because of the lack of a common representation a complex interfacing and data transfer verification process (to insure data integrity) must be implemented.

The system developed by Texas Instruments utilizes a common data representation and an object-oriented approach. Objects include, components, pins, routes, holes, boards, etc., each with associated specific electrical and mechanical attributes supporting a design from conception through production. This allows each designer to manipulate the representation from the user's own perspective. Communication across applications is much simpler as translators and data verification are eliminated along with the errors associated with them.

(b) Knowledge base

A central requirement of any successful concurrent engineering system is an accurate and complete knowledge base of information encompassing all relevant functions. This has more recently been carried out by physically applying teams of experts to the problem. While this is a significant improvement on the sequential approach, there is strong justification for automating this process. This will make the experts apparently available to the designer on an ongoing basis, allowing the application of expert knowledge continuously and during all stages in the design process.

Texas Instruments created a knowledge base by interviewing experts from the relevant functions, extracting their expertise and coding it. It is important to note that the inclusion of advice is ultimately at the discretion of the designer, not the tool. In using the system, the designer has complete access to and control over the entire set of constraints via menus that allow creation, browsing, and editing.

The degree of interaction between the knowledge base and the tool is controlled by the in-line, on-demand and batch mode options. In the in-line mode, relevant constraints are triggered interactively as an object is acted upon. All constraints for one specific object are triggered at the request of the user in the on-demand mode. Finally, for the batch mode, the entire set of constraints relating to all objects is evaluated at the request of the user. An additional feature allows the designer to choose specific objects which the constraint is to ignore. This may be used on a nonstandard component for which a waiver exists.

(c) Technology changes

Included in the knowledge base is technology information. Electronic distribution of this data disperses it to a wider audience in an expedient fashion. PWB component, interconnect and manufacturing technologies have and will continue to change rapidly. This system provides the designer with a menu driven change mechanism which allows the designer to alter or add new component and manufacturing technologies. Little or no software changes are necessary, minimizing support requirements, while allowing flexible expansion of technology.

(d) Manufacturing process models

Traditionally DFM has been a post-design process step. This has led to suboptimal solutions. Thermal analysis was carried out during prototype assembly to identify hot spots. This analysis may indicate a layout or component selection problem. To rectify the problem at this stage may involve a redesign which adds significant cost and lengthens time-to-market. Detailed research has made it feasible to create models which accurately predict performance parameters, such as cost, thermal characteristics, vibration profiles and solder joint reliability. While all parameters may not be available for input to the models during the early design phases, first order estimates can be generated with relatively high accuracy and used until real data becomes available. The Texas Instruments system utilizes several approximate and exact models developed by the defense industry. The designer can make tradeoff decisions by optimizing the performance in one area and monitoring the impact on others.

10.4 REFERENCES

Amundsen, M. and Hutchison, K.K. (1990) *Concurrent Engineering for Electronic Circuit Design*, Proceedings of Second National Symposium on Concurrent Engineering, Morgantown, WV, USA, February 7–10.
Boothroyd, G. and Dewhurst, P. (1987) *Design for Assembly – A Designers Handbook*, Boothroyd & Dewhurst Inc., Wakefield, RI, USA.
Boothroyd, G. (1988) *Estimating the Cost of Printed Circuit Board Assemblies*, University of Rhode Island, Dept. of Industrial and Manufacturing Engineering, Kingston, RI, USA.
Coffman, C. (1987) *Make Me a Match – Getting Design and Manufacturing Together, Simultaneously*, Automotive Industries, December.
Conradson, S.A., Barford, L.A., Fisher, W.D., Weinstein, M.J. and Wilker, J.D. (1988) *Manufacturability Tools: An Engineer's Use and Needs*, IEEE Engineering and Management.
Craig, M. (1989) *Managing Variation by Design using Simulation Methods*, Applied Computer Solutions, Inc., St. Clair Shores, Michigan, USA.
Cutkosky, M.R., Brown, D. and Tenebaum, J.M. (1989) *Extending Concurrent Product and Process Design Towards Earlier Design Stages*, Proceedings of the Symposium on Concurrent Engineering, ASME, December.
Digital Equipment Corporation (1990) *Archive Data Structure File Specification*, Internal Digital document, Document Identifier: A-SP-ELEN432-00-0.
Digital Equipment Corporation (1991) *Component Categories and Codes for Machine and Non-Machine Insertable Components*, Internal Digital Document, Document Identifier: A-SP-ELMF228-00-0000.
Druary, C. (1989) *Activity-Based Costing*, Management Accounting, September.
Fazio, T.L. (1990) *A Prototype for Feature-Based Design for Assembly*, Proceedings of Second National Symposium on Concurrent Engineering, Morgantown, WV, USA, February 7–10.
Jakiela, M. and Papalambros, P. (1984) *Concurrent Engineering with Suggestion-Making CAD Systems: Results of Initial User Tests*, Dept. of Mech. Eng., Mass. Institute of Technology, Cambridge, Mass., and University of Michigan, Ann Arbor, Michigan, USA.

Jakiela, M., Papalambros, P. and Ulsoy, A.G. (1984) *Programming Optimal Suggestions in the Concept Design Phase: Boothroyd Assembly Charts*, Proceeding of ASME Design Engineering Conference, Cambridge, Mass., USA.

Jeans, M. and Morrow, M. (1989) *The Practicalities of Using Activity-Based Costing*, Management Accounting, November.

Kroll, E., Lenz, E. and Wolberg, J.R. (1988) *Manufacturing Review*, 1(2) June.

Lu, S.C-Y., Subramanyam, S., Thompson, J.B. and Klein, M. (1988) *A Co-operative Product Development Environment to Realise the Simultaneous Engineering Concept*, Dept. of Mech., Ind., and Comp. Science, University of Illinois, Urbana, Il, USA.

Luby, S.C., Dixon, J.R. and Simmons, M.K. (1986) *Computers in Mechanical Engineering*, November, 25–33.

Miles, B.L. (1989) *Design For Assembly – A Key Element within Design For Manufacture*, Proceedings of the Institution of Mechanical Engineers, IMechE, vol. 20.

Miyakawa, S. and Ohashi, T. (1986) *The Hitachi Assemblability Evaluation Method (AEM)*, Proceedings of First International Conference on Product Design for Assembly.

Pugh, S. (1981) *Concept Selection – A Method that Works*, Proceedings of International Conference on Engineering Design (ICED), Rome, March 9–13.

Mazzullo, T. (1989) *The Transition from Design into Manufacturing*, Printed Circuit Design, July.

CHAPTER 11

Multiattribute design optimization and concurrent engineering

Deborah L. Thurston and Angela Locascio

11.1 INTRODUCTION

The approach to concurrent design described in this chapter is motivated by the observation that, despite the progress made by 'design for X' approaches, engineering designs are ultimately evaluated with respect to many criteria. We assert that the design process should be driven by simultaneous consideration of multiple criteria, rather than evaluation of a single criterion at each stage of the design process with modification along the way. Recognizing the need to satisfy the ultimate evaluator of designs, the customer, Cook defines quality as 'net value of product to society' (1991a). Cook (1991b) and Cook and DeVor (1991) construct a model for evaluating the quality of a product based on the value to the customer as well as cost. By listening to the 'voice of the customer', the 'house of quality' approach attempts to integrate customer desires into the design process (Hauser and Clausing, 1988). The *voice of the customer* defines certain design criteria, or attributes, to be considered. These attributes are related to the engineering domain as performance criteria through quality function deployment (Sullivan, 1986; Clausing and Pugh, 1991). Further analysis of these desired performance criteria leads to identification of relevant decision parameters over which designers have direct control. For example, the customer attribute of 'quieter car' may be translated into the performance goal of low engine noise. The designer then attempts to control variables that affect engine noise, thus indirectly controlling the customer attribute.

The 'house of quality' identifies the relationships between customer attributes and engineering characteristics, using symbols to indicate the degree of 'positive' or 'negative' influence. Similar symbols are used to describe the relative importance of attributes, often with numbers assigned to each attribute such that the total sum is fixed (say, at ten), in an effort to reflect the so-called 'imputed importance'. This information is sometimes used to infer appropriate tradeoffs between attributes.

The 'house of quality' and quality function deployment approach represent a major advance in engineering design methodology by making clear the connections between engineering design decisions and their impact on the customer. These tools have gained wide acceptance by breaking down some of the traditional barriers to communication between design, marketing and manufacturing interests within the organization. Now that these tools have been accepted by industry, we believe the next logical and necessary step is to put the information contained in the house of quality to better use. We propose two advances:

1. Rather than stopping with a designation of 'positive' or 'negative' to indicate the influence of engineering characteristics on customer attributes, quantify these relationships; and
2. Rather than using a weighted sum to represent imputed importance, use a more rigorously determined multiattribute evaluation function with strong theoretical foundations.

These advances allow a more general interpretation of the concurrent engineering problem and permit a more mathematically-based approach for its solution.

We interpret the house of quality approach as a general formulation of a multiattribute design optimization problem. Figure 11.1 shows a house of quality for the design of an automotive door panel. The ovals indicate the optimization interpretation for the part of the house labeled. The 'customer attributes' shown along the side of the house comprise the design goals or attributes. The 'engineering characteristics' shown along the top of the house represent the design decision variables. The 'relationship matrix' that links the voice of the customer to the engineering domain represents the constraint functions. The 'relative importance' is replaced with a multiattribute objective function for design optimization.

As with the house of quality, the approach presented in this chapter is also driven by the ultimate satisfaction of the end user. We exploit the observed relation between the house of quality representation and the design optimization formulation by constructing an optimization problem to concurrently design with respect to all customer attributes. We build on the quality-based approach described above by quantifying the relationship between design parameters and the customer attributes. Furthermore, the relationship between attributes and the trade-off between them is assessed through a formal procedure. These trade-offs are not constrained to be constant as they are in the house of quality; nor is the relation between performance level and its contribution to the overall design required to be linear.

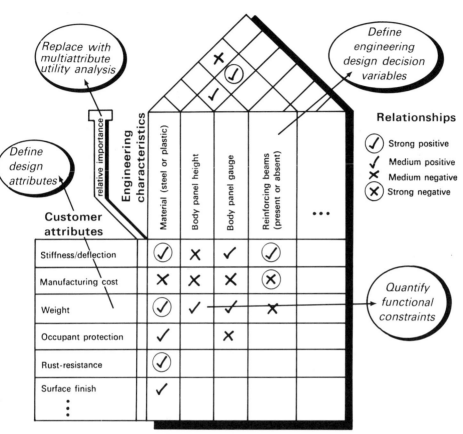

Fig. 11.1 Relation of house of quality to multiattribute design optimization for the design of a parking garage.

This approach provides a rigorous, analytically sound procedure for guiding the design process. Based on multiattribute utility theory (von Neumann and Morgenstern, 1947; Keeney and Raiffa, 1976) and optimization theory, the general methodology consists of determining these relationships and integrating all important design criteria into a design evaluation function. This function represents design attributes and the relation between them, as dictated by the end user. After constructing this function, formal optimization is performed to identify the best design alternative. The result is a numerical evaluation of the design with respect to all design goals, and a design that is optimal with respect to this combination of performance criteria.

Although the methodology we propose is motivated by the house of quality and quality function deployment methods, neither is necessary to use our methodology. They serve as a convenient representation of some of the concepts and relationships we present, but are not a necessary step in our problem formulation.

Section 11.2 describes our general methodology for concurrent design optimization. Section 11.3 presents an example of material selection and design optimization for an automotive door panel. Section 11.4 presents a second example from the field of structural dynamics to elucidate the modeling advantages of utility theory and computational procedure available for this type of formulation. Following the summary and conclusions in Section 11.5, some terminology is provided as a reference to clarify the discussion.

11.2 GENERAL METHODOLOGY FOR CONCURRENT DESIGN

Typically, certain factors are considered independently during the design process, while the customer will evaluate the final design simultaneously with respect to many criteria. For example, the design team might consider functional or performance requirements at first, then the manufacturing department might modify the design for ease of assembly, then management might verify that the cost of the final design is within the budget. This process is inefficient; designers are now attempting concurrently to design with respect to these criteria throughout the design process, letting the trade-offs between criteria lead to the optimal design. The method presented here uses a design evaluation function that considers all the design attributes and reflects the designers' willingness to make trade-offs between them. The multicriteria design evaluation function is derived from multiattribute utility analysis. The design model is based on optimization theory with the house of quality motivating a general formulation for the design optimization problem. The formulation and solution procedure for using utility theory and the house of quality to construct this multiattribute utility function for design evaluation is as follows:

11.2.1 Define attributes from customer attributes

Using the attributes that the customer has identified as important, the design attributes can be defined over an acceptable range. These are the relevant characteristics of the overall design that the decision-maker will consider. The imputed worth of varying levels of each attribute in isolation is expressed through the single attribute utility function for each attribute. The trade-off between attributes that the designer is willing to make is represented by the scaling factors. Determination of the single attribute utility functions and scaling constants are described below.

11.2.2 Replace 'relative importance' with multiattribute utility analysis

Utility is determined as a function of the levels of performance in each attribute that the design alternative exhibits.

Given conditions of preferential and utility independence of attributes, the overall multiattribute utility of an alternative is calculated from the multiplicative form shown in Equation (11.1a)

$$U(\mathbf{x}) = \frac{1}{K}\left[\left(\prod_{i=1}^{n} Ka_i U_i(x_i) + 1\right) - 1\right]$$ (11.1a)

where

$U(\mathbf{x})$ = the overall utility of a design alternative characterized by the vector of customer attributes $\mathbf{x} = (x_1, \ldots, x_n)$
x_i = the performance level of customer attribute i
$U_i(x_i)$ = the single attribute utility function for attribute i
i = 1, 2, ..., n attributes
a_i = the single attribute scaling constant
K = the normalizing constant, derived from

$$1 + K = \prod_{i=1}^{n} (1 + Ka_i).$$

If the more restrictive condition of additive independence of attributes holds, the overall multiattribute utility of an alternative is calculated from the additive form shown in Equation (11.1b):

$$U(\mathbf{x}) = \sum_{i=1}^{n} a_i U_i(x_i).$$ (11.1b)

Tests for the independence conditions are described in Keeney and Raiffa (1976).

The single attribute utility functions, $U_i(x_i)$, are assessed from the design decision maker(s) through a series of lottery questions. For group decision-making (e.g., members from marketing, engineering, and manufacturing), responses to lottery questions can be determined by group consensus. The single attribute scaling constants, a_i, which represent the trade-off between attributes the designer is willing to make, are assessed using a similar lottery technique. The certainty equivalence method realized by lottery questions is described in the Appendix. Figure 11.2 shows an example of a multiattribute utility function, $U(\mathbf{x})$, with two design attributes, x_1 and x_2, where x_1 is cost and x_2 is weight.

11.2.3 Define design decision variables

Identify the variables, $\mathbf{y}$, of the design about which the designer can make direct decisions in order to control the customer-defined attributes, $\mathbf{x}$. These decision variables are represented along the top of the house of quality. The designer cannot directly select an attribute level, for instance, but can control the variables which affect the attributes. Examples of design decision variables include material choice and component geometry.

Utility vs. Weight and Cost

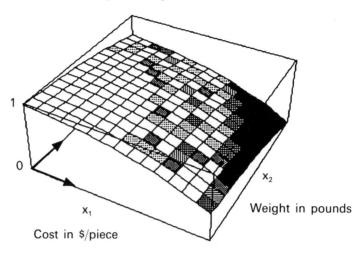

Fig. 11.2   Multiattribute design utility as a function of cost and weight.

11.2.4 Define constraint functions

The relation between the decision variables and the attribute levels is represented by the *constraint functions*, which are represented symbolically in the relationship matrices (top and bottom) of the house of quality. A constraint function may relate, for example, the changes in a component's dimensions to changes in the attributes of weight, stiffness, or manufacturing cost.

11.2.5 Determination of the attribute bounds

Because the customer attributes are each defined over a specific allowable range, a bound on one attribute may prevent another from reaching its bound. For example, an automotive design which achieves the lower bound on cost may have an unacceptable noise level, i.e., it exceeds the bound on the interior noise attribute. In the house of quality, only target values are estimated, without regard for feasibility or influence on other attributes. It is therefore necessary to determine the allowable ranges of all combinations of attributes so that the single attribute utility functions can be defined and assessed over a range that corresponds to a feasible design artifact. In our approach, this task is accomplished by solving a set of $2n$ optimization problems. See Thurston *et al.* (1991) for further discussion of the structure and solution of these optimization subproblems.

11.2.6 Structure of the optimization problem: design to maximize utility

The utility function is defined in terms of customer attributes, **x**, while the designer has direct control over the design variables, **y**. The goal of maximizing the multiattribute utility function $U(\mathbf{x})$ by choosing elements of the design vector can be written as a nonlinear program:

$$\text{maximize } U(\mathbf{x})$$
$$\mathbf{y}$$

subject to

$$\mathbf{x} = \mathbf{g}(\mathbf{y})$$

and

$$g_i(\mathbf{y}) \geqslant x_{il}$$
$$g_i(\mathbf{y}) \leqslant x_{iu} \qquad \text{for } i = 1, \ldots, n$$

where $\mathbf{g}$ is a function relating attributes $\mathbf{x}$ to the design variables $\mathbf{y}$ and x_{il} and x_{iu} are lower and upper bounds, respectively, on attribute i.

After the relationships between the attributes and the design variables are known, substitution of this representation $\mathbf{g}(\mathbf{y})$ into the objective function yields a new objective function, $V(\mathbf{y}) = U[\mathbf{g}(\mathbf{y})]$ and the maximization problem becomes

$$\text{maximize } V(\mathbf{y})$$
$$\mathbf{y}$$

subject to

$$g_i(\mathbf{y}) \geqslant x_{il}$$
$$g_i(\mathbf{y}) \leqslant x_{iu} \qquad \text{for } i = 1, \ldots, n$$

11.2.7 Computational approach

Depending on the specific design application at hand, the optimization problem might have a special structure which facilitates its solution. Design models often have constraint functions with special forms (e.g., monotonic in certain variables, quadratic, or simply bounded). In optimization practice, exploiting any special structure is strongly recommended (Gill *et al.*, 1981). For example, Section 11.4 presents a design problem with a simply-bounded feasible region. This structure permits a relatively straightforward approach for the solution procedure, as described in Section 11.5. For a general non-linear programming problem, a global optimization method is recommended.

11.3 EXAMPLE – MATERIAL SELECTION FOR DESIGN OPTIMIZATION

Manufacturing economics play a particularly significant role in decision-making for materials selection, since different materials are formed using different processing technologies. The manufacturing economics of stamping and injection molding, for example, are significantly different. The former requires a large capital investment in tooling machinery, while the latter typically does not. Conversely, material costs for steel are generally lower than for advanced polymer composites. The total manufacturing cost, therefore, is dependent on both material choice and production volume. The methodology is applied to the design and material selection of an automotive body panel, shown in Figure 11.3.

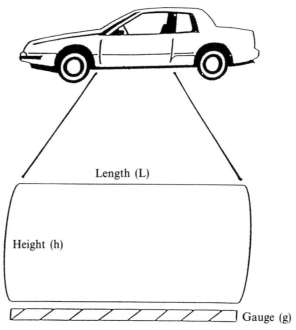

Fig. 11.3 Illustration of automotive door panel.

The end users of automobiles, the customers, desire that the automotive door be 'strong, rust-resistant, and safe' and contribute to the vehicle being 'inexpensive' and 'fuel-efficient'. Reflecting the voice of the customer, the house of quality is shown in Fig. 11.1. The customer attributes, shown along the side, are body panel stiffness, weight, manufacturing cost, surface finish, rust-resistance, and occupant protection. From the engineering characteristics, shown along the top, the design decision variables are identified as the material choice, reinforcing beams (present/absent), body panel gauge, and body panel height. The panel is modeled as a rectangular plate with length L, height h, and gauge g. Two materials are considered: steel and reaction injection molded (RIM) plastic. Using the relationship matrix of the house of quality as a basis, expressions for the customer attributes in terms of the design decision variables, or the constraint functions, are determined (see Fig. 11.1).

For this example, we consider a portion of the house of quality by examining three attributes and two design variables. The attributes are deflection, manufacturing cost, and weight; the design variables are material choice and body panel gauge.

11.3.1 Constraint functions

Choosing a material fixes material properties such as modulus of elasticity E and Poisson's ratio v. Modeled as a wide beam subjected to one-way bending due to a concentrated load at its center, the constraint function for deflection δ for a given material is

$$\delta = (1 - v^2)[FL^3/48EI] \tag{11.2}$$

where F is the concentrated force and I is the moment of inertia of the cross-section, $I = hg^3/12$.

The constraint function for weight, w, is simply the product of panel volume and material density, ρ, such that

$$w = (Lhg)\rho. \tag{11.3}$$

The constraint function for manufacturing cost, c, is the result of a multilinear regression on a data set to determine a reasonably simple relation between the design decision variables of gauge and the attribute of manufacturing cost as a function of production volume.

$$c = [C_m(Lgh) + C_n]V \tag{11.4}$$

where C_m is the material cost factor, C_n is the nonmaterial cost factor, and V is the volume sensitivity factor. The material cost factor relates unit cost to panel surface area; the nonmaterial cost includes costs which are independent of panel size; the volume sensitivity factor relates unit cost to a production volume function obtained for each material. Data was obtained from several sources including Douty *et al.* (1988), Ellerby and Nelson (1989), Hemphill (1988), Martinez and Sato (1989), O'Malley (1990), and Vesey and Abouzahr (1988). More accurate and precise cost estimate models can also be used (Boothroyd, 1988; Poli and Knight, 1984; Poli *et al.*, 1988).

To assist the design engineer in implementing the methodology, a computer aid to decision-making with a spreadsheet interface was previously developed (Thurston and Essington, 1991). This tool is used to determine the multiattribute utility function, compare the relative desirability of material alternatives, determine the optimal gauge for each material, and perform sensitivity analysis.

Assessment of the single attribute utility functions and scaling constants is performed using the procedure described in the appendix. Assembling these into the evaluation function, multiattribute utility for both material alternatives can be plotted as a function of panel gauge as shown in Fig. 11.4.

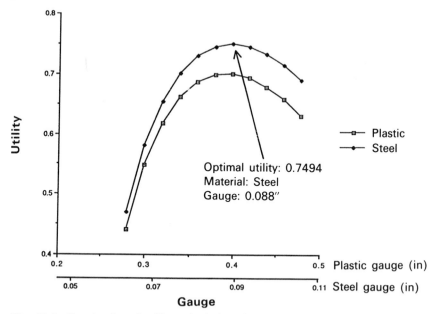

Total utility v. gauge

Fig. 11.4 Graph of total utility calculations for plastic and steel at a production volume of 500 000 units per year.

This figure clearly illustrates that the steel design is superior to polymer composite, as indicated by the higher utility value for steel. If the decision-making criterion is weight minimization alone, the material choice is plastic with a gauge of 0.28 inches (0.71 cm) (determined by substituting the density and minimum gauge for each material in Equation (11.3), then comparing the weight for each material). When all three attributes of weight, cost, and deflection are considered, however, the optimal combination of these attributes occurs in the steel panel with gauge of 0.088 inches (0.224 cm).

Sensitivity analysis can be used to evaluate the trade-offs between the attributes of cost, weight, and deflection. Iso-utility curves, shown in Fig. 11.5, are used to provide direction to the design engineer in the iterative design process. Indicating the combinations of key attribute levels for which total utility is constant, these curves illustrate the preferred trade-offs between attributes. Iso-utility curves for weight versus deflection are shown in Fig. 11.5. The solid lines represent combinations of weight and deflection where the total utility is constant, with utility increasing nearer the origin. The heavy dotted line represents the feasible combinations of weight and deflection as constrained by Equation (11.2). For example, consider a design with a gauge of 0.071 inches (0.180 cm) and a utility of 0.65. To improve utility, the decision-maker can determine from the graph that given the choice between increasing or decreasing gauge, the best choice is to increase

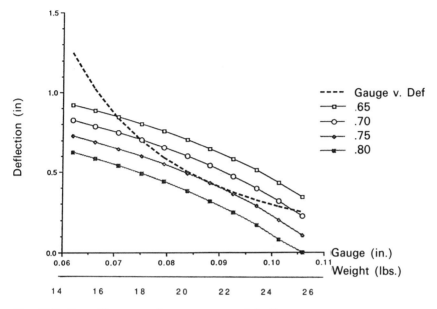

Fig. 11.5 Iso-utility curves for gauge versus deflection.

gauge to 0.088 (0.224 cm) inches, with a resulting increase in utility from 0.65 to 0.75.

Changes in production volume can affect the optimal material choice. Comparing the utility versus gauge functions for the panel at 250 000 units per year (Fig. 11.6) to that of 500 000 units (Fig. 11.4), it is observed that plastic is preferred to steel at low production volume while steel is preferred at the higher volume.

11.4 EXAMPLE – CONSIDERATION OF STRUCTURAL DYNAMICS IN DESIGN

Customer satisfaction is also essential when the product is a structure. We consider an office building which contains machinery (compressors and centrifuges for water filtration) that produces dynamic loading and thus motion in the structure. The clients and employees in the building, as well as their computing equipment, are adversely affected by this motion. The occupants, or customers, want to work in the building 'comfortably without incurring too much cost'. They also want their building to be 'constructed quickly and have a long service life'. Assume that the topology and general plan of the building are fixed as the two-story, one-bay frame structure shown in Fig. 11.7(a). For design purposes, the structure is modeled as the

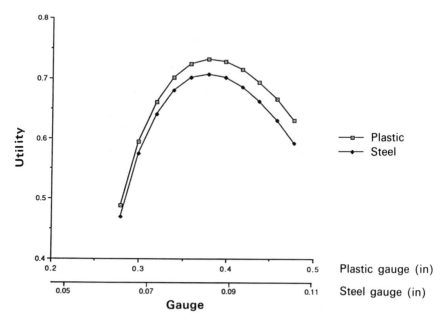

Fig. 11.6 Graph of total utility calculations for plastic and steel at a production volume of 250 000 units per year.

equivalent spring-mass system in Fig. 11.7(b). See Locascio and Thurston (1992) for a detailed development of the model.

The house of quality for this design task is shown in Fig. 11.8. The voice of the customer dictates the attributes of occupant discomfort and structural cost, as shown along the side. The engineering characteristics that relate to these attributes are the stiffness, damping, and mass of each floor, as shown along the top of the house of quality.

For this example, we consider two attributes: structural cost and occupant comfort. The multiattribute goal is to design the structure to yield optimum levels of the competing attributes of occupant discomfort and structural cost. Assigning typical values for the mass and damping of each floor (Berg, 1989), we define the design decision variables as the column stiffness of each story of the building, k_1 and k_2, which depend on column material and geometry.

11.4.1 Constraint functions

Using the relationship matrix of the house of quality as a starting point, we can determine the constraint functions: the relation between performance levels of customer attributes and the design variables.

We wish to define the total structural cost of the design as a function of the design variables. Selecting A36 steel as the column material and

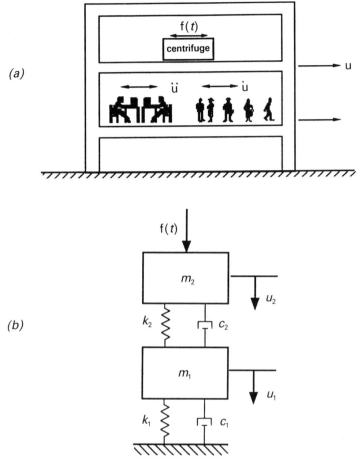

Fig. 11.7 (a) Two-story, one-bay frame structure; (b) equivalent spring-mass system.

standard I-beam sections for the geometry (AISC, 1980), fabrication and construction of the columns costs approximately $2 per pound (*Means*, 1991). If k_i is the stiffness (in *kips/in*) of floor i, the constraint function for the total cost of the design is

$$x_1 = \text{cost} = \$2.05 \ (k_1 + k_2) \tag{11.5}$$

and we have an expression for the cost attribute explicitly in terms of the design variables.

The allowable attribute bounds are determined from the largest and smallest standard W14 designations over the range of W14 × 132 to W14 × 22. The allowable range on cost is [$657.25, $5053.25]. For this example, there is a nonlinear relation between cost and utility such that

$$U_1(x_1) = 1.1249 - 0.0899e^{0.0005x_1} \tag{11.6}$$

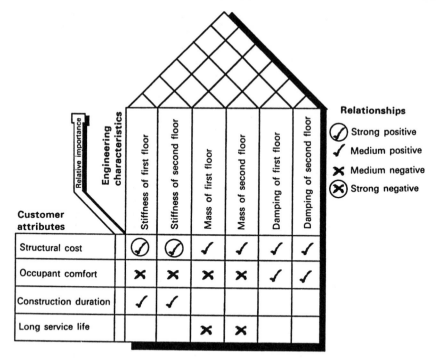

Fig. 11.8 House of quality for dynamically-loaded building design.

corresponding to $U_1(657.25) = 1$ and $U_1(5053.25) = 0$ as shown in Fig. 11.9.

Unlike the constraint function for cost, discomfort level cannot be defined directly in terms of the design variables. An implicit representation, however, is possible. Richart (1962) describes qualitatively how humans react to vibration, as shown in Fig. 11.10.

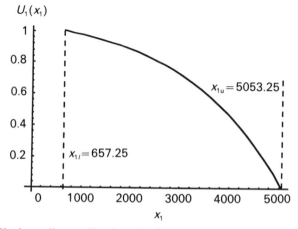

Fig. 11.9 Single-attribute utility function for cost.

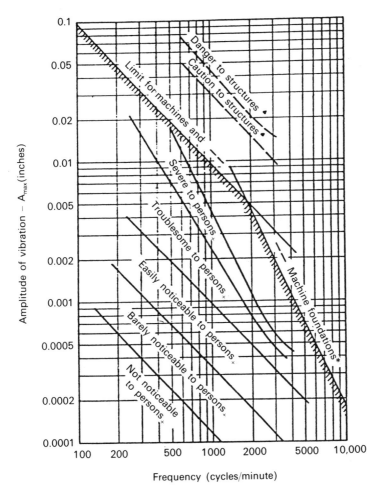

Fig. 11.10 Displacement limits of persons (from Richart, 1962).

These results aid in constructing the constraint function for occupant discomfort level. We approximate each level of discomfort defined by Richart as $\log f + \log u = c$, where f is the frequency of vibration in cycles per minute, u is the displacement in inches, and c is a constant which differs for each line. Defining this constant to be the discomfort level for each (f, u) pair on a given line, the expression for the discomfort level attribute as a function of frequency and displacement is

$$x_2 = \text{discomfort level} = \log f + \log u \qquad (11.7)$$

As described in detail by Locascio and Thurston (1992), the displacement spectra together with the transfer functions permit an evaluation of the worst-case frequency-displacement pair. This calculation is essentially an equality-constrained minimization subproblem to determine the maximum discomfort for any design alternative.

A linear relation is assumed between discomfort level and utility. The upper and lower limits on discomfort level are $x_{2u} = -0.2974$, *barely noticeable to persons*, and $x_{2l} = 0.6436$, *severe to persons*. The utility function for discomfort level is given by

$$U_2(x_2) = 0.683953 - 1.0626933x_2 \qquad (11.8)$$

11.4.2 Design evaluation function

Using the lottery approach described in the Appendix, the scaling constants for Equation (11.1) are determined as $a_1 = 0.6$ and $a_2 = 0.4$, representing the decision-maker's willingness to make trade-offs between cost and discomfort. The design evaluation function is the additive form of the multiattribute utility function, Equation (11.1b), given by

$$U(x_1, x_2) = 0.6 \, U_1(x_1) + 0.4 \, U_2(x_2)$$

The design optimization problem then becomes one of maximizing this function subject to bounds on the design variables of stiffness (in *kips/in*):

$$k_1 \geqslant 160.3$$
$$k_1 \leqslant 1232.5$$
$$k_2 \geqslant 160.3$$
$$k_2 \leqslant 1232.5$$

These bounds on stiffness correspond on W14 sections with moment of inertia of 199 in^4 (8 283 cm^4) for k_1 (W14 × 22) and 1 530 in^4 (63 683 cm^4) for k_2 (W14 × 132).

11.4.3 Computational procedure

Observe that this nonlinear programming formulation has a special structure: nonlinear objective function and simple bound constraints (linear). We exploit this structure and employ a procedure recommended by Gill *et al.* (1981) especially for this type of problem. The optimization algorithm, described in Fig. 11.11, consists of an active set strategy for the inequality-constrained bounds, and a gradient-based ascent method for the unconstrained optimization of the Lagrangian of the objective function. The search direction is a normalized steepest ascent direction (positive gradient); numerical gradients are computed using forward finite-differencing.

Using this optimization algorithm, the optimum design was found to be the case where the stiffness of both floors is

$$k_1 = k_2 = 160.3 \, kips/in$$

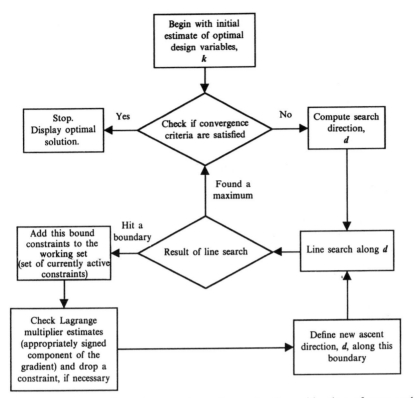

Fig. 11.11 Program flow chart to determine optimal combination of cost and occupant discomfort.

which corresponds to the minimum allowable level of structural cost, $x_1 = \$657.25$, and a discomfort level of $x_2 = -0.2560$ (easily noticeable to persons). This solution represents the design which gives the best combination of cost and discomfort level.

Sensitivity analysis may be performed for the decision-maker's scaling constants. Table 11.1 summarizes different optimal solutions for four different decision makers who have different values of these scaling constants.

For the first two decision-makers, the optimal design is also the minimum cost design. For designers 3 and 4, the optimal design yields neither minimum cost nor minimum discomfort, but the best combination of these attributes for their preference structure.

11.5 SUMMARY AND CONCLUSIONS

Manufacturing concerns have broken down the institutional barriers to communication between the customer, design engineers, manufacturing engineers, accounting and marketing personnel with popular 'house of

Table 11.1 Optimal solutions for different scaling constants

Decision-Maker # No.	Scaling constants		Optimal solution		Customer attributes	
	a_1	a_2	k_1	k_2	cost	discomfort
1	0.6	0.4	160.30	160.30	657.25	−0.256
2	0.5	0.5	160.30	160.30	657.25	−0.256
3	0.4	0.6	543.23	160.30	1442.24	−0.242
4	0.3	0.7	824.04	160.30	2017.90	−0.244

quality' or quality function deployment approaches. This chapter has presented a methodology for advancing beyond the benefits realized by these methods.

We move from qualitatively describing the relationships between customer satisfaction and design decisions to quantitatively describing them. In addition, we replace the *ad hoc* methods for judging the relative merits of a design with a rigorously determined multiattribute utility function with strong theoretical underpinnings. The methodology thus prescribes a concurrent multiattribute design optimization process, motivated by the desires of the customer, and driven by the subjective preferences of the designer.

Examples from two very different domains have been presented which demonstrate that the methodology can be used for both manufacturing and structural design decision-making. The examples also demonstrate that the explicit quantitative nature of the methodology not only identifies the optimal solution, but also provides explicit direction during the design process and enables sensitivity analysis on site-specific design parameters such as production volume and customer-specific preferences.

11.6 TERMINOLOGY

attribute: Any property or characteristic of a design alternative, such as cost, weight, stiffness.

certainty equivalent method: A technique for assessing a decision-maker's preferences over a single attribute range and assessing attribute trade-offs.

constraint function: Relates the design decision variables to the attributes, such as the relation between the geometry of a component and its weight.

design alternative: One of several solutions to a design problem which satisfies all functional design criteria.

design decision variable: The parameters that the designer may directly control, such as the choice of material or component geometry.

design evaluation: To determine the worth or value of a design alternative as a function of one or more attributes.

lottery: A situation where the outcome is uncertain; a construct used in assessing a decision-maker's preferences.

multiattribute utility function: A mathematical expression for the worth of a design alternative as a function of one or more attribute levels.

performance level: A quantitative measure of an attribute, such as 50 pounds, or $5 million.

single attribute utility function: A mathematical expression for the worth to a decision-maker of varying performance levels of one attribute.

utility, value, worth: The benefit or overall satisfaction derived by a design decision-maker, from either a single attribute or the entire design alternative. (These terms have distinct and specific technical meanings in formal decision theory. In this paper they are defined and used in a more general manner in the context of engineering design.)

11.7 ACKNOWLEDGMENT

The authors gratefully acknowledge the support of the National Science Foundation under PYI award DDM-8957420.

11.8 APPENDIX: ASSESSMENT OF UTILITY FUNCTION

11.8.1 Define attribute ranges

The decision maker is asked to define attribute ranges, based on his/her own estimates of the best or upper (u_i) and worst or lower (l_i) values of attribute levels that they anticipate being faced with or offered. The initial response for an attribute such as cost would normally be '0 cost to infinite cost'. This response is refined by asking the decision-maker to temper his or her estimate of the lower limit to that below which they could not tolerate going, despite highly desirable levels of other attributes. The upper limit is tempered to an optimistic yet realistic estimate of performance levels that alternative systems potentially offer and that they would be interested in. 'Interest' is defined as the willingness to pay in terms of performance in another attribute in order to achieve the upper limit. For example, plastic automobile parts can offer extremely high levels of corrosion resistance with optimistic estimates as high as fifty years. However, other constraints such as the designed service life of the vehicle (ten to fifteen years) limit the value of a fifty-year corrosion resistance guarantee, so the decision-maker places an upper limit of fifteen years on corrosion resistance. The decision-maker is not 'interested' in improving corrosion resistance to greater than fifteen years; any level of corrosion resistance above fifteen years has equal value in that application. He/she is not willing to 'pay' in terms of any other attribute in order to go from, say, sixteen to fifty years, since both values are outside the range. When attribute levels are greater than the upper limit of the defined range, the 'single attribute utility function' $U_i(x_i)$ for attribute x_i is assigned a value of 1. In a sense, beyond the defined range the attribute is a binary characteristic: below the minimum range the system is unacceptable and above it the decision maker is indifferent to changes.

After the attribute bounds (u_i, l_i) have been determined, the $2n$ optimization problems described in Section 11.2 are solved to obtain the (possibly) more restrictive bounds (x_{il}, x_{iu}) on the attribute levels which reflect the constraints.

11.8.2 Determine single attribute utility functions and scaling constants

Two types of valuations or preferences of the design decision maker must be assessed. One is the imputed worth of varying levels of each attribute in isolation, expressed in the single attribute utility function for each attribute $U_i(x_i)$. The other relates to the trade-off between attributes the designer is willing to make. This information takes the form of the scaling factors a_i, and as discussed in Kirkpatrick *et al.* (1983), should not be confused with concepts of relative importance of attributes or weighting factors.

Figure 11.12 is an example of the type of 'lottery' question used to determine the scaling constants a_i used in Equation 11.1. The value of a_i is equal to the utility where x_i is at its best level, x_{iu}, and all of the other attributes are at their worst levels; at this point $U(x_{1l}, ..., x_{iu}, ..., x_{nl}) = a_i$. The 'certain alternative' shown on the left in Fig. 11.13 represents an alternative with attributes at the levels shown for certain, and the lottery on the right represents an alternative in which there is uncertainty as to the attribute levels of an alternative. The lottery shows a probability p of 60% that weight will be 100 lb and cost will be \$10 000, and probability $1-p$ of 40% that weight will be 400 lb and cost will be \$90 000. When a user responds to the query 'Which do you prefer, the certainty, the lottery, or are you indifferent?' that he or she prefers the certainty, the value of p is increased to a more desirable value which is half-way between the previous level and the last level at which the decision maker preferred the lottery. The value of p at which the decision-maker is indifferent between the 'certain alternative' and the lottery is thus obtained by iteration between extreme values of p. The multiattribute utility of the situation where all attributes are at their best, or most desirable levels x_u is set equal to 1, and where they are at their worst or least desirable levels x_l set equal to 0. The value of a_i is then determined by

$$U(x_{1l}, ..., x_{iu}, ..., x_{nl}) = pU(x_u) + (1-p)U(x_l) \tag{11.A1}$$
$$U(x_{1l}, ..., x_{iu}, ..., x_{nl}) = p(1) + (1-p)(0) \tag{11.A2}$$
$$U(x_{1l}, ..., x_{iu}, ..., x_{nl}) = p \tag{11.A3}$$

Where $a_i = p$, since $U(x_{1l}, ..., x_{iu}, ..., x_{nl}) = a_i$.

Points which determine the single attribute utility function for each attribute are assessed using a similar type of lottery question, except

For each value of p indicated in Column 1, indicate the decision-maker's preference for the certainty or the lottery with an X.

Which do you prefer?

p	Certainty	Lottery
0%	X	
10%	X	
20%	X	
30%		
40%		
50%		
60%		
70%		
80%		
90%		
100%		

Check: For which value of p is the decision-maker indifferent between the certainty and the lottery?

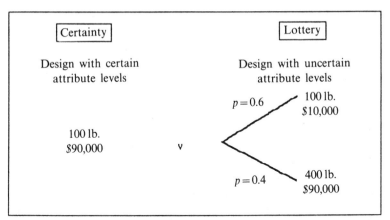

Fig. 11.12 Lottery question to assess scaling constant a_i for weight.

only one attribute is considered as shown in Fig. 11.13. The designer is asked to imagine that two alternative designs are being considered, each alike in every respect except that the 'certain' alternative's performance level for attribute x_i is known with certainty to be some value x, while the lottery alternative represents a design alternative in which there is uncertainty as to the attribute level. The lottery in Fig. 11.13 shows a probability p of 60% that weight will be 100 lb and probability $1-p$ of 40% that weight will be 400 lb. When the indifference point is reached, the relative value placed on the certainty equivalent as determined by the following

For each value of *w* indicated in column 1, indicate the decision-maker's preference for the certainty or the lottery with an X.

Which do you prefer?

w	Certainty	Lottery
100 lbs.	X	
150 lbs.	X	
200 lbs.		
250 lbs.		
300 lbs.		X
350 lbs.		X
400 lbs.		X

Check: For which value of *w* is the decision-maker indifferent between the certainty and the lottery?

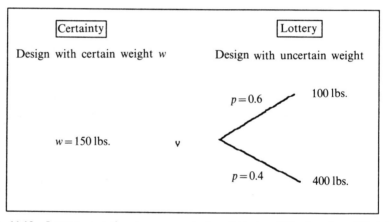

Fig. 11.13   Lottery question to assess single attribute utility function $U_i(x_i)$ for weight.

equation is one point on the single attribute utility function shown in Fig. 11.9.

$$U_i(x_i) = pU_i(x_{iu}) + (1-p)U_i(x_{il}) \qquad (11.A4)$$

$$U_i(x_i) = p(1) + (1-p)(0) \qquad (11.A5)$$

$$U_i(x_i) = p \qquad (11.A6)$$

The 'certainty equivalent' method just described to obtain indifference statements from the design decision-maker to determine the single attribute utility functions and scaling constants is discussed in greater detail by Dyer (1990) and Clausing and Pugh (1991). It can be performed by the analyst either manually or with the aid of a computer-based assessment program. When the assessment is performed manually, it is recommended that the analyst quickly sketch the single attribute utility function data points on a grid scaled from 0 to 1 during the survey to check for inconsistencies. The shape of the curve reflects the decision maker's non-linear valuation of changes in attribute levels.

11.9 REFERENCES

Berg, G.V. (1989) *Elements of Structural Dynamics*, Prentice-Hall, Inc., Englewood Cliffs, N.J.

Boothroyd, G. (1988) *Mechanical Engineering*, February, 28–31.

Clausing, D. and Pugh, S. (1991) *Design Productivity International Conference*, Honolulu, Hawaii, Feb. 6–9, 15–25.

Cook, H.E. (1991a) New avenues to total quality management, submitted to *Manufacturing Review*, November.

Cook, H.E. (1991b) *Manufacturing Review*, **39**(2) June, 106–14.

Cook, H.E. and DeVor, R.E. (1991) *Manufacturing Review*, **39**(2) June, 96–105.

Douty, D.L., McDermott, J.F. and Shipe, W.R. (1988) A steel-plastic hood concept for passenger cars, *SAE Technical Paper* 880355.

Dyer, J.S. (1990) *Management Science*, **36**(3), March.

Ellerby, G. and Nelson, Sr., D. (1989) Reaction injection molded composites–properties and economics, *SAE Technical Paper* 890197.

Gill, P.E., Murray, W. and Wright, M.H. (1981) *Practical Optimization*, Academic Press, Inc., San Diego, CA.

Hauser, J.R. and Clausing, D. (1988) *Harvard Business Review*, **66**(3) 63–73.

Hemphill, J. (1988) 'Polyurea RRim for automotive body panels, *SAE Technical Paper* 880361.

Keeney, R.L. and Raiffa, H. (1976) *Decisions with Multiple Objectives*: *Preferences and Value Tradeoffs*, John Wiley & Sons.

Kirkpatrick, S., Gelatt, C.D. and Vecchi, M.P. (1983) *Science*, **220**(4598) 671–80.

Locascio, A. and Thurston, D.L. (1992) Multiattribute optimal design of structural dynamic systems, submitted to *ASME Journal of Mechanical Design*, January.

Manual of Steel Construction (1980) American Institute of Steel Construction, Inc., Chicago, Illinois.

Martinez, E.C. and Sato, T.J. (1989) RIM – a process for the 90's, *SAE Technical Paper* 890698.

Means Building Construction Cost Data (1991) 49th Annual Edition, Construction Consultants & Publishers, Kingston, Mass., p. 120.

O'Malley, M. (1990) Plastic fenders: an emerging market? *SAE Technical Paper* 900289.

Poli, C. and Knight, W.A. (1984) *Design for forging handbook*, Technical Report, Department of Mechanical Engineering, University of Massachusetts.

Poli, C., Escudero, J. and Fernandez, R. (1988) *Machine Design*, November, 24.

Richart, Jr., F.E. (1962) *American Society of Civil Engineers Transactions*, **127**(I), 863–99.

Sullivan, L.P. (1986) *Quality Progress*, 39–50.

Thurston, D.L. (1991) *Research in Engineering Design*, **3**, 105–22.

Thurston, D.L. and Essington, S. (1991) *Decision Systems Laboratory Report*, November.

Thurston, D.L., Carnahan, J.V. and Liu, T. (1991) *Proceedings of ASME Design Theory and Methodology Conference*, Miami, Florida, 173–80.

Vesey, D. and Abouzahr, S. (1988) E-Coat Capable Plastics – A New Generation of Materials for Automotive Exterior Body Panels, *SAE Technical Paper* 880352.

von Neumann, J. and Morgenstern, O. (1947) *Theory of Games and Economic Behavior*, (2nd ed), Princeton University Press, Princeton, N.J.

CHAPTER 12

Concurrent cell design and cell control system configuration

F. Frank Chen

12.1 INTRODUCTION

The design and implementation of manufacturing systems have long been recognized as the essential efforts supporting the execution of business plans by providing a company with adequate production capability and capacity. The older generations of industrial/manufacturing engineers have been trained and challenged to design manufacturing systems which are both technically capable of executing desired manufacturing processes and economically justifiable. Past manufacturing system design practices have worked well for manufacturers in the era where mass production is the only way to cut cost and stay ahead of most competitors. Yet, the principles of just-in-time (JIT) production and group technology (GT) which gained popularity in the 1980s have imposed a totally new perspective on manufacturing system design. Production flexibility and rapid response to customers' needs became critical surviving factors. Moreover, with the advent of computer integrated manufacturing (CIM) technology, automation projects on the shopfloor require state-of-the-art control and management mechanisms to support factory integration efforts. Not only has the information technology been applied to facilitate islands of automation management, but also it serves to maintain the data integrity across both vertical and horizontal factory integration units. The new breed of industrial/manufacturing engineers is therefore expected to implement manufacturing cells with both production hardware (machine/equipment/layout) design and software (cell control system) design in mind such that the most viable cell configuration can be obtained.

This chapter first briefly describes the typical factory control architecture. A review of traditional approaches to manufacturing system (cell) design is provided in Section 12.3. The concurrent cell design and control system configuration is presented in Section 12.4. A case study in Section 12.5 is

finally employed to illustrate the use of this new cell engineering methodology.

12.2 FACTORY CONTROL SYSTEM HIERARCHY

Good factory managers are now characterized by realism, vision, and control
as the watchwords of the 1990s are innovation, speed, service and quality.
With the development of advanced factory control systems – particularly over
the past five years – the imperative to manage has finally triumphed over the
temptation to trust technological solutions somehow to work on their own.
The developments such as materials requirements planning (MRP) and
manufacturing resources planning (MRPII), for example, have led to some
promising manufacturing performance. MRP does enable valuable high level
planning, but on its own, MRP can never deliver the day-to-day, minute-
by-minute production cells (or lines) manufacturing efficiency needed by
world-class manufacturers. MRP systems were not designed to interface with
the shopfloor. Yet what the factory shopfloor has to tell the managers,
supervisors or line/machine operators is vital.

Factory management is about making things happen with optimum
efficiency in the context of the business plan; meeting customer requirements; controlling events; using capacity; understanding the capability of
machines and people; moving resources; integrating design, planning, control, distribution and financial systems. Figure 12.1 is a typical configuration

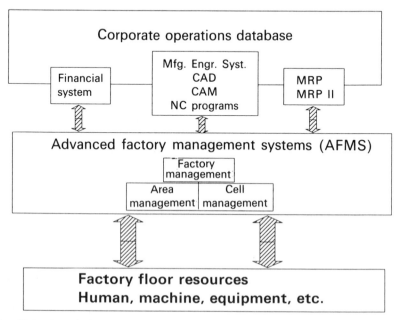

Fig. 12.1 A typical CIM architecture for manufacturing firms.

of a computer integrated manufacturing (CIM) architecture which illustrates the integration of advanced factory management systems with other vital functional components of a manufacturing firm. Figure 12.2 depicts the possible factory control stratifications, namely, the factory level, the area level, the cell level, and the station/equipment level. Table 12.1 summarizes typical software and hardware requirements of factory CIM solution hierarchy.

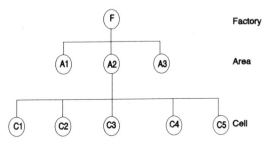

Fig. 12.2 Factory control system hierarchy.

Table 12.1 Software and hardware requirements of factory CIM solution hierarchy

Control hierarchy	CIM software	Running platform (hardware)
Factory	CAE, CAD, MRP, MRPII	Mainframe computer
Area	SPC/SQC, tool management maintenance management, CAPP	Industrial/minicomputer
Cell	Sequencing, monitoring, tracking DNC	PLC, industrial computer, CNC
Station/equipment	Machining/assembling/ handling	Microcomputer, PLC, CNC, Industrial computer, I/O Motion control

A factory control architecture such as the one shown in Fig. 12.2 is intended to offer complete control and total flexibility. For example, linking the CAD/CAM functions into day-to-day manufacturing enables the downloading of revised part programs to the production machinery. This ensures that engineering and design functions are more responsive to customers' needs. The long-term benefits of a factory control system are also clear. It allows for a more far-sighted business and production strategy, one based upon real-time operations, rather than upon studying factory performance data well after the event.

12.3 THE TRADITIONAL CELL ENGINEERING APPROACH

Cellular manufacturing systems have been identified as manufacturing systems which produce medium-volume/medium-variety parts more teconomically than other types of manufacturing systems (Black, 1983; Groover, 1980). The major components of cellular manufacturing systems are groups of machines/equipment and families of parts and associated tools. Partitioning of parts, tools and machines using group technology (GT) concepts has been a widely addressed research topic in the 1980s. There are some promising mathematical models and efficient solution approaches developed for planning manufacturing cells (King and Nakornchai, 1982; Ham *et al.*, 1985; Kusiak, 1985; Chen *et al.*, 1988; Choobineh, 1988; Ventura *et al.*, 1990, Harhalakis *et al.*, 1990). However, grouping parts and machines has become a less concerned issue mainly because the mission of designing a cell has already altered from 'rearranging' current factories to 'reinventing' new factories for support of factory CIM architecture. In other words, groups of parts for cellular manufacturing are always known (given) in many industrial settings.

Simulation modeling has been a popular technique for assessing the feasibility of physical layout and 'system dynamics' of manufacturing cells. There is abundant literature addressing the use of simulation in manufacturing system design. The major limitation of simulation is that it cannot prescribe system designs. It can only provide information about what is not advisable. Cell design using simulation is thus a trial and error process which is highly time consuming and sometimes cost prohibitive (Bullinger and Sauer, 1987; Caramanis, 1987; Kusiak, 1987). To supplement simulation deficiencies, Chryssolouris *et al.* (1990) proposed a neural network approach which sought to provide the inverse of the simulation function: given a set of desired cell performance measures, the neural network outputs a suitable cell design. However, the training of an artificial neural network calls for simulation runs which may have defeated the original initiative of such a seemingly attractive approach.

The real challenge for a cell designer becomes how to acquire machines, equipment, and a cell control system in such a way that the cell can support a threefold objective:

1. Capability to execute process plans of all parts;
2. Integration requirements of factory control hierarchy; and
3. Cost effectiveness of cell implementation.

12.3.1 The 'sequential' design methodology

Cell control systems are fundamental building blocks of a factory CIM hierarchy (see Fig. 12.2). The transition from a non-CIM to a total CIM environment of a factory has made cell engineering a two-step sequential process:

1. Machine/equipment acquisition; and
2. Cell control system implementation.

In this case, cell designers first select the appropriate sets of machinery/ equipment based on process plans, machine capability and machine capacity, and then ask for the system integrator (the cell controller suppliers) to provide cell control systems in accordance with the cell operating procedure and operator interface. Such a two-step cell engineering method is illustrated in Fig. 12.3. Depending on the cell complexity, simulation modeling may be performed before or after the implementation of cell control systems. This cell design methodology seems logical and, in fact, is the necessary painful step as a company seeks its journey to CIM through integrating existing factory cells.

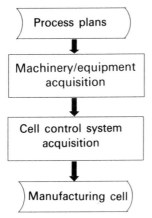

Fig. 12.3   The traditional two-step cell engineering method.

12.3.2 The high-cost dilemma

Implementation costs of a cell control system are highly related to the complexity of chosen cell control functionality and the connectivity of device drivers. Installing a cell control system on top of the machines and equipment already operating in the shop tends to be very costly due to vast efforts needed to build an interface (or adaptor) for communications to and from device drivers. The inconsistent operating procedures across different cells will also result in difficulties in choosing appropriate complexity levels of cell control functions. Both scenarios have been commonly seen in industry. This may have partially explained the ever-startling CIM start-up costs. The very likely result of the two-step sequential cell engineering practice is either a cell with a cost prohibitive cell controller or a cell with a mighty, yet consistently under-utilized, cell controller.

12.4 THE CONCURRENT CELL ENGINEERING METHODOLOGY

Coordination, communication, and control requirements of a manufacturing cell are always provided by a set of software functions which facilitate the system integration and manageability. Figure 12.4 depicts five typical modular control functions furnished by a cell control system. It is very difficult, if not impossible, for the cell designer to determine the appropriate functional configuration of each module unless the cell end-user pinpoints exactly how the cell is to be operated and managed. The central theme now becomes a journey in search of a methodology that will facilitate the acquisition of necessary cell control functions with appropriate levels of complexity while considering the inherent constraints imposed by the machines and equipment.

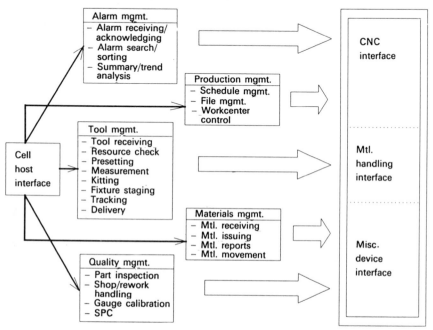

Fig. 12.4 Typical modular functions provided by a cell control system.

As the concept of concurrent product and process development (CPPD) is being widely employed to enhance the productivity and quality of prismatic part manufacturing, the idea of concurrent machine/equipment design and cell control system configuration is not difficult to comprehend. A basic framework illustrating implementation procedures of this proposed methodology is described in Fig. 12.5. Major components of this methodology are discussed in the following subsections.

12.4.1 Description of operations

Description of operations (DO) is a documentation of the definition of overall manufacturing operations to be performed within the cell. It defines individual operations, the intended process flow/sequence, and related quality control requirements. DO is normally prepared by the manufactur-

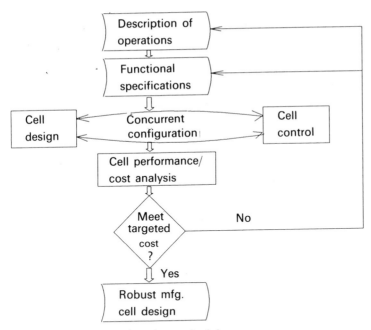

Fig. 12.5 Integrated cell engineering methodology.

ing engineers as a portion of the generic manufacturing plan for the set of parts selected to run in the cell. Tables 12.2 and 12.3 give two example production operations defined to process a family of mechanical parts. As shown in both tables, definition of operations starts with a general description followed by more detailed elaborations on process transformation, material handling, and operation sequencing issues.

12.4.2 Functional specifications

Functional specifications (FS) are the development and documentation of overall machine/equipment functions required to perform the manufacturing processes and supporting logistics (material handling/transport) defined in DO. They state detailed hardware functional requirements at the individual cell components level, such as the type of machine/equipment, and the inspection/probing/gauging mechanism. It also covers desirable software functions reflecting operating procedures at individual operation level. Table 12.4 shows an example list of types of machine/equipment required to perform manufacturing operations of a family of mechanical parts.

12.4.3 Concurrent design and control function configuration

This is the stage where detailed machine/equipment specifications and cell control functions are considered simultaneously in an effort to come up with an optimal cellular manufacturing system.

Table 12.2 Example description of operation (DO) – hard turn of mechanical parts

General description
  The turning of two bearing surfaces on the shaft and the facing of the shaft ends to length comprise the hard turn operation. One part at a time is machined with accuracy checked by automated in-process gauges, and manual post process gauges.

Process transformations

	Inputs	*Outputs*
Materials	Cooled parts	Hard turned parts
Tooling	Cutting tools	Worn/damaged tools
Controls	Manufacturing plan	None
Information	Work order	Gauging data
	NC part programs	
	Gauging data	
Alarms	None	Gauging system alarms
		Machine alarms

Material handling
  Operator 2 removes one part at a time from the cool down tank with the hoist, and transports the part to the load assist device on the hard turn machine (lathe). When the hard turn operation is complete, it the part is a #1234, then operate 2 uses the hoist to move it to the radial drilling station. Otherwise, Operator 2 hoists the part to the appropriate pickup area.

Operation sequence
  Operator 2 places part in a load assist device.
  Operator 2 verifies tool length gauge, in-process part gauge and the post-process part gauge.
  Operator 2 starts machine and at the end of the cycle, he (or she) manually gauges the part and enters the data to the quality system workstation.
  If part is within tolerance:
    If part is a #1234, Operator 2 hoists it to the radial drilling station.
    If part is not a #1234, Operator 2 hoist it to the appropriate pickup location – front or rear.
  Otherwise, the part must be examined for possible rework.

Backup procedures
  The soft turn machine should be used as a backup to the hard turn machine.

(a) Cell design

A complete cell simulation study should be performed to determine the number of each type of system component (e.g. machines, pallets, fixtures, gauging stations, and material handling equipment) and cell layout. The horsepower requirements, cutting feed/speed rates, size of batch processing (heat treat operations, etc.) are also to be determined at this time. Type of machine controllers and equipment control interface are to be selected in conjunction with desirable cell control functions as specified by the DO and FS. For instance, if both downloading and uploading NC part programs are desirable, selection of a proper machine controller is essential. However, the

Table 12.3 Example description of operation (DO) – hardening of mechanical parts

General description
    The hardening furnace maintains the temperature, pressure and gas concentration required to austenitize up to sixteen parts at one time. Via a PLC, the intelligent operator interface (IOI) directs the robots to load the furnace to the quench tank elevator at the proper time to insure the desired material hardness is attained. Under normal conditions, the robot handles all material transfer in the heat treat complex while Operator 2 is occupied with the hardness testing, hard turn, radial drill and stud press tasks.

Process transformations

	Inputs	Outputs
Materials	Machined parts	Hardened parts
Tooling	None	None
Controls	Manufacturing plan	None
Information	Temperature process set	Temperature process set
	Temperature idle set	Temperature idle set
		Temperature – actual
	Recipe 1–15 hard time	Recipe 1–15 hard time
	Idle start date and time	Idle start date and time
	Process start date and time	Process start date and time
Alarms	None	Over-temperature
		Cool water pressure low/high
		Methanol/nitrogen failure
		Natural gas failure
		Fan/motor/oil pump failure

Material handling
    Parts arrive on either of two inbound conveyors where a robot loads one part at a time into the hardening furnace. The robot later removes each part from the furnace and transports it to the quench tank elevator.

Operation sequence
    If the hardening furnace has an empty position, and a part is present on either inbound conveyor:
        Then the intelligent operator interface (IOI) checks the receipe of the part on the long inbound conveyor to verify that it can complete the entire heat treat cycle before the idle period begins.
        If the part can complete the heat treat cycle, the robot loads it into the furnace.
        If the part cannot complete the entire heat treat cycle, and a part is present in the short conveyor, IOI checks its recipe to verify that the part on the short conveyor can complete the heat treat cycle.
        If this other part can complete the heat treat cycle, the robot loads it into the furnace.
        Otherwise, no more parts can be loaded into the hardening furnace that day.
    When the IOI determines that a part is finished, it signals the robot via a PLC to remove the appropriate part from the furnace. The robot then removes the indicated part from the furnace and transports it to the quench tank elevator.

Backup procedures
    None

Table 12.4 Example machine/equipment functional specifications – a mechanical parts manufacturing cell

Machine/equipment	Operation(s) performed
CNC milling center	End and face milling; center and drill a hole for use by lathes
Lathe 1	Soft-turn to the desired dimensions
Spline roller	Form the splines on top and bottom of parts
Machining center	Drill and tap the parts and deburr the holes
Heat treat complex:	
Hardening furnace	Austenitize parts
Quench tank	Rapidly cools the hardened parts
Inspection station	Manual hardness testing of parts
Tempering furnace	Temper parts
Cool down tank	Cool parts to the same temperature
Lathe 2	Hard-turn of heat-treated parts
Radial drill press	Drill and ream holes in the parts
Stud press	Insert studs into each part
Support equipment:	
Gauges	Measure dimensional data of parts
Quality data systems	Enter and process gauging data
PLC and IOI	Coordinate robot operations in support of heat treat process

vendor cell control system may not currently support certain machine controllers or device drivers (for robots, AGV, etc.). Therefore the final decision on a machine controller, for example, should be based on the following two elements:

1. Machine tool builder's preference/specification; and
2. Connectivity to the cell control system being considered.

It must be noticed that most interface/adaptors built for a machine controller are reusable for machines using the same controller with minor programming changes. While the selection preference, simply for economic reasons, should be given to machine controllers and device drivers currently supported by the cell control system, an alternative would be to build the interface/adaptor which can be reused on other cells throughout a corporation for best cost leveraging.

(b) Cell control

Cell control functional specifications must be defined according to manufacturing engineering and operating requirements specified in the DO. Depending on the functionality and connectivity of machine controllers and device drivers, the cell control system should possess necessary functions to facilitate the operating philosophy defined by the cell management/operation personnel. Major cell control functions (see also Fig. 12.4) normally include manufacturing plan management, materials and resource management, production/labor reporting, perishable tool management, quality plan management, scheduling, etc. Each control function can be performed with different levels of complexity such as the two example functions presented in Tables 12.5 and 12.6. Most cell control levels are defined in such a way that more basic (less complex) levels are inclusive in more advanced (more complex) levels. This is to ensure that a manual backup method is available at the time that continued production is mandatory. The configuration of cell control functions must undertake two 'layers' of tasks:

1. The selection of appropriate set of functions; and
2. The selection of appropriate complexity level for each of the selected cell control functions.

Table 12.5 Example leveled cell control function – program management support

	Complexity level	*Required hardware/ software features*
Level 1 (L1)	Manually download program files to individual workstations through cell controller i.e., after receiving the producton command, the operator can retrieve and view all files and reliability status from the upper level systems and download to the correct station. (example files: NC programs, PLC data program files, CMM programs, *et al.*).	Cell control terminal. Connectivity to upper level system and to the individual workstation controller. Interface/adaptor between cell controller and machine controllers and device drivers.
Level 2 (L2)	L1 + the ability to manually upload program files to upper level systems through cell controller, e.g., uploading of robot programs (teach mode) and NC programs that have been edited on the shop floor.	Same as above.
Level 3 (L3)	L2 + full direct numerical control capabilities, i.e., uploading and downloading of programs and files between the workstation and the cell controller is initiated by the cell controller without human intervention (usually seen in flexible manufacturing systems).	Same as above + workstation controller must have full DNC capability.

The cost concern and/or machine controller interface can very well direct the selection of cell control functions and the complexity level of each control function. For example, the cell management may choose to select no program support (referring to Table 12.5) function for a particular cell. But the cell control terminal will be utilized to access existing download capabilities through its 'second session' capability. On the other

Table 12.6 Example leveled cell control function – perishable tool management

	Complexity level	Required hardware/ software features
Level 1 (L1)	View and print perishable tool requirements (tool ID #, type, name, quantity, *et al.*) for each operation in the manufacturing.	Cell control terminal.
Level 2 (L2)	L1 + view the quantity, location and inventory of all tools in the cell; provide graphic tool setup instructions contained in the manufacturing plan; and monitor tool life.	Cell control terminal with graphic capability. Microlog or barcode interface. Machine controllers with adaptive control function to monitor dull/broken tools.
Level 3 (L3)	L2 + order/replenish tools by the operator; download nominal tool set-up parameters to the tool preset station; upload actual values (offsets, *et al.*) to the cell control level for later downloading to machine controller (CNC).	Same as above + machine controllers with full DNC capabilities. Connectivity to higher level systems (e.g. a tool inventory management system).
Level 4 (L4)	L3 + monitor remaining tool life for all tools in the cell; predict what and when tools will need to be replenished; automatically issue reorder requests electronically to central tool crib and notify the automated delivered device (e.g., AGV) for handling services.	Same as above + proper control interface with material handling devices

hand, if cell workstations utilize a behind tape reader (BTR) interface, Level 2 program management functionality as defined in Table 12.5 is the highest level implementable.

The selection of complexity of tool management function has a lot to do with the machine controller. 'Intelligent' machine controllers will generate dull tool warnings and errors based on actual cut time and dull/broken errors based on prescribed adaptive control such as horsepower monitoring. To achieve the same Level 2 functionality as shown in Table 12.6, for example, with 'nonintelligent' machine controllers, the cell control system will have to estimate actual tool usage based on cycle count multiplied by cut time to generate dull tool warnings and errors. The cost trade-offs between the level of 'intelligence' of the machine controller and the complexity of cell control functions can be easily seen in this case.

12.4.4 Cell performance and cost analysis

The concurrent configuration of cell hardware (machine, equipment, and controllers) and cell software (cell control system) functions is primarily driven by DO and FS as specified in Fig. 12.5. In other words, the concept design up to this stage should have ensured the technical feasibility of the cell. However, the eventual project feasibility relies on its successful economic justification. To perform a thorough cost analysis, a detailed simulation study has to be done to capture the overall dynamics of a cell. This simulation study is intended to reassure the hardware requirements and layout plan identified in Section 12.4.3(a) after incorporating the integration effects resulting from application of the cell control system. Hence, the simulation model is a much more detailed one which encompasses not only the operator interface with machinery/equipment but also various information passages enabled by the cell control system. Such a simulation model can be considered as the closest to a real operating cell where the assessment of cell throughput rate can be used to calculate the part production cost. The piece part cost derived from a simulation performance study can then be compared with the targeted accounting cost to decide if revisions of DO and FS are needed. Options to reduce production costs include redesign of manufacturing processes and/or choice of a different set of production machinery/equipment. Since the cell design and control function election are driven by DO and FS, the cost justification may prompt an iterative cell engineering process (as shown in Fig. 12.5) finally to come up with a robust manufacturing cell design.

12.5 CASE ILLUSTRATION

Suppose a family of mechanical parts has the process flow diagram as shown in Fig. 12.6. Manufacturing engineers have defined a complete set of detailed operation descriptions for all processes. Tables 12.2 and 12.3 enumerate two representative operations – hard turn and hardening. After an initial simulation study, the machinery and equipment chosen to perform all necessary manufacturing operations are summarized in Table 12.4.

The selection of cell control functions has been based primarily on the DO. For instance, the DO has dictated that all machines except the machining center be equipped with an integrated process control station to provide for manual entry of post-processing gauging data, to calculate SPC data, to issue alarms and to communicate with the cell control system. Post-processing gauging data are used for tool wear compensation on the part to be processed next. This requires the cell controller to possess a quality management function which will support the manual downloading of quality instruction files to all process control stations and uploading of quality data from station level database. However, to facilitate such a quality management function, appropriate interface between the cell control system and device (process control station) drivers must be made available.

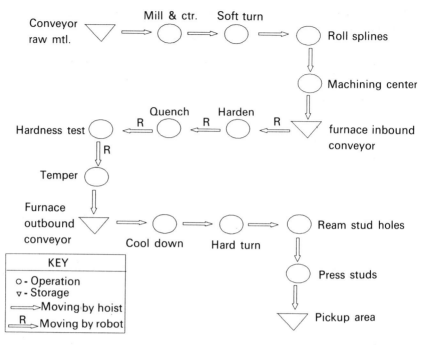

Fig. 12.6 Process flow diagram for the example mechanical parts manufacturing cell.

The DO also specifies that, at the start of the shift, the operator should refer to the cell control terminal or daily printout for any changes in part programs. If any part programs have been changed, the operator downloads the new programs to the appropriate machines from the cell control system. This demands that the cell control system should provide the program management support function which allows manual downloading of program files to individual workstations. Interfaces/adaptors between the cell control system and machine controllers must be made available. Similarly, the DO requires that alarms and parameters (recipes, temperatures, etc.) normally accessed from the intelligent operator interface (IOI) screen at the heat treat unit must be uploaded to the cell controller from the PLC each time an alarm arises or a parameter changes. During any combination of failures among the PLC, IOI, furnace, or tanks the cell control system will provide the operator with the capability of downloading PLC and/or IOI programs. These operation requirements again necessitate the interfaces between the cell control system and the PLC/IOI.

With desirable cell control functions in mind, a detailed cell simulation study has been performed to confirm the technical and economical feasibility of the initial cell configuration. Selection of machine controllers must refer to the list of currently supported interfaces by the cell control system

Table 12.7 Control hardware requirements – the example mechanical parts manufacturing cell

Device	Quantity
DEC Micro-VAX II with 16 MB RAM	1
DEC DELQA RA82 disk drive	2
DEC TK70 tape drive	1
DEC printer	1
DEC VT-220 terminal	2
A-B PLC-2 with KG-1771 module	1
C-M milling machine with GE2000 controller	1
C-M turning lathe with GE2000 controller	2
C-M T30 machining center with Acromatic850 controller	1
IOI for heat treat unit control	1
Process control station	5

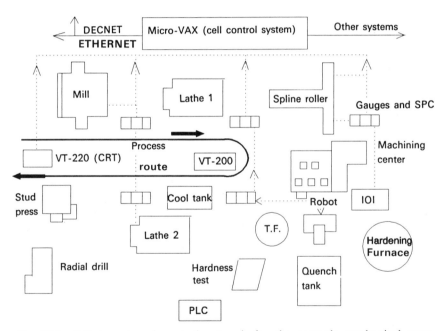

Fig. 12.7 Cell layout and control network for the example mechanical parts manufacturing cell.

vendor and the machine suppliers' specifications on station control. Table 12.7 presents a summary of control hardware requirements which specifies the use of two machine controllers – GE 2000, and Acromatic 850. The set of cell control (software) functions for this manufacturing cell is summarized in Table 12.8. Figure 12.7 depicts the final U-shape cell layout and the computer control network recommended through implementing the proposed concurrent cell engineering methodology.

Table 12.8 Cell control functions – the example mechanical parts manufacturing cell

1. Vertical and horizontal interface	The transfer of information to the external nodes is required in order for the cell to perform its task of producing finished parts from raw material. These external nodes include the corporate planning and scheduling system (running at a corporate mainframe IBM), the downstream assembly cell control systems, scrap bins/dunnage, gauge/tool crib, the area control system, and the operations data interface (ODI) required to have the access to the corporate operations database (ODB).
2. Scheduling and data maintaining	This function provides the ability to maintain the files used for the scheduling and production of parts.
3. Security data management	This function provides the support for adding entries to the 'security database' as well as changing and deleting these entries.
4. Process control and management	This function will generate hardcopy reports from quality control data collected during production. It also provides the ability to upload/download a program from/to the process control stations.
5. Material request and delivery	This function can receive a material request from either an operator or another function. It issues a request to the outside world and handles the reply.
6. Manufacturing operations control	This function controls the manufacturing operations within the cell. This includes providing instructions to the operator, downloading programs to SPC stations and machines, and issuing recovery data to a station.
7. Manufacturing operations monitoring	This function monitors the machines, processors, and terminals in the physical cell. If a significant event occurs within the cell, it may be 'signalled' to other cell functions for responding to this event. Quality data will also be monitored by this function via interface with SPC stations where manual gauging data are collected.

12.6 DISCUSSION AND CONCLUSIONS

The traditional cell design methodology (two-step, sequential method) has not worked well as companies are taking their long journeys to company-wide CIM from islands of automation to fully integrated production systems. Concurrent consideration of cell physical configuration and required cell control functions is one good way of ensuring economic implementation of factory cells. The concurrent cell engineering methodology has its strengths and several unique features:

- The new methodology uses intended system operating/management philosophy and desirable performance measures (cost etc.) to drive tasks of cell design and cell control function definition. This is a breakthrough from traditional thinking that performance of an engineering design must follow its design.
- This new methodology will help ensure that only necessary control functions are acquired, and those acquired functions are implemented with adequate levels of complexity. This is the key to economic justification of cellular manufacturing systems.

12.7 REFERENCES

Black, J.T. (1983) *Industrial Engineering*, November, 36–48.

Bullinger, H.J. and Sauer, H. (1987) *International Journal of Production Research*, **25**(11), 1625–34.

Caramanis, M. (1987) *International Journal of Production Research*, **25**(8), 1223–34.

Chen, F.F.,Ventura, J.A. and Leonard, M.S. (1988) *Proceedings of the 1988 International Industrial Engineering Conference*, 512–18.

Choobineh, F. (1988) *International Journal of Production Research*, **20**(7), 1161–72.

Chryssolouris, G., Lee, M., Pierce, J. and Domroese (1990) *Manufacturing Review*, **3**(3), 187–94.

Groover, M.P. (1980) *Automation, Production Systems, and Computer-Aided Manufacturing*, Prentice-Hall, New Jersey.

Ham, I., Hitomi, K. and Yoshida, T. (1985) *Group Technology*, Kluwer-Nijhoff Publishing, Boston.

Harhalakis, G., Proth, J.M. and Xie, X.L. (1990) *Journal of Intelligent Manufacturing*, **1**, 185–91.

King, J.R. and Nakornchai, V. (1982) *International Journal of Production Research*, **20**, 117–33.

Kusiak, A. (1985) *Annals of Operations Research*, **3**, 279–300.

Kusiak, A. (1987) *International Journal of Production Research*, **25**(3), 319–25.

Ventura, J.A., Chen, F.F. and Wu, E.H. (1990) *International Journal of Production Research*, **28**(6), 1039–56.

CHAPTER 13

A generalized methodology for evaluating manufacturability

Srinivasa R. Shankar and David G. Jansson

13.1 INTRODUCTION

Research in the design for manufacturing (DFM) area has evolved primarily as independent studies in different manufacturing domains. These studies have yielded many different techniques for evaluating how easy it is to manufacture a part. In most cases the techniques are valid for a specific manufacturing process. Since almost all products require several manufacturing operations, a designer needs to be familiar with several different DFM techniques in order to assess the manufacturability of a design. Manufacturability is defined here as the ability to manufacture a product to obtain the desired quality and rate of production while optimizing cost. The ease of manufacturing is an inherent part of this definition due to its influence on all of these factors. The methodology presented in this chapter provides designers with a structured approach to DFM and acts as an aid in identifying the critical parameters which affect manufacturability in any design. Further, the methodology yields quantitative metrics which can be used to evaluate and compare designs.

13.2 BACKGROUND

A number of different approaches to DFM exist in the literature. These include design rules which may be found in handbooks such as those by Bralla (1986) and Bolz (1977). These rules are the result of empirical knowledge gained over many years of experience. A comparison between different designs based on design rules is difficult since they do not provide a quantitative evaluation. Domain specific evaluation procedures such as those developed by Boothroyd and Dewhurst (1987) and Poli *et al.* (1988) typically provide a rough estimate of the cost of the part. Yannoulakis *et al.*

(1991) describe a method for obtaining quantitative measures of manufacturability for rotationally-symmetric machined parts. A drawback of these procedures is that they do not yield any feedback on those aspects of manufacturability, such as part quality, which are not directly related to cost. To estimate the quality of a part, one can use a simulation package such as Moldflow, which simulates flow and warpage in plastic injection molded parts. The methodological approaches suggested in the literature include those by Suh (1990) and Ishii *et al.* (1988). In Suh's method, manufacturability is determined by the relationships between design variables, process parameters, and functional requirements. These relationships are expressed in the form of design matrices. With hindsight it is often possible to examine these matrices and determine manufacturability. However, this approach does not provide any assistance to the designer in identifying the different manufacturability issues during the design process. The procedure developed by Ishii *et al.* employs fuzzy measures to evaluate the compatibility of a proposed design with the required specifications.

Although some of these approaches are quite general, they do not organize the manufacturability issues in a form which would help novice designers deal with the complexity of the problem. There is clearly a need for a uniform strategy which would allow a designer to obtain an initial estimate of manufacturability without having to be proficient in domain specific DFM techniques. A detailed discussion of such a methodology is presented in Shankar (1992). This paper highlights some of the important aspects of the methodology.

13.3 A METHODOLOGY FOR MANUFACTURABILITY

There are three basic objectives which the methodology must satisfy. These are:

1. Establish a uniform, domain-independent approach to manufacturability evaluation;
2. Facilitate a quantitative evaluation of manufacturability even during the early stages of design; and
3. Generate a number of measures which deal with different aspects of manufacturability.

Each of these points is further elaborated. A uniform or generalized methodology provides a framework within which manufacturability of any product and process combination may be examined. A domain-specific evaluation procedure, on the other hand, only aids the designer in designing for that one process. A uniform approach is by necessity one that is based on fundamental concepts rather than factual information. When faced with a design which requires novel manufacturing processes, the designer can extend these fundamental concepts to address manufacturability for the new problem. A generalized methodology, therefore, acts as a strong educational tool.

The need for an early evaluation of manufacturability is well recognized. The effective merger of manufacturing and other life-cycle issues into the early stages of design is one of the primary objectives of concurrent engineering efforts. Figure 13.1 depicts a view of the concurrent engineering process in which manufacturability evaluation is the key element that allows the designer to ensure compatibility between product and process descriptions.

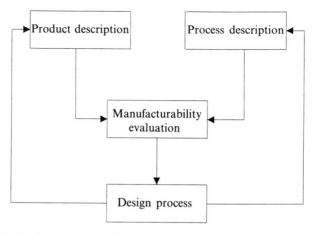

Fig. 13.1   Role of manufacturability evaluation in a concurrent engineering process.

The justification for a quantitative set of measures stems from a belief that the process of assigning numerical values to the evaluation promotes an in-depth analysis of the problem. It is important to keep in mind that at the early stages of the design, any quantitative metrics are likely to be less accurate than at the later stages and should therefore be interpreted with sound engineering judgment.

The main argument in favor of the approach in which a number of different measures are generated is that a single measure cannot satisfactorily cover all the issues which are encompassed within the scope of a DFM effort. The availability of multiple quantitative metrics to evaluate manufacturability will provide designers with a rich source of feedback and may lead to fresh insight on how to improve the design.

13.4 CORE MANUFACTURABILITY CONCEPTS

The common issues which arise in the study of manufacturability in a wide variety of manufacturing domains are presented here as a set of core manufacturability concepts, or CMCs. There are five core concepts namely, compatibility, complexity, quality, efficiency and coupling. Each core concept consists of several sub-categories which share the theme expressed in the core concept. Figure 13.2 shows the hierarchical structure of manufacturability which forms the basis for the evaluation methodology.

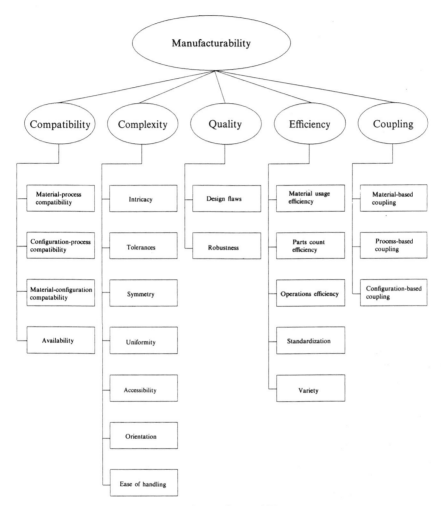

Fig. 13.2 Hierarchical structure of manufacturability.

The evaluation process for any product begins with an identification of the critical parameters which affect the manufacturability of that product, through a detailed examination of the CMCs. A discussion of the CMCs and the subcategories within each core concept is presented next. In this discussion, a reference is often made to examples from different manufacturing domains in order to demonstrate the generalized nature of the CMCs.

13.4.1 Compatibility

The purpose of a compatibility check is to ensure that the design requirements can be satisfied. A design requirement is any objective or constraint which the design must satisfy. Requirements may be classified as either functional requirements or production requirements. In a manufacturability evaluation effort, the focus is on production requirements such as cost, production rate, quality, and repeatability. The information content of a design changes as a result of activities such as selection of a material or process, addition of a feature, specification of a tolerance, or the design of a subsystem. When this occurs, it is important to check if compatibility has been affected. Frequent checks on compatibility will prevent the need for costly design modifications at later stages of the design. Compatibility may be classified into the following categories:

(a) Material-process compatibility

This category deals with material properties and their relationship with the characteristics of the manufacturing process. The question to be asked here is: How well can the material be processed? Machinability, extrudability, weldability, and forgeability for different materials and alloys are all different indicators of material-process compatibility.

(b) Configuration-process compatibility

Under this category the designer is mainly concerned with the ability of the process to generate the desired form and tolerances at the desired production rates. For processes such as injection molding or forging, where the geometry is created in a single step, or in multiple but discrete steps, the machine capacity determines the feasibility of creating the desired form. For processes in which the form is created continuously, as in machining, cycle time in the key variable which affects configuration-process compatibility.

(c) Material-configuration compatibility

The main issue here is the relationship between a specific material property and the final design configuration. An example of a situation where this form of compatibility is important, is the selection of a material for a high precision injection molded part. The material property which affects compatibility in this case is the difference between the shrinkage parallel and perpendicular to the flow. To increase compatibility, the designer should choose a material for which this difference is small.

(d) Availability

The last category of compatibility is aimed at the need to ensure that various resources required to create the product are available. These resources include processing technology, machines and equipment, skilled and unskilled labor, materials (in the desired form), standard components, and capital.

13.4.2 Complexity

The next CMC which the designer must address is complexity. Longer processing and set-up times, more expensive tooling, and higher labor costs are some of the effects of increased design complexity. Also, it is likely that more parts will be rejected due to problems which arise during manufacturing. Reducing the complexity of the design is therefore one of the primary objectives of any manufacturability evaluation effort. There are a number of factors which can affect the complexity of a manufacturing process. The focus here is on those factors over which the designer has considerable influence.

(a) Intricacy

Intricacy may be defined as the amount of detail in a part. Parts which have several features located over a small region are often difficult to manufacture. A rough estimate of the intricacy of a part can be obtained by counting the number of surfaces on the part.

(b) Tolerance and surface finish

Tight dimensional and geometric tolerances increase the amount of processing required. Specification of tight tolerances also results in the need for more careful measurement of dimensions, frequent regrinding of tools, and use of more expensive equipment. The costs rise very sharply as tolerance or surface finish is increased, an effect which emphasizes the importance of addressing this issue early in the design process.

(c) Symmetry

The concept of symmetry is important in many manufacturing domains. In most cases, making a part more symmetrical reduces the complexity of the associated manufacturing processes. For example, in assembly operations, an increase in the symmetry value tends to reduce the amount of reorientation that the part must undergo. In machining, a shaft which is symmetrical is much easier to turn as compared to a nonsymmetrical crank shaft. In injection molding, an asymmetrical layout of the mold cavities implies that the runners have to be sized individually, thus increasing the mold cost.

(d) Uniformity

Uniformity is similar to symmetry in the sense that it too is a concept which has different implications in different manufacturing domains and has an influence on the complexity of many manufacturing processes. In injection molding, uniform wall thicknesses are recommended in order to reduce the intensity of sink marks. Furthermore, nonuniform cooling of a part can result in high stresses and part warpage. In filament winding operations, if the diameter of the mandrel over which the filament is being wound is nonuniform, maintaining a constant thickness of material over the entire length becomes very difficult.

(e) Accessibility

It is important that the particular surface or region of the part which is affected by any operation is easily accessible. This is clearly true for manual and robotic assembly, most machining operations, and measurement and inspection procedures. If a particular surface in a part is difficult to access, there may be a need to construct a special tool in order to accomplish the desired task. This is viewed as an increase in the complexity of the task.

(f) Orientation

The orientation of a surface is defined in terms of the principal directions of the part. The principal directions are determined in the following way. The part is assumed to be enclosed within a rectangular box, such that the base surface of the part is aligned with one side of the box and the primary axis of the part is either parallel or perpendicular to each side of the box. One must identify a base surface and a primary axis for the part *a priori*. The principal directions then correspond to the three mutually perpendicular edges of the box. Surfaces which make compound angles with the perpendicular directions are said to be skewed. Such surfaces typically require the part to be reoriented during manufacturing, and therefore increase complexity.

(g) Ease of handling

It is important to remember that besides all the operations performed on a part to create its desired form, a part also passes through a number of handling operations. These include transferring the part from one workstation to the next, feeding, locating, orienting, and initial set-up of the machines. The complexity of these operations is influenced by size, weight, fragility, and shape.

13.4.3 Quality

The final quality of a product depends on both the design of the product and the ability of the manufacturing equipment to achieve the desired form and tolerances. In the current context, a high quality part is one which can be made with a high degree of repeatability and which conforms to the designer's specifications. The part designer can, through proper design, make it much easier to obtain high quality parts from any given production equipment. There are two main issues within this definition of quality which are important.

(a) Design flaws

The design should not contain features which can cause critical flaws. These features may often arise in processes such as injection molding, casting, and die casting where material orientation and cooling play important roles in determining part quality. Irrespective of the manufacturing process, features which are difficult to inspect can result in flaws which remain undetected and should be avoided.

(b) Robustness

The design should be robust with respect to minor variations in material properties, machine settings and other process parameters. With age, the repeatability of machines can deteriorate. A robust design will accommodate small variations in a manufacturing process without significantly affecting either subsequent manufacturing operations or the functional performance of the product.

13.4.4 Efficiency

The efficiency of a manufacturing process depends on both the design of the product as well as on the optimal selection of process parameters. The purpose of an efficiency check is to identify opportunities to improve the efficiency of the manufacturing processes through better design. The issues which the designer must address in order to improve the efficiency of the manufacturing process include:

(a) Material usage efficiency

Process selection should be such that material wastage during processing is kept at a minimum. In some processes such as machining, a lot of scrap is generated. In other processes such as casting, there is essentially no scrap, and the material usage efficiency is therefore high.

(b) Parts count

The use of fewer parts in a design leads to the elimination of several assembly and manufacturing operations. As long as the complexity of the resultant parts is not increased much, part reduction is a useful guideline to follow.

(c) Operations efficiency

The designer should always attempt to eliminate all unnecessary operations, both in manufacturing as well as in handling. In addition, the amount of processing associated with any expensive operations should be reduced as much as possible. Some of the design changes which will increase efficiency are the use of looser tolerances and elimination of features which cause excessive tool wear.

(d) Standardization

The use of standard components wherever possible in the design is well recognized as a fundamental design guideline. The advantages include fewer manufacturing operations and hence lower costs. Besides standard components, design for use of standard raw materials, operations, and tools leads to further savings.

(e) Variety

Another important factor closely related to standardization, is variety. The designer should specify as few a number of varieties for any component or feature as possible. The specification of fewer varieties of standard components implies that the parts can be bought in greater quantities which is usually accompanied by savings. When fewer feature varieties are specified, the manufacturing process becomes more efficient since the number of nonproduction operations such as machine set-up, tool change, tool regrind, measurement and gauging is reduced.

13.4.5 Coupling

In a concurrent engineering or DFM strategy, the impact of each design decision on both functional and production requirements must be assessed. A change in a design parameter will typically affect different requirements differently. Whereas one requirement shows an improvement, another may be adversely affected. When this happens, the design is said to be coupled and the parameter is referred to as the coupling variable. Unless it is handled properly, strong coupling between requirements can hinder a DFM effort. The designer should therefore identify coupling variables which affect functionality on the one hand and manufacturability on the other.

Coupling as it affects manufacturability may be classified into the following categories:

(a) Material-based coupling

Here, a material parameter is the primary coupling variable. An example of such coupling may be found in the processing of fiber-reinforced composites. Fiber orientation is the variable which affects functional requirements such as strength or stiffness. At the same time, manufacturability is influenced by the ability of the process to deliver the desired fiber orientation.

(b) Process-based coupling

In this case, a process parameter is the source of the coupling. An example is the number of cavities in an injection mold. From the point of view of increasing production rate, one would like to have a large number of cavities. However, there is a price to pay in terms of reduced quality and repeatability.

(c) Configuration-based coupling

In this case, the primary source of the coupling is a particular design feature or parameter. For instance, increasing the wall thickness in an injection molded part makes it easier to fill the mold, but increases cooling time and material cost.

It is important to keep in mind that the CMCs are aimed at obtaining manufacturability improvements through better design rather than through innovations in manufacturing processes. A comprehensive analysis will require the designer to visualize the entire process plan. This should include all manufacturing, assembly, handling, and set-up operations. The CMCs should then be used with this process plan to identify features in the design which adversely affect manufacturability.

Although the CMCs are presented here as a set of five different concepts, they are not entirely independent of each other. It may be noted that any attempt to reduce complexity will also improve efficiency, and reduction in coupling may improve the repeatability of the process. In a multidimensional approach to manufacturability, correlation between the different issues is unavoidable. It is more important to obtain adequate coverage of all major activities associated with a manufacturability study. Since the list of core concepts has been generated after examination of a number of designs in different domains, and after reviewing much of the literature on DFM, it is expected that coverage is good. It is also expected that the general nature of the concepts will allow one to fit new issues that may emerge in special situations into one of the existing categories.

13.5 QUANTITATIVE EVALUATION OF MANUFACTURABILITY

One of the guiding principles of this methodology is that the evaluation should be quantitative. For this purpose, the designer should use the CMCs to identify potential metrics which feed back information about manufacturability. At this stage, a distinction is made between the terms measure and index.

A *measure of manufacturability* is defined as a metric for manufacturability which is obtained primarily from an analysis of the manufacturing process. A *manufacturability index*, on the other hand, is defined as a metric which is obtained primarily from an analysis of the product.

13.5.1 Measures of manufacturability

The measures can be classified into three categories depending on the type of analysis required in each case. Cost estimates will be referred to as Type I measures of manufacturability. Manufacturing process costs can be broken down into material cost, equipment cost, labor cost, tooling cost, and overhead cost. In order to calculate these costs, a fairly detailed consideration of the manufacturing processes including the equipment to be used, the sequence of operations, and the process parameters is necessary. Much of this information is not known to the designer, a fact which makes cost estimation a difficult and time-consuming effort. The availability of life-cycle cost estimating techniques which are applicable during the early stages of design will no doubt be of great benefit to designers.

The Type II measures of manufacturability are based on estimates of the time required for different manufacturing operations and results which may be obtained through simulation of the manufacturing process. The measures are: time for primary operations; time for nonproductive operations such as handling and set-up; time for reworking and finishing operations; and process simulation results. An estimate of the time required to manufacture a product is a rich source of feedback to the designer. A time estimate can be converted into a cost estimate by taking the machine and labor rates into account. A detailed model for calculating process time should also enable the designer to identify those features which require excessive processing and are hence inefficient from a manufacturing point of view. The advantage of simulation tools is that they provide additional information which can also be used to improve manufacturability. The pressure required to fill an injection mold, the stress at various locations, and the amount of warpage to be expected in the part are some of the results which can be obtained from simulation packages such as Moldflow.

In order to generate Type I or Type II measures, the designer needs sophisticated models and tools which can predict cost, cycle time, and product quality from a description of the design. Precise calculation of these measures also depends on detailed information about various aspects of the design including part configuration, material properties, and process variables.

The Type III measures rely on the ability of the designer to visualize the sequence of processes and reason about them. In general, these measures may be obtained by first determining the process plan and then using it to identify those tasks which decrease process efficiency. Some of the measures which may be obtained from the process plan are: number of labor intensive operations; number of adjustments required during each operation; number of tool changes; number of reorientations; and number of finishing operations. The main advantage of Type III measures is that they force the designer to think explicitly about the different forms and features which produce the design configuration, and their effect on manufacturability.

13.5.2 Manufacturability indices

Whereas the measures of manufacturability are obtained by analyzing the manufacturing process, the manufacturability indices are directly related to the design variables. Design variables are variables which are obtained from the physical representation of the design. For instance, the specification of a tight surface finish (design variable) has an effect on the manufacturing cost. Cost is therefore a measure of manufacturability for that surface. However, the cost is affected by many other factors such as machine and labor rate, the orientation of the surface, and accessibility. Instead of calculating cost, the designer may choose to compare two or more designs based on the finish specified for different surfaces. The surface finish in this case is an index of manufacturability.

The main advantage of utilizing manufacturability indices as metrics is that it is much easier to calculate indices than measures. Indices require only partial information about the design and can therefore be calculated very early in the design process.

There are two types of indices, generalized manufacturability indices (GMI) and domain-specific manufacturability indices (DSMI). The GMI are not related to any particular manufacturing process. They can be computed for any product and represent a very coarse set of metrics for estimating manufacturability. The DSMI, on the other hand, utilize information which is specific to a particular manufacturing process and are therefore not generalized in nature.

13.5.3 Accessibility: an example of a GMI

One of the issues mentioned in the discussion on CMCs was accessibility. An index for the difficulty of access for a surface can be constructed purely from a consideration of geometry. For any feature in the part, one must first identify a reference plane with respect to which accessibility is to be calculated. Typically, the plane is chosen such that all sections (of the feature) parallel to this plane are identical. A few terms are now defined to simplify the discussion. A primary access line for any surface is a straight line drawn from a point on the surface to any point outside the part such that the line is as close as possible to the surface normal and also lies in a plane parallel to the reference plane. α is the angle in degrees between the primary access line and the surface normal. A secondary access line is similarly defined as a line from the surface to a point outside the part, such that the line is as close as possible to a line parallel to the surface (or tangential for curved surfaces) and also lies in a plane parallel to the reference plane. β is the angle subtended between the secondary access line and the line parallel to the surface. The index is then calculated as

$$\text{Index value} = \frac{(2\alpha + \beta)}{270} + 1$$

For convenience, the index has been normalized so that the resultant value lies in the range [1, 2]. If a primary or secondary access line cannot be drawn for a surface, the value of α or β is taken to be 90° respectively. Thus, for a surface which cannot be accessed at all by a straight line, the index value is 2. Since different points on the same surface can have different values for α and β, the index is calculated by first looking for the point with the largest value of α. If a unique point is found, the index for this point is taken to be the value of the index for the surface. If multiple points are found, then the one with the largest β is taken as the representative point. Figure 13.3 shows a part in which two different features have been identified.

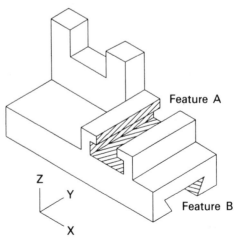

Fig. 13.3 Example part showing two features.

Accessibility is to be determined with the XZ plane being the reference plane for feature A and the YZ plane being the reference plane for feature B. The surfaces associated with each feature are indicated by the hatch lines. Figures 13.4 and 13.5 show the primary and secondary access lines for some of these surfaces.

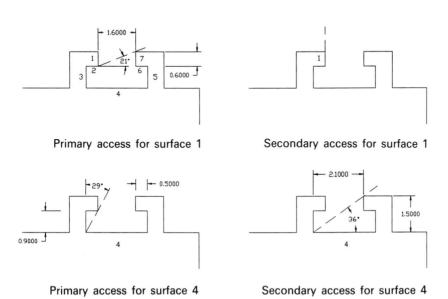

Primary access for surface 1   Secondary access for surface 1

Primary access for surface 4   Secondary access for surface 4

Fig. 13.4 Primary and secondary access lines for surfaces in feature A.

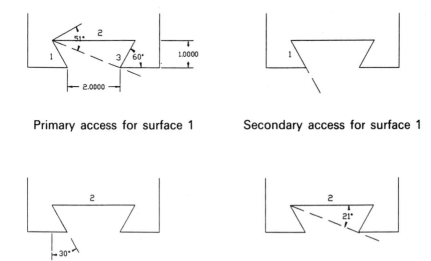

Primary access for surface 1   Secondary access for surface 1

Primary access for surface 2   Secondary access for surface 2

Fig. 13.5 Primary and secondary access lines for surfaces in feature B.

Access lines for surfaces 2, 3, 5, and 6 in feature A are not shown because a straight line from the surface to a point outside the part cannot be drawn. The values of the index for the different surfaces and a cumulative value for each feature are provided in Tables 13.1 and 13.2.

Table 13.1 Value of α, β and index of accessibility for each surface in feature A

Surface number	α	β	Index value
1	21	0	1.16
2	90	90	2.00
3	90	90	2.00
4	29	36	1.35
5	90	90	2.00
6	90	90	2.00
7	21	0	1.16
Cumulative			11.67

Table 13.2 Value of α, β and index of accessibility for each surface in feature B

Surface number	α	β	Index value
1	51	0	1.38
2	30	21	1.30
3	51	0	1.38
Cumulative			4.06

From the tables, the designer can quickly compare the manufacturability of different surfaces and features. Feature A contains several surfaces with very poor accessibility as is indicated by the value of 2 in the last column. Also, since feature A has more surfaces, the cumulative value of the index is much higher than that for feature B.

One must keep in mind that this example is very simple and that the comparison is based on only one aspect of manufacturability, which is the difficulty of accessing the surface. For the analysis of complex designs, the availability of several manufacturability indices dealing with different issues will be very useful to the designer. The benefit of this approach is enhanced if it is possible to calculate values for different indices automatically. For this purpose, it is recommended that the indices be defined using information which can be easily obtained from a geometric model of the part. Whenever the value of a particular index for a feature or surface exceeds a threshold, the system can warn the designer. The designer then has the option to modify the design or analyze it more accurately using measures of manufacturability rather than indices.

13.6 CONCLUSION

A generalized methodology for the early evaluation of manufacturability has been presented. The methodology utilizes quantitative metrics for evaluating manufacturability of designs. To aid the designer in identifying suitable metrics, a hierarchical model of the various aspects of manufacturability has been constructed. A distinction has been made between measures of manufacturability and manufacturability indices. An important aspect of the methodology is that the designer can interact with the evaluation process. Depending on the particular product being evaluated, the designer can define and use new measures and indices which provide relevant feedback. Since manufacturability indices can be calculated with incomplete information, it is appropriate to conduct a preliminary investigation of manufacturability using indices and use specific measures during the later stages of design. The methodology thus establishes a framework within which the role played by different DFM techniques is clearly defined.

13.7 REFERENCES

Bolz, R.W. (1977) *Production Processes: The Productivity Handbook*, Conquest Publications, Winston-Salem NC.

Boothroyd, G. and Dewhurst, P. (1987) *The Design for Assembly Software Toolkit*, Boothroyd-Dewhurst Inc., Wakefield RI.

Bralla, J.G. (ed) (1986) *Handbook of Product Design for Manufacturing*, McGraw-Hill Book Company, New York, NY.

Ishii, K., Adler, R. and Barkan, P. (1988) *AI EDAM*, **2**(1), 53–65.

Poli, C., Esudero, J. and Fernandez, J.E. (1988) *Machine Design*, **60**, 101–4.

Shankar, S.R. (1992) *A Generalized Methodology for Manufacturability Evaluation*, Ph.D. Dissertation, Texas A&M University.

Suh, N.P. (1990) *The Principles of Design*, Oxford University Press, New York, NY.

Yannoulakis, N.J., Joshi, S.B. and Wysk, R.A. (1991) *Proceedings of ASME Design Theory and Methodology Conference*, ASME Publication DE **31**, 217–26.

CHAPTER 14

Evaluating product machinability for concurrent engineering*

Dana S. Nau, Guangming Zhang,

Satyandra K. Gupta and Raghu R. Karinthi

14.1 INTRODUCTION

Decisions made during the design of a product can have significant effects on product cost, quality, and lead time. Such considerations have led to the idea of identifying design elements that pose problems for manufacturing and quality control, and providing feedback to the designer so that the designer can change the design to improve its manufacturability (Vann and Cutkosky; 1990, Cutkosky and Tenenbaum, 1991).

In the case of machined parts, a part is often considered as a collection of machinable features (Butterfield *et al.*, 1986; Hummel and Brooks, 1986; Shah *et al.*, 1988; Rogers, 1989; Hummel, 1990; Shah, 1990). If we can evaluate the machinability[1] of these features, then the information produced by such an analysis can be used to provide feedback to the designer identifying problems that may arise with the machining. For example, if it is not possible to produce some feature to within the desired tolerances, then the designer may need to change the design accordingly. Thus some of the goals of concurrent engineering can be addressed through the following steps:

1. Generate a feature-based model for the object, i.e. an interpretation of the object as a collection of machinable features; and
2. For these feaures, select appropriate machining processes and process parameters, and evaluate whether the selected processes and parameters will be capable of producing the object to within the desired tolerances, and if so, what the associated machining costs and times will be.

We will now discuss two major issues that arise in performing the above steps.

First, what tolerances and surfaces finish a machining process can create will depend on the feature geometry and the machine tool (Machinability Data Center, 1972; DeVor *et al.*, 1980, Kline *et al.*, 1982; DeVor *et al.*, 1983; Sutherland and DeVor, 1986). But in addition, variations in hardness in the material being machined cause random vibration during machining (Zhang and Hwang, 1990a, 1990b, 1990c; Zhang and Kapoor, 1991a, 1991b; Zhang *et al.*, 1991), and this vibration is one of the major factors affecting the quality of the resulting surface.

Second, existing approaches for obtaining machinable features from a CAD model (Henderson, 1984; Woodwark, 1986; Brown and Ray, 1987; Ide, 1987; Shah and Rogers, 1988; Turner and Anderson, 1988; Pinilla *et al.*, 1989; Shah *et al.*, 1989a, 1989b; Sakuri and Gossard, 1990; Vandenbrande, 1990) normally produce a single interpretation of the part as a collection of machinable features. However, there can be several different interpretations of the same part as different collections of machinable features – or equivalently, different sequences of machining operations for creating the same part (Hayes and Wright, 1989; Vandenbrande, 1990; Karinthi and Nau, 1992a, 1992b). To determine the machinability of the part, all of the alternative interpretations should be generated and examined.

For example, in the machined part shown in Fig. 14.1, there are several different ways to interpret the hole h_1 and its relation to the slot s_2 and the shoulders s_1 and s_3. These interpretations correspond to different sequences of machining operations. For example, interpretation (a) corresponds to making h_1 after s_1 and s_2, interpretation (b) corresponds to making h_1 after s_1 but before s_2, and so forth. Depending on the feature geometry, tolerance requirements, surface finish requirements, and process capabilities, one or

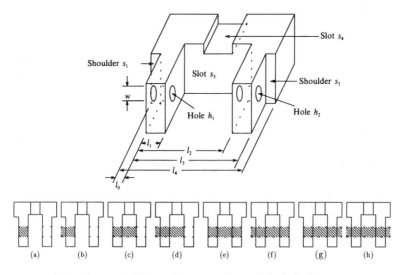

Fig. 14.1 A bracket, and different interpretations of the hole h_1.

another of these interpretations will be preferred. Here are a few of the trade-offs involved:

- Interpretations (e)–(h), in which the two holes are made in a single step, produce the tightest concentricity on h_1 and h_2. Thus, one of these interpretations may be necessary if the concentricity tolerance is tight.
- If l_3 is large, then interpretations (e)–(h) may require special tooling. Thus, if the concentricity is not tight and l_3 is large, then one of (a)–(d) may be preferable.
- If interpretations (e)–(h) do not require special tooling, then they have the advantage that they minimize the number of setups. Among these interpretations, (h) has the smallest number of tool changes; but (e) has the smallest tool travel distance.
- Interpretation (a) has the advantage that it incurs the smallest amount of wear on the drilling tool. But if l_1 is small, then interpretation (a) may cause excessive workpiece vibration.

To address the issues described above, we are developing a system to generate and evaluate machining alternatives. Section 14.2 describes our work on how to generate feature interpretations, and Section 14.3.2 describes our work on how to evaluate their machinability. Section 14.4 describes research issues that are currently being addressed, and Section 14.5 contains concluding remarks.

14.2 GENERATING FEATURE INTERPRETATIONS

For obtaining machinable features from a CAD model (such as a boundary representation), there are three primary approaches. In human-supervised feature recognition, a human user examines an existing CAD model to determine what the manufacturing features are (Brown and Ray, 1987). In automatic feature recognition, the same feature recognition rask is performed by a computer system (Henderson, 1984; Woodwark, 1986; Pinilla *et al.*, 1989; Hiroshi and Gossard, 1990; Vandenbrande, 1990). In design by features, the designer specifies the initial CAD model in terms of various form features which translate directly into the relevant manufacturing features (Ide, 1987; Shah and Rogers, 1988; Turner and Anderson, 1988; Shah *et al.*, 1989a, 1989b).

All three of these approaches will normally produce a single interpretation of the part as a collection of machinable features. However, because of interactions among features, there can be several different interpretations of the same part as different collections of machinable features – or equivalently, different sequences of machining operations for creating the same part (Hayes and Wright, 1989; Karinthi, 1990; Vandenbrande, 1990; Karinthi and Nau, 1992a, 1992b). To determine the machinability of the part, all of the alternative interpretations should be generated and examined.

To generate alternate interpretations of a part as a collection of machinable features, we have been developing an algebra of feature interactions (Karinthi and Nau, 1989a, 1989b, 1989c, 1992a, 1992b; Nau and Karinthi, 1990). Given a set of features describing a machinable part, other equally valid interpretations of the part can be produced by performing operations in the algebra.

Mathematically, an algebra consists of a domain D, along with binary operations defined on members of D. The domain of the feature algebra consists of the set of 'all possible machinable features', along with operations such as truncation and maximal extension as described below.

Since each machining operation removes a volume of material, we want a 'machinable feature' to be a three-dimensional solid corresponding to the volume of material that is removed. In addition, some portions of the surface of a feature are *blocked*, i.e., they separate air from metal, and some are *unblocked*, i.e., they separate air from air. Thus, we define a *feature* to be a pair $F = (S, P)$, where S is any compact, regular, semianalytic set (Requicha and Voelcker, 1985) and P is any partition of $b(S)$ into regular semianalytic subsets, each of which is labeled as 'blocked' or 'unblocked'. The domain $\mathscr{F}$ of the feature algebra is the set of all pairs (S, P) satisfying the above definition.

When two features $x, y \in \mathscr{F}$ are adjacent, we can define two operations:

$E(x, y)$: the maximal extension of x into y; and

$T(x, y)$: the truncation of x by y.

In the general case, the definitions of these operations are mathematically complex, so for brevity we omit them here (the reader is referred to Karinthi and Nau (1992a) for the details). However, Fig. 14.2 illustrates the definition

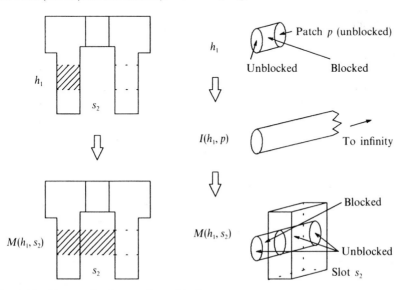

Fig. 14.2 The maximal extension of h_1 into s_2.

of the maximal extension operator on a simple example: the maximal extension $E(h_1, s_2)$ of the hole h_1 into the slot s_2, where h_1 and s_2 are as given in Fig. 14.1. As shown in Fig. 14.2, one takes the 'infinite extension' $I(h_1, p)$, where p is the patch bounding one end of h_1; and truncates $I(h_1, p)$ where it hits the far end of s_2.

From the definitions of the algebraic operators, we have proved various mathematical properties (associativity, etc.), which can be used to predict that various combinations of operators will produce the same feature. We have developed a prototype version of a feature interpretation system making use of these properties which, given an interpretation of an object (i.e., a set of features), uses state-space search techniques (Nilsson, 1980) to generate all of other interpretations of the same object that result from applications of the algebraic operators. For example, Fig. 14.3 shows the state space produced for operations on h_1, h_2, s_1, s_2, and s_3.

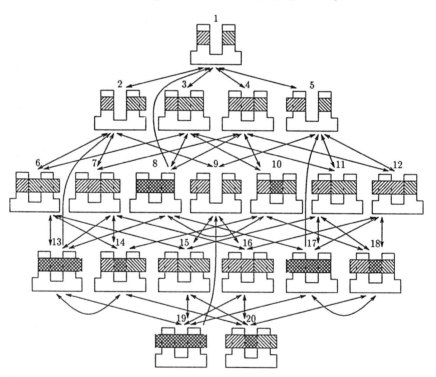

Fig. 14.3   State space for operations on h_1, h_2, s_1, s_2 and s_3.

14.3 PROCESS SELECTION AND MACHINABILITY EVALUATION

Given an interpretation of the object as a collection of machining features, we need to evaluate the various possible machining operation sequences for producing these features, to see whether or not any of them can satisfactorily achieve the design specifications. As described below, this involves two steps:

1. select candidate operation sequences; and
2. evaluate them.

14.3.1 Process selection

For each feature, we need to select machining operations and associated cutting parameters. Sometimes a feature can be created by a single machining operations (e.g., drilling or face milling), and other times it will require a sequence of operations (e.g., drill then bore then hone, or rough-face-mill then finish-face-mill). Also, in some cases there can be more than one sequence of machining operations that can create the feature geometry and also satisfy the tolerance requirements. Cutting parameters are selected based on past experience or handbook recommendations. Sometimes available machining facilities also affect the choice of cutting parameters.

Due to accessibility (Nau *et al.*, 1992) and setup constraints (Hayes and Wright, 1989), the set of features that comprise an object cannot necessarily be machined in any arbitrary sequence. Instead, these constraints will require that some features be machined before or after other features. However, for a given set of features, usually there will be more than one machining operation sequence capable of creating it. For example, in the bracket shown in Fig. 14.1, consider interpretation 1 of Fig. 14.3. In this interpretation, the two holes h_1 and h_2 must be made after the two shoulders s_1 and s_2 and the slot s_3. But there are two possible orderings for making h_1 and h_2, and six possible orderings for making s_1, s_2, and s_3, so interpretation 1 gives us twelve possible orderings in which to make the features.

Most knowledge-based systems for automated process selection have been rule-based (e.g., see Chang and Wysk, 1985; Brooks and Hummel, 1987; Nau, 1987a; Ham and Lu, 1988). Our process selection system, although knowledge-based, is based on a different approach. To represent and use problem-solving information for process selection, we use a hierarchical abstraction technique which we call hierarchical knowledge clustering. This approach has implemented in a system called SIPS, and later in a more sophisticated system called EFHA (Nau and Gray, 1986; Nau, 1987a, 1987b; Nau and Gray, 1987; Nau and Luce, 1987). These systems have been used both in the AMRF project at the National Institute for Standards and Technology (Brown and McLean, 1986; Brown and Ray, 1987), and at Texas Instruments (Nau, 1987a; Nau and Luce, 1987).

In SIPS and EFHA, knowledge about machining processes is organized in a taxonomic hierarchy. As shown in Fig. 14.4, each node of the hierarchy is a frame which represents a class of machining processes such as 'milling' or 'hole-create-process'. These frames contain knowledge about the intrinsic capabilities of various machining processes. Given the description of a

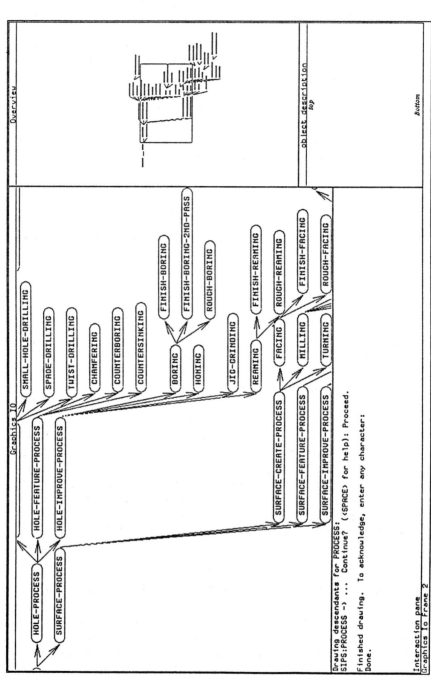

Fig. 14.4 A portion of SIPS' frame hierarchy.

machinable feature, SIPS and EFHA use the information in the frame hierarchy to guide a state-space search to find sequences of processes capable of creating the feature.

For example, in Fig. 14.5, SIPS has been given the task of making a hole. By doing a state-space search, it has determined that the best sequence of machining operations for making the hole consists of a twist-drilling operation followed by a rough-boring operation. If asked to continue, SIPS would eventually find each of the sequences of machining processes capable of creating the hole.

Researchers at Texas Instruments have extended SIPS and EFHA's knowledge bases to include information which enables them to select cutting tools and compute feed rates and cutting speeds.

14.3.2 Machinability evaluation

Given a candidate operation sequence, the machining data for that sequence, the feature's dimensions and tolerances, and the workpiece material, we want to evaluate whether or not it can satisfactorily achieve the design specifications. The capabilities of the machining process depend on the following factors:

1. The machining system parameters, such as the feed rate, cutting speed, depth of cut, and structural dynamics. Their effects on the process capabilities can be modeled and evaluated deterministically (Machinability Data Center, 1972; Wu, 1977; DeVor *et al.*, 1980, 1983; Kline *et al.*, 1982; Sutherland and DeVor, 1986; Chryssolouris *et al.*, 1988).
2. The natural and external variations in the machining process. For example, variations in hardness in the material being machined cause random vibration, which is one of the major factors affecting the surface quality. Such variations are unavoidable in practice, and are best dealt with statistically. This introduces a margin of error into our calculations of the process capabilities. The margin of error needs to be large enough that product quality is maintained, and yet small enough that the cost of the machining process is manageable (Lu and Zhang, 1990; Zhang *et al.*, 1990, 1991; Zhang and Hwang, 1990a, 1990b, 1990c; Zhang and Kapoor, 1990a, 1990b; Zhang and Lu, 1990).

We have developed a computer-based system for machinability evaluation, which is capable of determining the achievable machining accuracy such as surface finish, variation of dimensional sizes, and roundness and straightness of rotational surfaces. This system is built on an integration of machining science, materials science, and metrology. It produces a model of the surface texture formed during machining, and displays a graphic image to aid in visualization. Currently, it can estimate the achievable machining accuracy of four machining processes: turning, boring, drilling, and end milling.

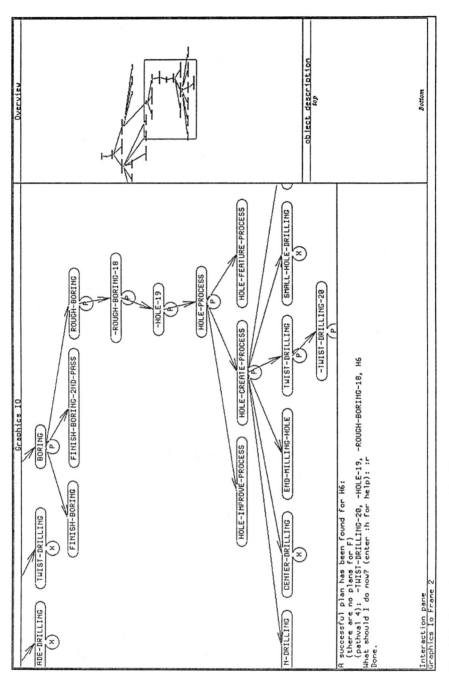

Fig. 14.5 A search space generated by SIPS.

The basic methodology of the evaluation system is shown in Fig. 14.6. The input consists of the cutting parameters for the selected machining process, and the basic properties of the workpiece material. Through simulation of the variations in cutting force based on the cutting mechanics and prediction of the tool vibratory motion during machining, the system produces a simulation of the topography of a machined surface, such as the one shown in Fig. 14.7.

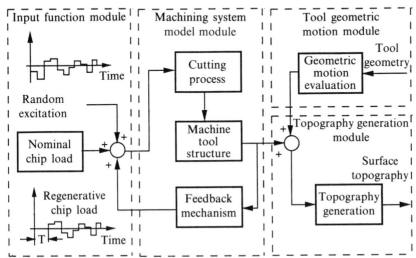

Fig. 14.6 Methodology to simulate the topography of a machined surface.

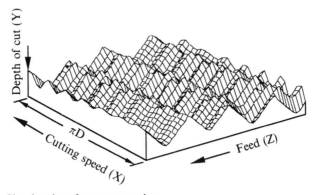

Fig. 14.7 Simulated surface topography.

Based on this information, the system assesses the machinability of the feature to be produced by the machining process. For example, as shown in Fig. 14.8, from the simulated surface topography of a hole, the system can take a cross-section perpendicular to the hole's axis and calculate the maximum and minimum diameters, in order to determine the hole's dimensional tolerances. As shown in Fig. 14.9, it can take a cross-section parallel to the hole's axis in order to calculate the hole's straightness. In Fig. 14.9, the

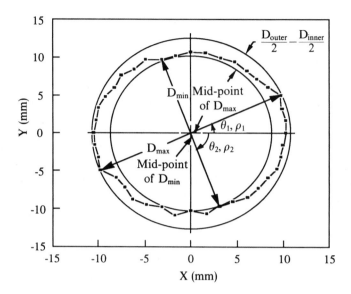

Fig. 14.8 Contour of the machined surface at a certain cross-section.

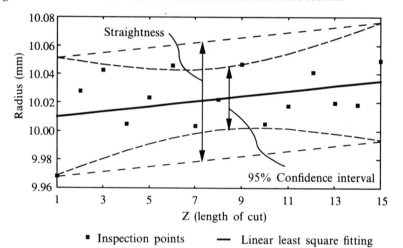

■ Inspection points   — Linear least square fitting

Fig. 14.9 Straightness estimation based on confidence band.

confidence band explicitly defines the achievable tolerance for the cylinder being machined.

This system has been tested by research institutions such as the National Institute of Standards and Technology (in the Precision Engineering Division and the Ceramics Division). It is being used by several industries, including the Ford Center for Research and Development and Allied Signal Corp., for evaluating the dimensional accuracy and surface finish quality during the machining of cylindrical surfaces (Zhang and Hwang, 1990a, 1990b; Zang and Kapoor 1991a, 1991b).

14.4 RESEARCH ISSUES

14.4.1 Generating alternative feature interpretations

In order to produce all of the alternative feature interpretations relevant for machining, some additional operators are needed in the feature algebra. We are developing definitions of these operators. In addition, some of the interpretations currently produced by the current operators are useless in terms of actual manufacturing practice, and we are modifying the feature interpretation system to discard such interpretations. For example, in Fig. 14.1, suppose the ratio $l_3/w < 2$. Then we will never drill from both sides, so any interpretations that require drilling from both sides should be eliminated, unless some very specific manufacturing requirements dictate otherwise.

We are developing methods for assigning tolerance requirements for each new feature produced by the feature algebra. For example, in Fig. 14.1, suppose the diameter specification for interpretation (a) is $\phi 10 + 0.20$, and that the length of interpretation (b) is twice that of interpretation (a). During the manufacturing process, in most cases a looser diameter specification for interpretation (b) (such as $\phi 10 + 0.5$) would be sufficient to achieve the diameter specification of $\phi 10 + 0.20$ for interpretation (a). If we use $\phi 10 + 0.20$ for interpretation (b), then in most cases we are using a tighter tolerance specification than is actually required, resulting in an unnecessarily high machining cost.

The dimensions of the feature to be machined sometimes depend on the direction from which the tool will be approaching. For example, consider interpretations (b) and (c) of Fig. 14.1, which are reproduced in Fig. 14.10. In interpretation (b), we must machine a hole of length $l_0 + l_1$. However, in interpretation (c), we do not need to machine a hole of length l_2. Instead, as shown in Fig. 14.10, the length may be between l_1 and l_2. We are working out the mathematics governing the relationships between the features and the machining operations.

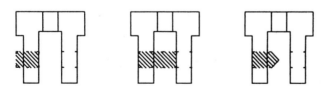

Fig. 14.10   Interpretations (b) and (c), and how to machine interpretation (c).

14.4.2 Machinability evaluation

Evaluating the machinability of alternative interpretations of an object will require repeated calls to the machinability evaluation module. To reduce the total time required for that task, we intend to augment the machinability evaluation system to provide means for fast approximate estimation of machining economics indices. This capability will quickly eliminate those feature interpretations that are infeasible in view of common manufacturing practice.

We intend to extend the system to make it capable of evaluating additional machining processes, such as face milling and grinding, and additional geometric tolerance parameters, such as the cylindricity of rotational surfaces and the flatness of planar surfaces.

Currently, the machinability evaluation is based on a model of tool deflection but not workpiece deflection. To make the machinability evaluation more sophisticated, we intend to incorporate into the machinability evaluation the effects of the static and dynamic deflection of the workpiece during machining, as well as the structural dynamics of the machine tool.

14.5 CONCLUSIONS

Decisions made during the design of a product can have significant effects on product cost, quality, and lead time. This has led to the evolution of the philosophy of concurrent engineering, which involves identifying design elements that pose problems for maufacturing and quality control, and changing the design, if possible, to overcome these problems during the design stage.

Our research directly addresses the above issue. The analysis performed by our system will enable us to provide feedback to the designer by identifying what problems will arise with the machining. By comparing the tolerances achievable by various machining operations with the designer's tolerance reqirements, we should be able to suggest to the designer how much the design tolerances should be loosened in order to make the feature easier (or possible) to machine.

In addition, for features whose tolerance and surface finish requirements are more easily achieved, the analysis will typically provide several alternative machining operations capable of achieving them. Such information will be useful to the manufacturing engineer in developing alternative plans for machining the part if the preferred machine tools or cutting tools are unavailable.

We anticipate that this work will provide a way to evaluate new product designs quickly in order to decide how or whether to manufacture them. Such a capability will be especially useful in flexible manufacturing systems, which need to respond quickly to changing demands and opportunities in the marketplace.

14.6 NOTES

*This work was supported in part by NSF Grants NSFD CDR-88003012, IRI-8907890 and DDM-9201779 to the University of Maryland and Defense Advanced Research Projects

Agency (DARPA) Contract No. MDA972-88-C-0047 for the DARPA Initiative in Concurrent Engineering (DICE).

1. By the machinability of a part, we mean how easy it will be to achieve the required machining accuracy. This is somewhat broader than the usual usage of 'machinability'.

14.7 REFERENCES

Brooks, S.L. and Hummel, K.E. (1987) XCUT: A rule-based expert system for the automated process planning of machined parts, *Technical Report BDX-613-3768*, Bendix Kansas City Division.

Brown, P. and Ray, S. (1987) *Proceedings 19th CIRP International Seminar on Manufacturing Systems*, **June**, 111–19.

Brown, P.F. and McLean, C.R. (1986) *Symposium on Knowledge-Based Expert Systems for Manufacturing at ASME Winter Annual Meeting* **December** 245–62. Anaheim, C.A.

Butterfield, W.R., Green, M.K., Scott, P.C. and Stoker, W.J. (1986) Part features for process planning, *Technical Report R-86-PPP-01, Computer Aided Manufacturing International*, November.

Chang, T.C. and Wysk, R.A. (1985) *An Introduction to Automated Process Planning Systems*, Prentice-Hall, Englewood Cliffs, NJ.

Chryssolouris, G., Guillot, M. and Domroese, M. (1988) *Transactions of the ASME, Journal of Engineering for Industry*, **110**(4); 397–8, November.

Cutkosky, M.R. and Tenenbaum, J.M. (1991) *International Journal of Systems Automation, Research and Applications*, **7**(3), 239–61.

DeVor, R.E., Kline, W. and Zdeblick, W. (1980) *Proceedings of NAMRC*, **8**, 297–303.

DeVor, R.E., Sutherland, J. and Kline, W. (1983) *Proceedings of the NAMRC*, **11**, 356–62.

Ham, I. and Lu, S. (1988) *Annals of CIRP*, **37**(2), 1–11.

Hayes, C. and Wright, P. (1989) *Journal of Manufacturing Systems*, **8**(1), 1–15.

Henderson, M.R. (1984) *Extraction of Feature Information from Three Dimensional Cad Data*, PhD thesis, West Lafayette, IN, May.

Hummel, K.E. (1990) *Proceedings CAMI Features Symposium, P-90-PM-02*, 9–10 August, 285–320.

Hummel, K.E. and Brooks, S.L. *Knowledge-Based Expert Systems for Manufacturing Symposium, BDX-613-3580*, December.

Ide, N.C. (1987) Integration of process planning and solid modeling through design by features, Master's thesis, University of Maryland, College Park, MD.

Karinthi, R. and Nau, D. (1989a) *IJCAI-89*, August.

Karinthi, R. and Nau, D. (1989b) *AAAI Spring Symposium*, Stanford.

Karinthi, R. and Nau, D. (1992a) *IEEE Trans. Pattern Analysis and Machine Intelligence*, **14**(4), 469–84, April.

Karinthi, R. and Nau, D. (1989c) *Workshop on Concurrent Engineering Design*, Detroit, Michigan, August.

Karinthi, R. and Nau, D. (1992b) In *Artificial Intelligence Applications in Manufacturing* (eds F. Famili, S. Kim, and D.S. Nau) AAAI Press/MIT Press, Menlo Park, CA, pp. 41–59.

Karinthi, R. (1990) *An Algebraic Approach to Feature Interactions*, PhD thesis, Computer Science Department, University of Maryland, December.

Kline, W., DeVor, R. and Shareef, I. (1982) *Transactions of the ASME, Journal of Engineering for Industry*, **104**, 272–8.

Lu, S.C-Y. and Zhang, G.M. (1990) *Journal of Manufacturing Systems*, **9**(2), 103–15.

278 *Evaluating product machinability*

Machinability Data Center (1972) *Machining Data Handbook* (2nd edn) Metcut Research Associates, Cincinnati, Ohio.

Nau, D.S. (1987a) *TI Technical Journal*, **Winter**, 39–46. Award winner, Texas Instruments 1987 Call for Papers on AI for Industrial Automation.

Nau, D.S. (1987b) *Second Internat. Conf. Applications of Artificial Intelligence in Engineering*, SRC Tech Report 87–105.

Nau, D.S. and Gray, M. (1986) *Symposium on Integrated and Intelligent Manufacturing at ASME Winter Annual Meeting*, **December**, 219–25, Anaheim, CA.

Nau, D.S. and Gray, M. (1987) In *Expert Systems: The User Interface* (ed. J. Hendler), 81–98, Ablex, Norwood, NJ.

Nau, D.S. and Luce, M. (1987) *19th CIRP International Seminar on Manufacturing Systems*, SRC Tech Report, 87–106.

Nau, D.S., Zhang, G. and Gupta, S.K. (1992) Generation and evaluation of alternative operation sequences, ASME Winter Annual Meeting, November, ASME, New York.

Nau, D.S. and Karinthi, R.R. (1990) *Proceedings Manufacturing International 1990*, **August**.

Nilsson, N.J. (1980) *Principles of Artificial Intelligence*, Tioga, Palo Alto.

Pinilla, J.M., Finger, S. and Prinz, F.B. (1989) *Proceedings of the NSF Engineering Design Research Conference*, **June**.

Requicha, A.G. and Voelcker, H.B. (1985) *Proceedings of the IEEE*, **73**(1), 30–44.

Rogers, M. (1989) *Technical Report R-89-GM-02, CAM-I*, July.

Sakurai, H. and Gossard, D.C. (1990) *IEEE Computer Graphics and Appications*, **September**.

Shah, J., Sreevalsan, P., Rogers, M., Billo, R. and Mathew, A. (1988) *Technical Report R-88-GM-04.1*, CAM-I Inc.

Shah, J.J. and Rogers, M.T. (1988) *Computer Aided Engineering Journal*, **7**(2), 9–15, February.

Shah, J. (1990) *Proceedings of Feature Symposium*, number P-90-PM-02, Woburn, Boston, MA, August.

Shah, J., Rogers, M. and Sreevalsan, P. (1989a) *Technical Report R-90-PM-01*, CAM-I Inc.

Shah, J., Rogers, M., Sreevalsan, P. and Mathew, A. (1989b) *Technical Report R-89-GM-01*, CAM-I Inc.

Sutherland, J. and DeVor, R. (1986) *Transactions of the ASME, Journal of Engineering for Industry*, **108**, 269–79.

Turner, G.P. and Anderson, D.C. *ASME-computers in Engineering Conference*, **July–August**, San Francisco, CA.

Vandenbrande, J.H. (1990) *Automatic Recognition of Machinable Features in Solid Models*. PhD thesis, Computer Science Department, UCLA.

Vann, C. and Cutkosky, M. *ASME Symposium on Advances in Integrated Product and Process Design*, **November**.

Woodwark, J.R. (1986) *ACM* **18**(6), July/August.

Wu, S.M. (1977) Dynamic data system: A new modeling approach, *Transactions of the ASME, Journal of Engineering for Industry*.

Zhang, G.M. and Hwang, T.W. (1990a) *Symposium on Automation of Manufacturing Processes, 1990 ASME Annual Meeting*, **DSC-Vol. 22**, 31–3.

Zhang, G.M. and Hwang, T.W. (1990b) *Symposium on Fundamental Issues in Machining, 1990 ASME Winter Annual Meeting*, **PED-Vol. 43**, 25–37.

Zhang, G.M. and Hwang, T.W. (1990c) *Proceedings of the Fourth International Laboratory Information Management Systems Conference*, Pittsburgh, PA, June.

Zhang, G.M., Hwang, T.W. and Harhalakis, G. (1990) *Proceedings of the Second International Conference on Computer Integrated Manufacturing*, 339–45, Troy, NY, May.

Zhang, G.M. and Kapoor, S.G. (1991a) *Journal of Engineering for Industry, Transactions of the ASME*, **May**.

Zhang, G.M. and Kapoor, S.G. (1991b) *Journal of Engineering for Industry, Transactions of the ASME*, **May**.

Zhang, G.M. and Lu, S.C-Y. (1990) *Journal of the Operational Research Society*, **41**(5) 391–404.

Zhang, G.M., Yerramareddy, S., Lee, S.M. and Lu, S.C-Y. (1991) *Journal of Dynamic Systems, Measurement and Control, Transactions of the ASME*, **June**.

CHAPTER 15

Concurrent optimization of design and manufacturing tolerances

Chun Zhang and Hsu-Pin (Ben) Wang

15.1 INTRODUCTION

Mechanical engineering design is a complex process which involves creative thinking, experience, intuition, and quantitative analysis. The requirements for designing a complicated mechanical system are diverse and often contain conflicting goals. Stated in more detail, a design process can be divided into levels of activities that include functional design (product design), manufacturing (process design), and life-cycle considerations. Designers must consider not only the functional requirements of a product but also life-cycle issues, such as manufacturing, testing, assembly and maintenance.

Product design begins with a concept which specifies the capabilities and performance requirements and sets goals for production cost and annual volume. Next, the technical concepts for the product's function are established. Major decisions regarding configuration, materials, energy, size, and so on, are made at least tentatively. Following the initial specification, designers break the design into parts and subassemblies and then perform the design of single parts. The main consideration here is performance of the product, and designers consider such matters as materials, strength, weight, and tolerances on the dimensions of functional surfaces. In this phase, engineering analyses may be used to verify the parameters of the design. It is at this point that the design has traditionally been passed onto the manufacturing department, which must then figure out how to realize the design.

Process design (known as process planning), the conversion of design specifications into manufacturing instructions, is a key function in manufacturing. At this stage, process engineers determine the processes and their parameters to make parts. The major decisions involved in process planning include: process selection, process sequencing, machining dimension and tolerance calculation, machining parameter (speed, feedrate and depth of cut) selection, and machine selection, etc.

As product requirements become more stringent and manufacturing technologies become more sophisticated, process planning becomes more difficult and time-consuming. Current practice, an experience-based approach, may be characterized as qualitative, non-scientific and imprecise. A survey of related literature indicates that little, if any, has been done in the field of process planning optimization utilizing quantitative and scientific methodologies.

Products and processes traditionally evolved over a period of time with engineers extrapolating from their experience, then building and testing new products and processes in a time-consuming, iterative process. This worked well for small improvements. In modern industries, however, the situation is quite different. The development team is faced with totally new products to design, manufacture and introduce in much shorter time with little help from their experience. The time to build and test ideas in hardware has been greatly shortened as product changes are more frequent. Materials are more difficult to work with; complex components are required as near net shapes with tighter tolerances. Both experience-based process planning and sequential, iterative product and process design procedures need to be improved to meet these challenges.

In order to obtain an economically viable design solution for industrial production, product and process design problems must be considered concurrently. A new approach, called concurrent engineering, has recently emerged in which design, engineering analysis, and manufacturing planning can be carried out simultaneously. The objectives of concurrent engineering are to consider simultaneously all elements of product development cycle from conception through delivery, and to focus on reduction of cost and development time while maintaining high production quality through continuous improvement of manufacturing processes.

Tolerance is one of the most important parameters in design and manufacturing. Tolerance design and synthesis is a major task of product and process design and is in a period of extensive study due both to increased demands for quality products and to increasing automation of machining and assembly. Optimum tolerance synthesis ensures good quality product at low cost. Therefore, tolerancing plays a key role in concurrent engineering.

Two types of tolerances are often used in practice. They are design tolerances and manufacturing tolerances. Design tolerances are related to the functional requirements of a mechanical assembly or of a component; whereas manufacturing tolerances are mainly devised for a process plan for manufacturing a part. The key is that manufacturing tolerances must ensure the realization of design tolerances (Sfantsikopoulos, 1990). Traditionally, design tolerance synthesis and manufacturing tolerance allocation are performed separately by the designer and the process engineer, respectively. At the design stage, by considering assembly functions, the designer allocates design tolerances to different assembly component parts or part features based on his experience and knowledge, whereas at the manufacturing stage the process engineer determines manufacturing tolerances in an effort to realize the design tolerances. Unfortunately, in such disparate efforts, designers allocating design tolerances are often not aware of manufacturing processes and their capabilities to produce the product. This may be due to either lack of communications between the designer and the process engineer, or the designer's lack of knowledge on the manufacturing processes. Such a design often results in a process plan that cannot be effectively executed, or that can only be executed at unnecessarily high cost. The concurrent optimization of design and manufacturing tolerances is highly desired to achieve optimal design and economical manufacturing.

Considerable literature has emerged in the field of concurrent engineering in general, but not much of it addresses systematic methodologies to help designers conduct the discipline (Adler, 1989). Most of the research focuses on the development of strategies or frameworks for concurrent engineering systems. Analytical models are severely short for its implementation (Nevins *et al.*, 1989). Furthermore, most of the currently developed concurrent engineering strategies seek to meet targets of the costs in various stages of design rather than to minimize those costs. In our research, an analytical model has been developed and solved for concurrent optimization of design and manufacturing tolerance synthesis. Through the tolerance optimization, product functionality, manufacturability, assemblability and cost effectiveness are enhanced simultaneously. This modeling technique provides a tool for implementation of concurrent engineering.

15.2 PROBLEM STATEMENT

A product is usually composed of a number of component parts. The manufacturing cost of a product thus consists of the costs of fabricating those parts and others. Tolerance, including design tolerance (the final component tolerance) and manufacturing tolerance (intermediate component tolerance during fabrication), is one of the most important parameters which affect product quality and cost. Therefore, it is of great importance to concurrently optimize the design and manufacturing tolerances. An analytical model for concurrent tolerance optimization will greatly improve and expand the foundations of design and manufacturing for computer integrated manufacturing.

The concurrent optimization of both design and manufacturing tolerance syntheses is explained as follows. In general design practice, product functional requirements impose specifications on structural performance (such as displacement, stress on critical points) and assemblability (such as clearance). These specifications are translated to product design parameters, such as dimension and tolerance of a feature. From a design viewpoint, tolerances on design parameters result in variations in structural perform-ance. From an assembly viewpoint, an assembly dimension is dominated by a set of component dimensions, and thus its tolerance is stacked up by the tolerances on those components. From a manufacturing viewpoint, a component dimension may require several stages of machining. For example, a hole feature with high precision requirements may require machining by a chain of manufacturing processes, such as: drill→bore→ finishing bore→grind. At each stage, one process (or machine) will be chosen for the component fabrication from a number of alternative processes (or machines). Figure 15.1 shows the cost versus tolerance curves of intermedi-ate manufacturing processes for producing the cylinder bore of an engine assembly. In the figure, C_{d_i}, C_{b_i}, C_{f_i} and C_{g_i} denote cost-tolerance curves for drill, bore, finishing bore and grind operations, respectively. The numbers of alternative processes for these stages are 2, 3, 3 and 2, respectively. For a specific process, the tighter the processing tolerance, the higher the manu-facturing cost. The problem here is to determine the optimal combination of processes and manufacturing tolerances for component fabrication which will meet requirements on product functionality and assemblability. By solving the problem, design and manufacturing tolerances are simultaneous-ly optimized in terms of the total manufacturing cost of the product.

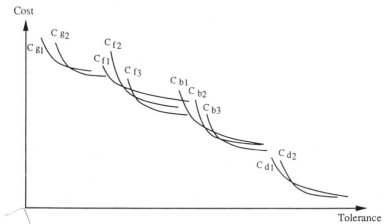

Fig. 15.1 Manufacturing cost–tolerance curves for producing a cylinder bore.

The objective of this research is two fold:

1. Formulation of a mathematical model for optimizing design and manu-facturing tolerance simultaneously based on the least manufacturing cost criterion; and
2. Development of a solution procedure to solve this model.

The symbols and terminology used in this paper are introduced in Section 15.3.1. Section 15.3.2 formulates the problem. In Section 15.3.3, the optimization model for concurrent design and manufacturing tolerance syntheses is presented. A simulated annealing solution procedure for this optimization problem is described in Section 15.4. Section 15.5 provides an example to illustrate the optimization model. Conclusions are made in Section 15.6.

15.3 OPTIMIZATION MODEL FOR CONCURRENT TOLERANCE SYNTHESIS

15.3.1 Notation

- δ_Σ – prespecified resultant assembly tolerance;
- δ_{im_ik} – design tolerance for component i;
- ψ – performance measure in component design;
- Δb – prespecified allowable variation of the performance measure;
- x_u – the uth controllable design parameter;
- q_v – the vth uncontrollable design parameter;
- $\dfrac{\partial \psi}{\partial x_u}$ – design sensitivity of ψ with respect to x_u;
- $\dfrac{\partial \psi}{\partial q_v}$ – design sensitivity of ψ with respect to q_v;
- l_c – number of controllable parameters;
- l_{nc} – number of uncontrollable parameters;
- Δ_i – upper bound of design tolerance for component i with respect to design performance measures;
- δ_{ijk} – manufacturing tolerance on component i in operation j by performing process k;
- G – total manufacturing cost of the assembly;
- A, α, δ_h, g_0 – constants for cost-tolerance functions;
- $g_{ijk}(\delta)$ – manufacturing cost of performing the kth alternative process in the jth operation (stage) for producing tolerance δ on component i;
- $\delta_{z_{ij}}$ – maximum variation of the stock removal of the jth operation to produce component i;
- δ_{ijk}^l – lower bound of δ_{ijk};
- δ_{ijk}^u – upper bound of δ_{ijk};
- n – number of components in the assembly;
- m_i – number of operations (stages) for producing component i;
- p_{ij} – number of alternative processes for the jth operation to produce component i;
- X_{ijk} – process selection coefficient; 1, if process k is chosen for the jth operation to produce component i; 0, otherwise.

15.3.2 Problem formation

The concurrent optimization of design tolerance synthesis and manufacturing tolerance allocation can be formulated as follows:

Given:

- A prespecified resultant assembly tolerance (δ_Σ);
- Assembly tolerance stackup relationship;
- Upper bounds of design tolerances on components (Δ_i);
- Process route and alternative processes (machines) for each component fabrication;
- Cost-tolerance functions for each machining process involved in the product manufacturing $(g_{ijk}(\delta))$.

Find:

- The optimal process (machine) to be selected at each manufacturing stage to produce components (X_{ijk}^*);
- Optimal design tolerances on components $(\delta_{im_ik}^*)$ and manufacturing tolerances for each process to produce components (δ_{ijk}^*).

15.3.3 Optimization model

(a) Manufacturing cost-tolerance model

It is known that in manufacturing the tighter the tolerance, the higher the manufacturing cost. Several modeling methods have been reported to describe this cost-tolerance relationship of different production processes. These models include reciprocal square function, reciprocal function, exponential function, the Sutherland function and Michael-Siddall function, etc. (Wu *et al.*, 1988).

In this study, an individual curve is employed for each manufacturing process. The total manufacturing cost for a design tolerance is modeled by assembling all the curves for the manufacturing processes involved in the tolerance machining. An exponential function $g(\delta)$ is used to model the machining data (although exponential functions are used in this study, the solution procedure is applicable to any other cost-tolerance models). For a particular manufacturing process, the following relationship is found.

$$g(\delta) = Ae^{-\alpha(\delta - \delta_h)} + g_0, \quad \delta_a < \delta < \delta_b$$

In this model, $g = f(A, \alpha, \delta_h, g_0, \delta)$. A, δ_h and g_0 determine the position of the cost-precision curve whereas α controls the curvature of it. These parameters can be derived using a curve fitting approach based on experimental data. δ_a and δ_b define a region in which the tolerance is economically feasible.

(b) Design constraints

As mentioned previously, product operational requirements often impose certain constraints on structural performance measures. These performance measures are usually expressed as stress, displacement and frequency on some critical points in components. For example, such a design requirement can be stated as: the stress at a point on the top edge of an engine piston should be less than 1×10^6 psi. In order to enhance product quality, variations on the performance measures should be controlled in certain ranges as well. For the above example, we may define an allowable variation of the stress as $\pm 5\,000$ psi. A large number of stress and displacement performance measures are normally considered for complete definition in a design problem.

In an engineering design problem, there are two types of design parameters: controllable parameters and uncontrollable parameters (Parkinson *et al.*, 1990). The controllable parameters are those variables whose values can be selected by the designer within the range of the upper and lower bounds. Design dimensions of components are controllable parameters. Tolerances are used to define the variations of component dimensions. The uncontrollable parameters denote those variables whose values are fixed as part of the problem specifications. We cannot adjust those parameters during design synthesis. Material density, Young's modulus and magnitudes of loads are examples of uncontrollable parameters. In an engineering design, both fluctuations (tolerances) of controllable parameters and those of uncontrollable parameters have impact on performance measures.

In engineering analysis discipline, design sensitivity analysis provides a valuable tool to assess the impact of design changes on performance measures (Godse *et al.*, 1991). Design sensitivity is defined as the change in a performance measure caused by a perturbation of a design parameter. The calculation of design sensitivities can be easily obtained by running existing engineering analysis packages. For a component, the simultaneous evaluation of changes in structural or dynamic performance measures due to tolerances on design parameters can be calculated using a Taylor series first order approximation, as follows:

$$\Delta\psi = \sum_{i=u}^{l_c} \left| \frac{\partial\psi}{\partial x_u} \cdot \Delta x_u \right| + \sum_{v=1}^{l_{nc}} \left| \frac{\partial\psi}{\partial q_v} \cdot \Delta q_v \right|$$

For an engineering problem, design constraints can be formulated as:

$$\sum_{u=1}^{l_c} \left| \frac{\partial\psi}{\partial x_u} \cdot \Delta x_u \right| + \sum_{v=1}^{l_{nc}} \left| \frac{\partial\psi}{\partial q_v} \cdot \Delta q_v \right| \leqslant \Delta b$$

where Δb is a prespecified value to control the variation of the performance measure. For a practical design problem, there may be a set of design constraints for stress, displacement and frequency, respectively.

In our tolerance optimization problem, assumptions are made so that only one controllable parameter (dimensional tolerance) is considered for a component and that tolerances on uncontrollable parameters are known. By simplifying the above inequality, the design constraints for a component can be expressed as follows:

$$\delta_i \leqslant \Delta_i$$

where Δ_i sets the upper bound for tolerance on component i with respect to design performance measures.

With consideration of process selection, this constraint can be rewritten as:

$$\sum_{k=1}^{p_{im_i}} X_{im_ik} \cdot \delta_{im_ik} \leqslant \Delta_i$$

where X_{im_ik} is the process selection coefficient for the last manufacturing process to produce component i. It will be discussed in Section 15.3.3(e).

(c) Assembly constraints

In a product design, dimensions of some critical elements may directly affect the product functionality. Thus, tolerances on these design dimensions are important parameters toward product design. For a product or an assembly, a design dimension may be affected by a set of interrelated dimensions. These dimensions form a close loop which is often called a dimension chain. The dimension which is obtained indirectly is called the resultant dimension and others are called component (or constituting) dimensions in a dimension chain.

The following are some examples of resultant design elements:

- The clearance between the piston and the cylinder bore in an engine assembly;
- The gap between the brake shoes and the caliper in a car disk brake system; and
- The gap between the bushing and the shaft shoulder in a gear box assembly.

Considering the worst case of tolerance stack-up, the tolerance on the resultant dimension is the summation of those on component dimensions. It can be expressed as follows:

$$\sum_{i=1}^{n} \delta_{i_d} \leqslant \delta_{\Sigma}$$

Here δ_{i_d} is the design tolerance on component i and δ_{Σ} is the critical tolerance which is stacked up by the component tolerances. This relationship forms a necessary constraint set for the concurrent tolerance optimization model. For example, in the engine assembly shown in Fig. 15.2, the

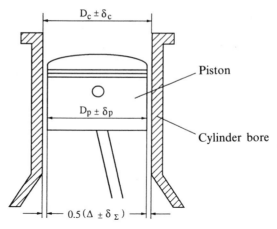

Fig. 15.2 The engine assembly for the example problem.

clearance between piston and cylinder bore (Δ) is the resultant dimension, and diameters of piston and cylinder bore (D_p and D_c) are constituting dimensions. The tolerance on the clearance (δ_Σ) is the summation of tolerances on piston and cylinder bore diameters (δ_p and δ_c).

Considering process selection, the design tolerance constraint can be represented as:

$$\sum_{i=1}^{n} \sum_{k=1}^{p_{im_i}} X_{im_ik} \cdot \delta_{im_ik} \leqslant \delta_\Sigma$$

(d) Manufacturing constraints

In concurrent tolerance syntheses, consideration should be given not only in ensuring design tolerances but also in providing appropriate amount of stock removal for operations. The stock removal is the layer of material which is to be removed from the surface of a workpiece in machining, in order to obtain the required accuracy and surface quality. The determination of stock removal greatly influences the quality and the production efficiency of the machined part. Excessive stock removal will increase the consumption of material, machining time, tool and power, and thus increase the manufacturing cost. On the other hand, if there is not sufficient stock removal, the surface roughness and defective surface layer caused by the preceding operation cannot be removed completely from the workpiece surface. Thus it will influence the surface quality of the part (Wang and Li, 1991).

The amount of a stock removal is the difference between the machining dimension obtained in the preceding operation and that in the current operation. Since the machining dimensions are not fixed and each of them is associated with a tolerance, the actual stock removals cut from workpiece surfaces vary in a certain range. The so-called stock removal is its nominal value. For a machining dimension, the variation of its stock removal is the sum of the manufacturing tolerances in the current and preceding operations. In practice, an appropriate stock removal should be provided for each process operation. It forms necessary constraints for the concurrent tolerance synthesis model. It can be described as follows:

$$\delta_{ij} + \delta_{ij-1} \leqslant \delta_{Z_{ij}}$$

where δ_{ij} and δ_{ij-1} denote respectively the manufacturing tolerances in the jth and $(j-1)$th operations for producing component i, and Z_{ij} denotes the allowable variation of the stock removal in the jth operation to produce component i. Z_{ij} is usually found in manuals, handbooks, etc.

Considering the process selection issue, this constraint set can be rewritten as:

$$\sum_{k=1}^{p_{ij}} X_{ijk} \cdot \delta_{ijk} + \sum_{k=1}^{p_{ij-1}} X_{ij-1k} \cdot \delta_{ij-1k} \leqslant \delta_{Z_{ij}}$$

In addition to critical design tolerance, stock removal, process selection and bounds on tolerance of each process form the necessary constraints for the optimization model. They are expressed as follows:

Process selection:

$$\sum_{k=1}^{p_{ij}} X_{ijk} = 1 \quad (i=1,\ldots,n;\ j=1,\ldots,m_i)$$

Here X_{ijk} is the process selection coefficient. $X_{ijk}=1$, if process k is chosen to produce the jth manufacturing tolerance on component i; 0, otherwise. This constraint set ensures that one and only one process be selected for each manufacturing tolerance fabrication.

Process bounds:

$$\delta^l_{ijk} \leqslant \delta_{ijk} \leqslant \delta^u_{ijk}$$

This constraint set defines the upper and lower boundaries on manufacturing tolerances for each process.

(e) Optimization model

The concurrent design and manufacturing tolerance synthesis model represents a well-defined optimization problem. Requirements on critical tolerance, stock removal, process selection and process capabilities form the necessary constraints for the problem.

A mathematical representation of the optimization model is given below:

$$\text{Min } G = \text{Min} \sum_{i=1}^{n} \sum_{j=1}^{m_i} \sum_{k=1}^{p_{ij}} X_{ijk} \cdot g_{ijk}(\delta_{ijk})$$

subject to:

$$\sum_{k=1}^{p_{im_i}} X_{im_ik} \cdot \delta_{im_ik} \leq \Delta_i$$

$$\sum_{i=1}^{n} \sum_{k=1}^{p_{im_i}} X_{im_ik} \cdot \delta_{im_ik} \leq \delta_\Sigma$$

$$\sum_{k=1}^{p_{ij}} X_{ijk} \cdot \delta_{ijk} + \sum_{k=1}^{p_{ij-1}} X_{ij-1k} \cdot \delta_{ij-1k} \leq \delta_{Z_{ij}}$$

$$\sum_{k=1}^{p_{ij}} X_{ijk} = 1$$

$$\delta_{ijk}^l \leq \delta_{ijk} \leq \delta_{ijk}^u$$

A nonlinear mixed binary-continuous programming model is formulated for simultaneously determining the optimal design and manufacturing tolerances. The objective function is the total manufacturing cost of the assembly. The design tolerances on components (δ_{im_ik}), manufacturing tolerances on those components created by intermediate manufacturing process (δ_{ijk}) and process selection coefficients (X_{ijk}) are to be determined (decision variables). The number of variables in the optimization model depends on the number of components, the number of manufacturing stages to produce those components and the number of alternative processes in each stage.

15.4 A SOLUTION PROCEDURE

In this concurrent tolerance optimization model, since binary variables are involved, combinatorial nature exists. For each combination of processes, a tolerance allocation must be performed to find the minimum cost tolerances. A change from one process to another causes a jump in cost, which may result in a pronounced shift in distribution of the tolerances. The surface of the solution may be in a very complicated form and many local minima are likely to exist. As the number of components and processes increases, the solution surface may become even more complex. In order to solve such a complex optimization problem, considerations on efficiency and reliability for finding the global optimum should be given.

Several solution algorithms have been reported for general design tolerance optimization problems, such as branch and bound (Lee and Woo, 1987), SQP and univariate search (Chase *et al.*, 1990). However, they either suffered low efficiency or lacked wide range of applicability. Therefore, an optimization algorithm with high efficiency and reliability for finding global optimum is highly desired for optimal tolerancing problems. In this study, simulated annealing algorithm is adopted as a tool for this purpose.

15.4.1 Optimization by simulated annealing

In the early 1980s, a global optimization algorithm, called simulated annealing, was introduced to the combinatorial optimization community (Kirkpatrick *et al.*, 1983). As was stated in many reviews, simulated annealing algorithm has been employed in VLSI layout, traveling salesman, image processing, and many other manufacturing related problems and had shown much effectiveness (Collins *et al.*, 1988). It is reported that the algorithm performs well in the presence of large number of variables.

As the name implies, simulated annealing is a simulation of the physical annealing process. It differs from iterative improvement algorithms in that the latter accept only those moves which will lead to the improvements of the results. The inherent problems with this kind of algorithms are that:

1. It can be easily trapped in a local optimum; and
2. It heavily depends on the initial solution.

By allowing perturbations to move to a worse solution in a controlled fashion, simulated annealing is able to jump out from a local optimum and potentially finds a more promising downhill path. Furthermore, the uphill moves (acceptance of worse solutions) are carefully controlled. Although the global optimum solution is not guaranteed, the annealing algorithm consistently provides solutions which are very close to the optimum. Moreover, by accepting uphill moves, this algorithm appears to be independent of the initial solution.

The basic elements of a simulated annealing algorithm include: configuration, move, neighboring configuration, objective function and cooling schedule. The components of the solution algorithm for concurrent tolerance optimization are provided as follows:

1. Configuration (or a solution) – any feasible allocation of component tolerances (δ_{ijk}) and combination of process selection coefficients (X_{ijk});
2. Move – a change of a component tolerance and a new combination of process selection coefficients;
3. Neighboring configuration – a new allocation of component tolerances and process selection combination after one move;
4. Objective function – total manufacturing cost of producing component tolerances using the selected processes; and
5. Cooling schedule – the way the algorithm progresses. The starting temperature should be high enough so that a large number of configurations can be explored before the configuration reaches the final one in order not to be trapped in a local optimum. It should not be too high to form a purely random search, resulting in wasting computer time. The simulated annealing process is terminated whichever of the following conditions hold:
   (a) the temperature reaches the freezing temperature, a very small t_0;
   (b) no significant changes at the last N_l consecutive temperature levels; and

(c) the objective function has been evaluated for certain number of times, say N_t.

Originally, simulated annealing algorithm was developed for combinatorial problems. Several pioneering studies reported the applications of the annealing algorithm in areas where variables are continuous in nature (Brooks and Verdini, 1988; Coranam *et al.*, 1987). The difference of the applications in continuous variables from those in combinatorial problems is that the former is similar to the random walks with bias. In other words, the determination of a move involves the selection of a direction along which a move is made, and the determination of a step size which decides how far a new solution is away from the current one. These requirements cause the difficulty of applying the simulated annealing algorithm over continuous variables. On the other hand, in the combinatorial problems, the moves are clearly defined as to switch from one point to another in the domain without any intermediate feasible solutions in between. Intuitively, in solving problems over continuous variables, the step size is desired to be small when the temperature is near freezing, i.e., the temperature to terminate the annealing process, in order to stay close to the final result, and large enough when the temperature is very high so that it is possible to jump out from a local optimum quickly. In additions, it is generally desired that the number of accepted and rejected moves are balanced, i.e., the ratio of accepted moves to rejected moves is close to 1. From an optimization point of view, a large number of acceptance with respect to rejection means that the new configuration is too close to the previous one. On the contrary, havng too many rejected moves represents that the move step is so large that the new configuration is too far from the current one. Therefore, the step size should be adjusted based on the ratio of accepted to rejected moves during the algorithm executes in a dynamic fashion.

Based on the understanding of the mechanism of simulated annealing algorithm, a solution procedure is developed for solving the mixed nonlinear tolerance optimization problem. In this algorithm, discrete and continuous variables are generated and handled separately. This helps reduce the computational time for checking the intermediate solutions against constraints and thus improves the efficiency of the algorithm. Figure 15.3 shows the procedure of simulated annealing algorithm for the concurrent tolerance optimization.

In this study, simulated annealing algorithm has been tested on the concurrent tolerance optimization problem. The computational results for an example problem are given in the next section.

15.5 AN EXAMPLE

In this section, an example is provided to illustrate the concurrent tolerance optimization model and the simulated annealing solution procedure. The

Begin with an initial solution $S(s_1(\delta_{ijk}), s_2(X_{ijk}))$ and temperature t
Determine the initial direction and step size of the move (for s_1)
Repeat
   Perform the following loop M times
     Pick a neighboring configuration s_1' of s_1 by a random move $(\delta_{ijk} \rightarrow \delta_{ijk}')$
     If the new solution does not fall in the feasible region, go back to generate a
     new move
     Generate a random combination of process selection coefficients (s_2')
     Let $\Delta = (f(s_1', s_2') - f(s_1, s_2))/f(s_1, s_2)$
     If $\Delta \leqslant 0$ (downhill move)
       Set $s_1 = s_1'$ and $s_2 = s_2'$
     Else (uphill move)
       Set $s_1 = s_1'$ and $s_2 = s_2'$ with the probability $e^{-\Delta/t}$
     Update the direction and step size of moves (for s_1) based on the ratio of
     accepted to rejected moves
   Lower t
Until one of the terminating conditions is true
Return S

Fig. 15.3 The solution algorithm for concurrent tolerance optimization.

example problem is to determine the optimal design and manufacturing tolerances on piston and cylinder bore diameters in an engine assembly (see Fig. 15.2). A description of manufacturing processes for producing the assembly is given in Table 15.1. The design specification of the assembly is that the clearance between the piston and cylinder bore should be kept within 0.0022 ± 0.001 in. The upper bounds for design tolerances on the piston and cylinder bore are: $\Delta_1 = 0.0006$ in and $\Delta_2 = 0.00053$ in, respectively.

Table 15.1 Description of the example problem

Piston		Cylinder bore	
Process route	*No. of alternative processes*	*Process route*	*No. of alternative processes*
1 rough turn	2	1 drill	2
2 finish turn	3	2 bore	3
3 rough grind	3	3 finish bore	3
4 finish grind	3	4 grind	2

By the terminology defined in Section 15.3, in this problem, the number of components is $n=2$ and the numbers of manufacturing stages for these two components are $n_{m_1} = 4$ and $n_{m_2} = 4$, respectively. For each stage of fabrication, there are two to three alternative processes, as shown in Table 15.1.

This example problem was solved by the simulated annealing algorithm over a number of runs. A tolerance allocation result is shown in Table 15.2.

Table 15.2 shows the result of design tolerance synthesis and manufacturing tolerance allocation for the piston and cylinder bore fabrication. For

Table 15.2 A tolerance allocation result by simulated annealing algorithm

Piston ($i = 1$)

Optimal manufacturing tolerances

Mafg. stage (j)	δ^*_{ijk} Alternative process (k)			X^*_{ijk} k		
	1	2	3	1	2	3
1	0.016	–	–	1	0	–
2	–	–	0.0035	0	0	1
3	0.0012	–	–	1	0	0
4	–	0.00047	–	0	1	0

Optimal design tolerance
0.00047

Cylinder bore ($i = 2$)

Optimal manufacturing tolerances

Mafg. stage (j)	δ^*_{ijk} Alternative process (k)			X^*_{ijk} k		
	1	2	3	1	2	3
1	–	0.018	–	0	1	–
2	–	–	0.0036	0	0	1
3	–	0.0012	–	0	1	0
4	0.00053	–	–	1	0	–

Optimal design tolerance
0.00053

example, for piston fabrication, process (or machine) 1 is selected for stage 1, with manufacturing tolerance allocated to 0.016; while for stage 2, process (or machine) 3 is selected with manufacturing tolerance allocated to 0.0035, and so forth. The optimal design tolerance on piston diameter is 0.00047.

The example problem is solved with the solution procedure starting from four different initial solutions. Table 15.3 shows the result of the computational experiments. It can be seen from the result that the minimum objective function values obtained through those different four runs are very close. The simulated annealing algorithm appears to be independent of initial solutions which is consistent with other researchers' findings (Lee and Wang, 1991).

Table 15.3 Minimum costs (G_{min}) obtained in experiments with different initial solutions

	Experiment			
	1	2	3	4
Minimum cost G_{min}	162.11	161.87	161.92	161.81

15.6 CONCLUSIONS

This paper has presented an analytical model and a solution procedure for concurrent tolerance optimization problem. A nonlinear mixed binary-continuous optimization model was formulated for design and manufacturing tolerancing with consideration of process (machine) selection. By solving the problem, design tolerance synthesis, manufacturing tolerances allocation and process selection are optimized simultaneously based on the least manufacturing cost criterion. Simulated annealing algorithm is employed as the optimizer for the problem. Based on our observations, simulated annealing appears to be an effective and robust algorithm for solving the concurrent tolerance optimization problem. Tolerance is an important parameter in design and manufacturing. This study provides a tool towards implementation of concurrent engineering.

15.7 REFERENCES

Adler, R.E. and Ishii, K. (1989) *Proceedings of the 1989 ASME International Computers in Engineering Conference*, 19–26.
Brooks, D.G. and Verdini, W.A. (1988) *American Journal of Mathematical and Management Sciences*, **8**(3/4) 425–49.
Chase, K.W., Greenwood, W.H., Loosli, B.G. and Hauglund, L.F. (1990) *Manufacturing Review*, **3**(1), 49–59.
Collins, N.E., Eglese, R.W. and Golden, B.L. (1988) *American Journal of Mathematical and Management Sciences*, **8**(3/4), 209–307.
Coranam, A., Marchesi, M., Martini, C. and Ridella, S. (1987) *ACM Transactions on Mathematical Software*, **13**(3), 262–80.
Godse, M.M., Haug, E.J. and Choi, K.K. (1991) A Parametric Design Methodology for Concurrent Engineering, *Technical Report R-109*, CCAD, The University of Iowa.
Kirkpatrick, S., Gelatt Jr., C.D. and Vecchi M.P. (1983) *Science*, **220**, 671–80.
Lee, S. and Wang, H.P. (1991) A Simulated Annealing Approach for Optimum Cam Design, Working Paper, No. 91-14, Department of Industrial Engineering, The University of Iowa.
Lee, W. and Woo, T.C. (1987) Optimum Tolerance Selection of Discrete Tolerances, Technical Report 87-34, Department of Industrial and Operations Engineering, University of Michigan, Ann Arbor, MI.
Nevins, J.L., Whitney, D.E. and De Fazio, T.L. (eds) (1989) *Concurrent Design of Products and Processes: a Strategy for the Next Generation in Manufacturing*, McGraw-Hill, New York.
Parkinson, A., Sorensen, C., Free, J. and Canfield, B. (1990) *Proceedings of the 16th Design Automation Conference*.
Sfantsikopoulos, M.M. (1990) *International Journal of Advanced Manufacturing Technology*, **5**(1), 126–34.
Spotts, M.F. (1973) *Transactions of ASME, Journal of Engineering for Industry*, **95**, 762–4.
Trucks, H.E. (1989) *Design for Economical Production*, (2nd edn), SME.
Wang, H.P. and Li, J.K. (1991) *Computer-aided Process Planning*, Elsevier, New York.

Wu, Z., ElMaraghy, W.H. and ElMaraghy, H.A. (1988) *Manufacturing Review*, **1**(3), 168–79.

Zhang, C. and Wang, H.P. (1991) The Discrete Tolerance Optimization Problem. To appear in the Manufacturing Review.

Zhang, C. and Wang, H.P. (1992) Integrated Tolerance Optimization with Simulated Annealing. To appear in the *International Journal of Advanced Manufacturing Technology*.

CHAPTER 16

Design for human factors

Fariborz Tayyari

16.1 INTRODUCTION

Nevins and Whitney (1991) define concurrent engineering (CE) as 'Design of the entire life cycle of the product simultaneously using a product design team and automated engineering and production tools'. Its objective is to 'design it right in the first place'. A product is manufactured, handled, installed and used by humans. Since the human is an inherent part of any engineering project and affected by the project design, the human factors must be considered in the design of the project, right at the beginning. Therefore, it is crucial for designers to understand and consider the capabilities and limitations of the targeted users population for whom a product or process is designed.

Automation has eliminated many of the heavy physical activities that used to be performed manually by human beings. Despite a tremendous amount of technological advancement and automation there are, however, still many physically demanding jobs to be performed by individuals. The physical demands can place stress on the worker while he/she is performing his/her job. This stress can affect the worker's job performance and his/her health and safety as well. Modernization of the industrial environment has, however, increased *mental demands* placed on the worker. In the past, physical demands of the job used to cause fatigue, but today mental demands are causing fatigue. This chapter introduces the concepts of ergonomics/human factors engineering and its principles.

16.1.1 Ergonomics

The main objective of ergonomics is to achieve an optimal relationship between man and his work environment. The two conflicting factors in this optimization process are the worker's productivity and his/her health and physical well-being. That is, while the worker should perform his/her job in the most efficient manner possible, he/she must also be protected against undue physical, biological, and psychological strains that may occur as a result of performing the required tasks. The ergonomics objective is achieved by application or ergonomic principles, which have been developed on its two basic principles:

- Fitting the job to the worker; and
- Fitting the worker to the job.

Industrial engineers (or work designers) put a great deal of effort into incorporation of the first basic principle in the early stage of the design by considering the human aspects of the work systems. The second basic principle should only be considered in existing work systems, when the incorporation of the first one may not be economically feasible. This is accomplished by selecting the right person for the job and training. Ergonomic principles are developed through the utilization of the following primary areas of expertise:

- Biomechanics: It deals with the assessment of the human's capabilities, and static and dynamic forces on the body structure.
- Anthropometry: It deals with the assessment of the human's limitations and variability among individuals due to body dimensions and muscular strength. The anthropometric procedures and population's dimensional and strength measurements are used to design workstations and tools that can be effectively used by people.
- Work physiology: It studies the physiological responses to work and environmental stresses for establishing a 'fair-day' work regimen.
- Industrial psychology: It studies the psychological effects of work and environmental stresses, and mental capacity and workload for improving employee safety and productivity.

Other areas (e.g., engineering, mathematics and statistics) are also heavily utilized in ergonomic research, design, simulation and training.

16.1.2 Occupational stress factors in the workplace

Among factors affecting the worker and his/her job performance, the following are most important to be considered in optimizing the relationship between the worker and his/her work environment:

1. Unnatural environmental conditions (i.e., thermal, illumination, noise, vibration, toxic materials, etc.) in the workplace;
2. Physical and mental requirements of the job;
3. Complexity and number of tools to be used in the workplace that makes the interaction between the worker and the work equipment difficult;
4. Inefficient workspace and workplace designs; and
5. Worker's exposure to hazardous materials.

The stress from these factors placed on the worker is significant. Ergonomic principles can be applied to design of the workplace so that the strain placed on the worker is minimized.

16.2 CONTROLS AND DISPLAYS, AND THEIR COMPATIBILITY

In a 'man-machine' system the human operator interacts with the machine. The system is a closed cycle (Fig. 16.1) in which the operator plays the key role because he/she is the decision-maker. The operator visually receives (perceives) information from displays on an instrument panel which he/she must understand and assess (interpretation). Based on the interpretation of the perceived information and his/her previous knowledge, the operator makes a decision on the action to be taken. The operator communicates his/her decision with the machine through manual controls. The machine displays the result of the operator's action which might require the operator's use of control for confirmation. The machine, then, carries out the process as programmed. A proper design of the man-machine interface is of a great importance for enhancing the operator's productivity, safety and well-being.

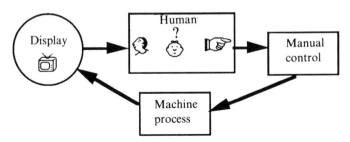

Fig. 16.1   An illustration of the man-machine system.

16.2.1 Display and control problems

Errors are the major problems associated with displays and controls in man-machine systems. They are caused by poorly designed displays and controls as well as display-control incompatibility and inappropriate layouts of the instruments.

Displays include gauges, lights, CRT monitors, dials, etc. Controls include valves, toggle switches, push-bottons, pedals, keyboards, levers, rotary knobs, thumb-wheels, hand wheels, and cranks.

There are three common types of display instruments (Van Cott and Kinkade, 1972; Alexander, 1986; Grandjean, 1988):

1. A window with digital read-out (counters) in which figures can be read directly (Fig. 16.2(a));

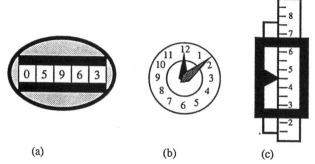

(a)   (b)   (c)

Fig. 16.2   Three types of display: (a) counter; (b) fixed scale with moving pointer; and (c) fixed pointer with moving scale.

2. A (fixed) circular scale with moving pointer (Fig. 16.2(b)); and
3. A fixed pointer over a moving scale (Fig. 16.2(c)).

16.2.2 Guidelines for design of displays and controls

Since controls and displays are usually interrelated, they should be designed to be compatible. They should be properly grouped together so that the relationship between each pair of control and display can easily be understood by the operator. The following can be used as general guidelines for design of controls and displays, their relationship, and their layouts (Van Cott and Kinkade, 1972; Kvålseth, 1985; Alexander, 1986; Grandjean, 1988).

(a) Guidelines for displays

- A right type of display must be used for the information needed.
- Use digital read-out (counters) displays when quick and accurate readings are to be made, provided that the displayed data do not change too quickly to be missed and that only one figure is visible at a time.
- Use the display with moving pointer on a fixed, circular scale which provides more information if it is necessary to monitor how a process is going on, or to note the amplitude of some change.
- Avoid the use of displays with moving scale and fixed pointer since it is not easy to memorize the previous reading, nor to assess the extent of movement. They appear to display poorly qualitative readings.
- The major scale markers should be numbered in a natural and most recognizable fashion (e.g., $(1, 2, 3, \ldots)$; $(5, 10, 15, \ldots)$; $(10, 20, 30, \ldots)$; $(100, 200, 300, \ldots)$; etc. rather than $(2, 4, 6, \ldots)$ or $(3, 6, 9, \ldots)$).
- Avoid multi-scale, multiple-pointer and nonlinear-scale displays, whenever possible.
- Install displays in the direct line of sight and avoid any obstructions.
- Avoid glare on the displays.
- Avoid small-size or illegible displays.

- The display must be readable from the operator's normal location.
- The size of letters and figures, thickness of lines and their distance apart must be determined based on the viewing distance between the eye and the display. Use the values given in Table 16.1 (Van Cott and Kinkade, 1972) for determining the heights of letters and numbers on displays.

Table 16.1 Standards for determining heights of letters and numbers

	Low light levels	Adequate light levels
Critical labels (information, data, emergency labels)	0.25 in (0.635 cm)	0.15 in (0.381 cm)
Noncritical labels (identification labels, instructions, nonemergency labels)	0.10 in (0.254 cm)	0.10 in (0.254 cm)

The values are given for a 28 in (71-cm) standard viewing distance. The values will increase proportional to the increase in the viewing distance (after Van Cott and Kinkade, 1972).

(b) Guidelines for controls

- Design controls based on the anatomy and functional characteristics of the body limbs (e.g., fingers and hands should be used for quick, precise movements; arms and feet for operations requiring force).
- Install hand-operated controls such that they can easily be reached and grasped, that is, between elbow and shoulder height.
- Install controls in full view.
- The distance between neighboring controls must be sufficient to avoid accidental activation.
- Install foot controls in such a way that accidental activation, which may lead to potential hazards and undesirable consequences, can be avoided. Foot controls may be necessary to guard.
- Use appropriate methods of control coding to enable the operator to identify each control correctly and quickly. Examples of common methods of coding are color, shape, size, texture, labeling and location.
- Consider the human anatomy in specifying distance between controls, for example:
  - finger-operated knobs or switches should be, at least, 1.5 cm (0.6 in) apart;
  - whole-hand operated controls should be, at least, 5 cm (2 in) apart.
- Use push-buttons, tumbler switches and rotating knobs for operations requiring little movement or muscular effort, small travel and high precision, and for either continuous or stepped operation (click-stops).
- Use long-armed levers, cranks, hand-wheels and pedals for operations requiring muscular effort over a long travel, and relatively little precision.
- Controls should be near the operator and usable in the normal sitting or standing position.

(c) Guidelines for control-display relationship

- Install controls near the displays which are affected by them.
- Install the control either below its corresponding display, or, if necessary, to the right of it.
- Place identification labels above the controls and identical labels above the corresponding display instruments.
- If the controls have to be in one panel and the displays in another, the two sets should be laid out in the same order and arrangement.
- If a series of controls are normally operated in sequence, they and their corresponding display instruments should be arranged on the panel in the same order, from left to right.
- When arranging a group of controls and displays, the controls and displays that are used more often should be located directly in front of the operator, and others to the sides.
- When a control is moved or turned to the right (clockwise), the pointer on the display must also move to the right on a horizontal scale or clockwise over a circular scale; on a vertical scale the pointer must move upwards.
- When a control is moved upwards or forwards, the display pointer must move either upwards, or to the right, or clockwise.
- A right-handed or clockwise rotation implies an increase, so the display instrument should also record an increase.
- A moving scale with a fixed indicator should move to the right when the control is moved to the right (clockwise), while the scale values should increase from right to left (counterclockwise), so that the rotation of the scale to the right (clockwise) shows increased readings.
- When a hand lever (control) is moved upwards, forwards, or to the right, the display reading should increase or a switch should move to the 'ON' position. To reduce the reading, or to switch to 'OFF' position, the lever is moved downwards, or towards the body, or to the left.

16.3 USE OF ANTHROPOMETRY IN WORKSTATION DESIGN

Work performance becomes more efficient when body postures and movements are natural. A natural posture is a position of body in which the trunk, arms and/or legs are not involved with static work (Grandjean, 1988). A static work, in which the muscles are loaded without movement, increases tension within the muscles, blocking the natural blood circulation in them and therefore results in muscular fatigue. To facilitate natural postures and movements, the workstation should be suited to the body dimensions of the worker. However, a major problem in workstation design is the existence of a wide variation in body dimensions among individuals. This problem to a significant extent is solved by engineering anthropometry, that has a great

77777777

role in equipment and workplace/work-space designs. Availability of anthropometric data has enabled the designers to accommodate a major proportion of the potential users' population in their designs. The desired information can be extracted from the anthropometric reference manuals and books (e.g., Van Cott and Kinkade, 1972; Roebuck *et al.*, 1975; NASA, 1978; Bailey, 1982; Jürgens *et al.*, 1990). Table 16.2 presents some anthropometric data which are commonly used in designing workstations to accommodate a predetermined proportion of the workforce population.

Table 16.2 US civilian anthropometric dimensions for workstation design, in cm (in), for ages 18 to 79 years

Measurement	Male/Female Mean($\bar{x}$)	S.D.* (s)
Sample size	3,091/3,581	—
Stature (standing height)	173.2/160.3 (68.2/63.1)	6.9/6.6 (2.7/2.6)
Sitting height	90.5/84.7 (35.6/33.3)	3.7/3.6 (1.5/1.4)
Popliteal height	44.0/39.7 (17.3/15.6)	2.7/2.6 (1.1/1.0)
Elbow rest height (from seat)	24.1/23.1 (9.5/9.1)	3.0/2.9 (1.2/1.1)
Knee height, sitting	54.2/49.7 (21.3/19.6)	2.9/2.7 (1.1/1.1)
Thigh clearance (thickness), sitting	14.3/13.7 (5.5/5.4)	1.7/1.8 (0.7/0.7)
Buttock-to-knee length, sitting	59.1/56.8 (23.3/22.4)	2.9/3.1 (1.1/1.2)
Buttock-to-popliteal length, sitting	49.4/48.0 (19.4/18.9)	3.1/3.1 (1.2/1.2)
Elbow-to-elbow breadth	42.0/39.0 (16.5/15.4)	4.7/5.4 (1.9/2.1)
Hip breadth, sitting	35.4/36.7 (13.9/14.4)	2.8/3.7 (1.1/1.5)
Shoulder-to-shoulder breadth (biacromial breadth)	39.6/35.4 (15.6/13.9)	2.1/1.9 (0.8/0.7)

* S.D. is the sample standard deviation (s).
Source: NASA, 1978.

It is assumed that the human anthropometric measures are normally distributed. That is, we can state that the anthropometric measure X has a normal distribution with mean μ and standard deviation σ. Hence, $Z=(X-\mu)/\sigma$ has a standard normal distribution and we can write:

$$P(X \leqslant x_p) = P[(X-\mu)/\sigma \leqslant (x_p-\bar{x})/s] = p \tag{16.1}$$

or

$$P(Z \leqslant z_p) = p \qquad (16.2)$$

where:

$z_p = (x_p - \bar{x})/s;$
$\bar{x}$ = the sample mean (average), obtained from anthropometric references;
s = the sample standard deviation, also from anthropometric references; and
z_p = values are given in standard normal probability (i.e., Z-distribution) tables for values of p (note that the size of the sample for which $\bar{x}$ is calculated should be $n \geqslant 30$; otherwise, t_p and the t-distribution table must be used instead of z_p, however the published anthropometric data are usually based on large sample sizes).

We can use $x_p = \bar{x} + z_p \cdot s$ to determine the pth percentile value for the anthropometric measure X.

The most commonly used percentile (%ile) and p values along with their corresponding z_p values are given in Table 16.3 for determining the size of an anthropometric measure. For example, based on an anthropometric measurements given in Table 16.2, the popliteal height of the US male individuals has a mean 44.0 cm and a standard deviation 2.7 cm. The popliteal height of the 95th percentile of males is $X_{0.95} = 44 + 1.64(2.7) = 48.4$ cm; and of the 5th percentile of females is $X_{0.05} = 39.7 - 1.64(2.6) = 35.4$ cm. Hence, when considering a heel height of 2.5 cm, an adjustable seat height ranging from 37.9 cm to 50.9 cm with 13 cm adjustability (i.e. 50.9 − 37.9) is suitable for 90% of the US industrial workforce (i.e. 5th percentile females to 95th percentile males).

Table 16.3 Standard normal z values for the commonly used percentiles

%ile	1st	2.5th	5th	10th	17th	50th	83rd	90th	95th	97.5th	99th
p	0.01	0.025	0.05	0.10	0.17	0.50	0.83	0.90	0.95	0.975	0.99
z_p	−2.33	−1.96	−1.64	−1.28	−0.955	0.0	0.955	1.28	1.64	1.96	2.33

16.3.1 Ergonomics of workstation design

The fundamental ergonomic principle in workplace design is aimed at designing the workstation to fit the operators. However, it may not be economical to fit the workplace to all individual operators due to a wide variability among them. Then, the workstation should be designed in such a way that it will minimally inconvenience a minimum number of operators. Therefore, a compromise must be made to accommodate a majority (e.g. 90% to 95%) of the population. There are three basic design strategies for this purpose:

1. Designing for the average;

2. Designing for the extreme; and
3. Designing for adjustability.

Each strategy may be suitable for a certain situation.

It is not usually appropriate to design a workplace to suit an average individual since the existence of such a person being average in all dimensions is extremely rare. Designers, who use the first design strategy, may assume that the design will at least fit a half of the workforce. They should keep in mind that the clearances may be too small for the larger half of the population and/or the reaches may be too long for the smaller half of the population, and as a result the workplace may be suitable for nobody in the entire population.

If possible, any piece of equipment or component of the workplace should be made adjustable (the third strategy) so that it can be efficiently used by a predetermined proportion of the workforce population. Such an adjustability feature can be used in some items (e.g. seats and tables) while it is difficult, if not impossible, to be incorporated in many other items (e.g. the location of a control, the height of a conveyor, and the width of an aisle). Under such conditions, the smallest person's reach should be used when a 'reach' is involved and the largest individual must be considered when a 'clearance' is involved (the second strategy). For example, if the 5th percentile females' reach is considered in determination of the location of parts bins or a control switch on a wall, only five out of one hundred female workers will become inconvenienced, if not excluded. In such a situation, the 5th percentile males and others with longer reaches will have no trouble with reaching the parts or control. Also, larger clearances (e.g. 95th percentile of male population) will fit smaller individuals.

16.3.2 Examples of workstation design using anthropometric data

The application of anthropometric data in workstations is best explained by examples. This is done by types of workstation designs accommodating a certain range of percentiles: (1) adjustable table and seat; and (2) fixed table, but adjustable seat. The heights of seat, table and footrest for each case are formulated as shown below:

(a) When both table and seat are adjustable
(fully adjustable workstation)

Footrest is not used since the workstation is fully adjustable and the feet rest on the floor.

Seat height = Popliteal height + Heel height (about 2.5 cm or 1 in).
Table height (undertable) = Seat height + Thigh clearance (thickness).
Tabletop height = Seat height + Elbow-rest height − 5 cm or 2 in (below elbow level).

Calculations of dimensions of fully adjustable workstation (i.e., when both the table height and seat height are adjustable) to fit 66% of the population are made as follows:

Footrest $=0$.
Seat Height:
    Minimum (for the shortest female) $= 39.7 - 0.955(2.6) + 2.5$ (Heel height) $= 39.7$ cm
    Maximum (for the tallest male) $= 44.0 + 0.955(2.7) + 2.5$ (Heel height) $= 49.1$ cm
Table height (undertable):
    Minimum (for the shortest female) $= 39.7 + [13.7 - 0.955(1.8)] = 51.7$ cm
    Maximum (for the tallest male) $= 49.1 + [14.3 + 0.955(1.7)] = 65.0$ cm
Tabletop height:
    Minimum (for the shortest female) $= 39.7 + [23.1 - 0.955(2.9)] - 5 = 55.0$ cm
    Maximum (for the tallest male) $= 49.1 + [24.1 + 0.955(3.0)] - 5 = 71.1$ cm

(b) When table is fixed, but seat is adjustable:

Footrest height $=$ the adjustability of the table height when both table and seat are adjustable; that is, footrest replaces the table adjustability.
Table height (undertable) $=$ Footrest height $+$ Heel height $+$ Popliteal height $+$ Thigh clearance
Tabletop height $=$ Table height (undertable) $+2.5$ cm or 1 in (table thickness).
Seat height $=$ Popliteal height $+$ Heel height (about 2.5 cm or 1 in) $+$ Footrest height

To fit the workstation to 66% of the population when the table height is fixed, but the seat is adjustable, dimensions are calculated as follows:

Calculations of dimensions for fitting 66% of the population:
Footrest:
    Minimum (for the tallest male) $=0$
    Maximum (for the shortest female) $= 13.3$ cm
Table height (undertable):
    Minimum (for females) $= 13.3 + 2.5 + 37.2 + 12.0 = 65.0$ cm
    Maximum (for males) $= 0 + 2.5 + 46.6 + 15.9 = 65.0$ cm
Tabletop height $= 65.0 + 2.5 = 67.5$ cm
Seat height:
    Minimum (for the tallest male) $= 44.0 + 0.955(2.7) + 2.5 + 0 = 49.1$ cm, or $= 65.0 - 15.9 = 49.1$ cm

Maximum (for the shortest female) $= 39.7 - 0.955(2.6) + 2.5 + 13.3 =$
53.0 cm, or $= 65.0 - 12.0 = 53.0$ cm

The results of calculations for the two workstation options for accommodation of various desired ranges of percentiles are given in Table 16.4.

16.4 MANUAL MATERIALS HANDLING

Manual materials handling (MMH) is one of the occupational factors that
can cause significant damage to the human body if performed improperly. It
is referred to the acts of manually lifting, lowering, holding and carrying an
object without mechanical aids (i.e., block-and-tackles, conveyors, hoists,
etc.), and also pushing and pulling an object with or without mechanical
aids (i.e., dollies, carts, etc.).

MMH places strains on the cardiovascular system and the musculoskeletal system. The strain problem associated with the cardiovascular
system is revealed by increase in oxygen consumption and heart rate
which is not usually so critical and can be resolved by resting. However, a musculoskeletal strain, which can be injurious, is of greater
concern.

The subject of MMH concentrates on the identification and control of
injury-causing conditions associated with manual materials handling, and
minimizing the health hazards by administrative controls (e.g., employee
training, rotations and selection) and engineering techniques (e.g., job
redesign, mechanical assists). The application of biomechanics (specifically
statics and dynamics) in MMH is tremendous.

16.4.1 A manual lifting task

A manual lifting task is the act of manually grasping and raising an object of
a definable size without mechanical aids (NIOSH, 1981). Since the time
duration of such an act is less than two seconds, the sustained exertion
required (as encountered in holding and carrying) is not significant to be
considered in the analysis of a lifting task. The lifting models and limits
developed and recommended by NIOSH in its Work Practices Guides in
1981 are intended to apply only for:

1. Smooth lifting;
2. Two-handed, symmetric lifting in the sagittal plane (directly in front of
   the body, no twisting during lift);
3. Moderate width (i.e., maximum 75 cm or 30 in);
4. Unrestricted lifting posture;
5. Good couplings (handles, shoes, floor surface); and
6. Favorable ambient environments.

Table 16.4 Workstation heights, in cm, for accommodation of various ranges of the population

Percentiles fitted		Adjustable table and seat				Fixed table/adjustable seat			
		17–83	5–95	2.5–97.5	1–99	17–83	5–95	2.5–97.5	1–99
Per cent fitted		66	90	95	98	66	90	95	98
Seat height	Min	39.7	37.9	37.1	36.1	49.1	50.9	51.8	52.8
	Max	49.1	50.9	51.8	52.8	53.0	57.3	59.2	61.6
	Range of Adjustability	9.4	13.0	14.7	16.7	3.9	6.4	7.4	8.8
Table height (undertable)	Min	51.7	48.6	47.3	45.6	65.0	68.0	69.4	71.1
	Max	65.0	68.0	69.4	71.1	65.0	68.0	69.4	71.1
	Range of Adjustability	13.3	19.4	22.1	25.5	0	0	0	0
Task or tabletop height	Min	55.0	51.2	49.5	47.4	—	—	—	—
	Max	71.1	74.9	76.8	78.9	67.5	70.5	71.9	73.6
	Range of Adjustability	16.1	23.7	27.3	31.5	0	0	0	0
Footrest height	Min	0	0	0	0	0	0	0	0
	Max	0	0	0	0	13.3	19.4	22.1	25.5
	Range of Adjustability	0	0	0	0	13.3	19.4	22.1	25.5

In using the NIOSH Work Practices Guides in real-world situations, the following considerations should be taken into account:

1. Other MMH activities (e.g., holding, carrying, pushing and pulling) are assumed to be minimal;
2. When lifting activities are not performed, the individual is assumed at rest;
3. The work force is physically fit and accustomed to physical labor; and
4. 'Safety factors' commonly used by engineers to account for the unexpected conditions are not included.

16.4.2 NIOSH lifting model and its variables

NIOSH has developed a mathematical model for estimating an action limit (AL, Equation 16.3) and a maximum permissible limit (MPL, Equation 16.4).

(a) Maximum permissible limit (MPL)

Lifting loads beyond the MPL should not be permitted, but engineering controls must be applied to lower the load within the acceptable range. This limit has been set based on four criteria (epidemiological, biomechanical, physiological, and psychological) as summarized below:

1. Musculoskeletal injury and severity rates increase significantly when a lifting task is performed above the MPL;
2. Biomechanical compression forces on the L5/S1 intervertebral disk above 650 kg (6 370 N or 1 430 lb) are not tolerable by most people;
3. Metabolic rates would exceed 5.0 Kcal/minute for most individuals performing a lifting task above the MPL; and
4. Only about 25% of male and less than 1% of female workers are capable of performing the lifting task above the MPL.

(b) Action limit (AL)

Because of the large variability in capacity of individuals, loads falling between the AL and MPL may be lifted if administrative controls (e.g. personnel selection and training) are applied since:

1. Musculoskeletal injury and severity rates increase moderately when workers perform a lifting task up to the AL;
2. A 350 kg (3 430 N or 770 lb) biomechanical compression force on the L5/S1 intervertebral disc imposed by the conditions described by the AL can be tolerated by most young, healthy workers;
3. Metabolic rates would exceed 3.5 Kcal/minute for most individuals performing a lifting task above the AL; and

4. Lifting loads up to the AL can be performed by over 99% of male and over 75% of female populations with a nominal risk of back injury.

(c) Lifting task variables

Primary lifting task variables that affect the AL and MPL are as follows:

1. Horizontal location (H) of the hands at origin of lift, measured from the midpoint between ankles (in centimeters or inches). H must be between 15 and 80 cm (6 to 32 in). The minimum 15 cm (6 in) is due to body interference.
2. Vertical location (V) of the hands at origin of lift, measured from floor level (in centimeters or inches). V must be between 0 and 175 cm (0 and 70 in), which is the range of vertical reach for most individuals.
3. Vertical travel distance (D) from origin to destination of lift (in centimeters or inches). D must be between 25 cm (10 in) and $(200 - V)$ cm $[(80 - V)$ in]. If the travel distance is less than 25 cm (10 in), then $D = 25$ cm (10 in) must be used.
4. Frequency of lifting (F), average number of lifts per minute. F must be between 0.2 (one lift every five minutes) and F_{max} (shown in Table 16.5). If the frequency of lift is less than once per 5 minutes, then set $F = 0$.
5. Maximum frequency of lifting (F_{max}) which is determined based on the duration or period of the task during the work shift. Lifting is assumed to be occasional (less than one hour) or continuous (8 hours). Table 16.5 provides the F_{max} values.

Table 16.5 Maximum lifts per minute (F_{max})

Period	Average vertical location cm (ins)	
	$V > 75$ (30) Standing	$V \leqslant 75$ (30) Stooping
1 hour	18	15
8 hours	15	12

NIOSH has categorized lifting frequencies as follows:

1. Infrequent lifting: either occasional or continuous lifting less than once per 3 minutes.
2. Occasional high frequency: lifting one or more times per 3 minutes for a period of one hour.
3. Continuous high frequency: lifting one or more times per 3 minutes continuously for 8 hours:

$$\text{AL(lb)} = 90x\left(\frac{6}{H}\right) \times (1 - 0.01\,|V - 30|) \times \left(0.7 + \frac{3}{D}\right) \times \left(1 - \frac{F}{F_{max}}\right)$$

(US customary units)   (16.3)

or

$$AL(kg) = 40x\left(\frac{15}{H}\right) \times (1 - 0.004|V - 75|) \times \left(0.7 + \frac{7.5}{D}\right) \times \left(1 - \frac{F}{F_{max}}\right)$$
(Metric units)

$$MPL = 3\,AL \hspace{4cm} (16.4)$$

where:

H = Horizontal location of lift centerline;
V = Vertical location of the hands at origin of lift;
D = Vertical travel distance from origin to destination of lift;
F = Frequency of lifting, average number of lifts per minute;
F_{max} = Maximum frequency of lifting which can be sustained (from Table 16.5);
AL = Action limit; and
MPL = Maximum permissible limit = $3 \times AL$.

The H, D and V factors are shown in Fig. 16.3, while Fig. 16.4 illustrates the NIOSH's recommended maximum lift weights at various horizontal distances for infrequent lifts from floor to knuckle height.

(d) Evaluation of lifting task variables

As shown in Equation 16.1, the NIOSH model for the action limit (AL) consists of four multiplicative factors. These factors are explained for the

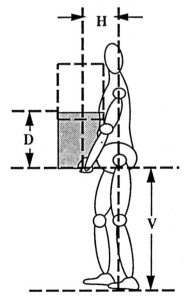

Fig. 16.3   Illustration of the independent variables in the NIOSH lifting model.

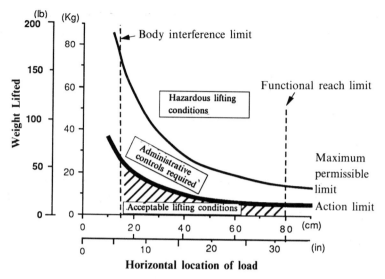

Fig. 16.4 NIOSH's recommended maximum lift weights at various horizontal distances for infrequent lifts from floor to knuckle height.

equation given for the use of metric units as follows:

H factor $=(15/H)$

This factor indicates the amount of adjustment necessary as a function of the horizontal location (H) and ranges from 0.1875 (for H = 80 cm) to 1 (for H = 15 cm). If H = 15 cm, this factor is equal to 1 and no adjustment for the horizontal location is necessary. When H = 60 cm, this factor is $(15/60) = 0.25$, which means that the AL is reduced from 40 kg to $40(0.25) = 10$ kg.

The factor for vertical location (V) deals with the absolute deviation of V from 75 cm (which is approximately equal to the knuckle height) ranges from 0.6 (for V = 175 cm) to 1 (for V = 75 cm). If V = 75 cm, then

V factor $=(1-0.004|V-75|)$
$\qquad =(1-0.004|75-75|)=1$

and, hence, no adjustment for the vertical location is necessary. If V = 25 cm:

V factor $=(1-0.004|25-75|)$
$\qquad =1-0.004(50)=0.8$

Also, for V = 125 cm,

V factor $=(1-0.004|125-75|)$
$\qquad =1-0.004(50)=0.8$

Thus, when V = 25 cm or 125 cm, the absolute deviation of V from 75 cm is 50 cm and the V factor will reduce the AL from 40 kg to $40(0.8) = 32$ kg.

The vertical travel distance (D) ranges from 0.7375 (for $D = 200$ cm or lifting from the floor to 200 cm above the floor) to 1 (for $D = 25$ cm, the minimum allowed value). If $D = 25$ cm,

D Factor $= (0.7 + 7.5/D)$
$= 0.7 + 7.5/25 = 1$

and, hence, no adjustment is necessary. If $D = 50$ cm,

D Factor $= (0.7 + 7.5/50) = 0.85$,

which reduces the AL from 40 kg to $40(0.85) = 34$ kg.

The frequency factor is slightly more complicated than the other three factors. If the observed frequency is nine lifts per minute (i.e. $F = 9$) and the lifting point is below 75 cm (i.e. $V \leqslant 75$ cm), then:

F Factor $= (1 - F/F_{max})$
$= 1 - 9/12 = 0.25$

which reduces the AL from 40 kg to $40(0.25) = 10$ kg. The F factor ranges from 0 (for $F = F_{max}$) to 1 (for $F = 0$, which is set for frequencies $\leqslant 0.2$ lifts per minute or 1 lift every 5 minutes).

To illustrate the combined effects of these factors, let's assume that a continuous (all-day long) lifting is performed above the knuckle height ($V = 90$ cm) with an average $H = 30$ cm and an average rate of six lifts per minute. The load is placed on shelves of an average height of 120 cm above the floor. This information is summarized as follows:

$H = 30$ cm,
$V = 90$ cm,
$D = 120 - 90 = 30$ cm,
$F = 6$ lifts/minute, and
$F_{max} = 15$ lifts/minute (for 8 hours in standing position).

Then,

AL $= 40 \times (15/30) \times (1 - 0.004|90 - 75|) \times (0.7 + 7.5/30) \times (1 - 6/15)$
$= 40 \ (0.5) \ (0.94) \ (0.95) \ (0.6) = 10.7$ kg
MPL $= 3 \times 10.7 = 32.1$ kg

As it can be seen, the H factor, followed by the F factor, has had the greatest effect on reducing the AL value.

16.5 ENVIRONMENTAL STRESSES AND THEIR CONTROLS

The effects of industrial environmental factors (e.g., thermal components, illumination, sound/noise, vibration) are so important that each should be studied in detail. However, each topic is briefly discussed in this section.

16.5.1 Illumination (lighting)

In order for personnel to perform their jobs effectively and safely, each task and area must be adequately illuminated. An adequate illumination (lighting) is not just adding light to increase its quantity. It rather consists of both quantity (i.e, the amount of light) and quality (i.e., contrast, color, glare free, etc.) of light.

Different tasks require different levels of illumination. For example, fine assembly tasks require more light than general tasks. An efficient use of light can be achieved by providing light at a level appropriate for at each task, where it is actually needed, and reducing light levels in general areas. ANSI Standard RP-7-1979 should be consulted for industrial lighting.

Excessive light causes glare, which obscures vision and is eye fatiguing. Visual display panels and monitors should be positioned in relation to light sources in such a way that reflected glare on them can be avoided. Windows, naked light sources and highly reflective surfaces can present glare. Thus, curtains and blinds should be used to control light from windows, naked light sources should be screened or relocated, and shiny surfaces should be avoided (e.g., opaque paints can be used for this purpose).

Poor illumination can deteriorate the worker's job performance and be a potential strain to his/her vision. It can also cause accidents, resulting in personal injuries.

16.5.2 Sound/noise

In occupational environments, workers may be exposed to various kinds or levels of noise or sound. Some consider noise and sound to be as two different concepts for the same phenomenon (Alexander, 1986), where the word noise is used for unwanted sound. To a certain level, noise may not be harmful but annoying. However, exposure to loud noise can cause hearing loss, especially if it lasts for a long period of time. Hearing loss may be either temporary or permanent. Loss of hearing can occur quickly (i.e. in one short exposure to extremely loud noise) or so slowly that the person would hardly notice it. Temporary hearing loss can occur when a loud noise is endured for a few minutes or hours, but it will return after the person is away from loud noises for a few hours. Permanent hearing loss can happen by long exposures to loud noise over months or years or short exposure to extremely loud noise. In the latter case, the hearing cannot be recovered regardless of the length of rest. Permanent hearing loss can also occur by the natural aging process which cannot be prevented by limiting noise exposure.

Noise is described by its two components:

1. Intensity (or loudness) of the sound energy (or sound pressure); and
2. Frequency.

Standard sound-level meters usually have three weighting scales (A, B, and C). Noise pressure intensity (or loudness) readings are indicated by the letter representing the scale used. For example, 78 dB(A) (or dBA) is the noise level measured using the 'A' scale. Since the human ear is less sensitive to low frequencies, the A-scale circuit of the sound-level meter attenuates very low frequencies to approximate the human ear's response. It is represented by a logarithmic scale ranging from about zero (0) dBA (hearing perception threshold) to about 120–130 dBA (pain threshold) (Ramsey, 1985). Noise frequency is measured in cycles per second (cps) or Hertz (Hz). Different frequencies are sensed as different tones or pitches. The audible frequency range for humans is 20 to 20 000 Hz (Ramsey). The human ear is able to respond to sound pressure ranging from 0.0002 to 2 000 dynes per square centimeter (microbars).

Octave-band analyzers are used to find at which frequency bands the noise is generated, or to detect the noise source. When an octave-band analyzer's switch is set for a certain frequency band, it will block out all other noise frequencies which are outside the octave range in question. Octave-band sound pressure levels can be converted into their equivalent A-weighted sound levels (i.e., dBA) using Fig. 16.5.

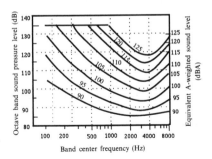

Fig. 16.5 OSHA's standards: dBA equivalent to octave-band sound pressure in dB.

The decibel (dB) is defined as ten times the logarithm of the ratio of a measured intensity to the reference intensity (Bels). That is,

$$L_i(\text{Bels}) = \log\left[\frac{\text{Measured intensity}}{\text{Reference intensity}}\right] \tag{16.5}$$

and

$$L_i(\text{dB}) = 10 \log\left[\frac{\text{Measured intensity}}{\text{Reference intensity}}\right] \tag{16.6}$$

where the reference intensity is 0.0002 microbars that corresponds to the hearing threshold.

The Occupational Safety and Health Administration (OSHA) of the US Department of Labor has set an action level (AL) of 85 dBA and a (maximum) permissible exposure limit (PEL) of 90 dBA to prevent noise hazards. Both levels are based on a time-weighted average (TWA) over an eight-hour exposure. OSHA has also specified 140 dBA as a ceiling (or C) value that should be considered as a limit for acute exposures only and is a safety hazard.

Noise level and exposure duration are two critical factors which industrial engineers and safety professionals must be concerned with. In reality, noise level varies from time to time during the work shift. The noise exposure under such variable conditions is assessed by the total exposure dose (D). This is performed by measuring the worker's actual exposure duration (C_i) to each noise level (L_i) during the work shift, then the total exposure dose, D, for the entire work shift is calculated using the following equation:

$$D = \sum_{i=1}^{n} \frac{C_i}{T_i} \qquad (16.7)$$

where:

D = Dose, total shift noise exposure as a proportion of PEL;
C_i = actual exposure time at noise level i, that is, L_i;
$T_i = 8/2^{(L_i - 90)/5}$ = maximum permissible exposure time at noise level i;
L_i = the noise intensity at level i, measure by a sound level meter using 'A' scale; and
n = number of all noise levels observed.

The total dose, D, must not exceed unity. If it exceeds one, appropriate corrective measures must be used to reduce the dose to acceptable levels; that is, 1 or less.

OSHA requires the employer to take the necessary action if noise levels reach the action level (AL) of 85 dBA for an eight-hour, time weighted average (TWA). The action includes:

1. A hearing conservation program;
2. Noise surveys;
3. Training; and
4. Recordkeeping.

(a) Noise control

There are three general noise control techniques:

1. Engineering controls;
2. Administrative controls; and
3. Personal protective equipment.

Engineering controls should be considered as a more adequate and permanent solution to the noise problem. Examples of engineering and administrative controls (NIOSH, 1973; Woodson, 1956) are given below.

Examples of engineering controls

- Maintaining and lubricating machines and equipment to eliminate rattles and squeaks.
- Replacing loose and worn parts of machines.
- Isolating noisy equipment, by moving the operator's station away from the source of noise. The intensity of noise from a source varies inversely with the square distance from the source, if the walls are not reflective.
- Modifying the plant layout to spread noisy machines out over the plant.
- Modifying or eliminating the process.
- Using plastic and wooden materials instead of noisy metal materials.
- Installing acoustic materials (sound-absorbing barriers) between machines.
- Padding inside sound booths or enclosures to prevent reflected noise from adding to the overall noise level.
- Placing equipment on rubber mountings to reduce its vibration.
- Acquiring new equipment with low noise levels.
- Using larger, low-speed fans instead of smaller, high-speed fans. The fan casing should be rigid and damped. Fans with silent motors should be selected.
- Using asphalt, cork or carpets on floors made of bare boards and concrete, that magnify impact noise.
- Staggering doors along a hallway. Opposite doors transmit noise more freely.
- Using double-panel windows with air space between panels.
- Covering walls with perforated acoustic tiles or rough plaster.
- Constructing irregular wall patterns.
- Avoiding domed ceilings in small rooms and enclosures which concentrate noises above the head of the occupant.
- Using boards suspended from the ceiling at different points.
- Using presses instead of drop hammers.
- Using large, slow machines instead of small, high-speed ones which have the same output capacity.
- Mounting vibrating machines on firm, solid foundations. Foundation bolts must be kept tight.
- Avoiding sharp directional and velocity changes in pipelines and ducts.
- Minimizing the velocity of air and liquids flowing through ducts and pipes.
- Securing pipes firmly to prevent rattling actions.

Examples of administrative controls

- Worker rotation to maintain his/her noise exposure dosage in an acceptable range.
- Scheduling production runs to split between shifts.
- Performing high-noise maintenance after the regular shift hours to minimize the number of exposed people.

Design for human factors

- Warning workers of noise hazards and the necessity for using protective devices by posting proper posters where such devices must be worn.
- Making sure that workers wear their protective devices at all times while in high-noise areas.
- Periodically checking the noise levels to identify areas in which the noise levels exceed the OSHA's TWA 85 dBA action limit.

Feasible engineering or administrative solutions must be applied first. If these remedies are found to be insufficient to reduce the noise exposure to an acceptable level, effective personal protective equipment must then be provided and used to reduce noise levels. Personal hearing protective equipment includes ear muffs, ear-canal caps and ear plugs. This is the most common, but least reliable, method of noise reduction. The major problem with the use of hearing protective devices is that they are not comfortable and the workers may not use them. They also make communication more difficult. Hence, they should be the last remedy resorted to for noise reduction.

16.5.3 Vibration

Vibration, at a high level, may be physiologically and/or psychologically harmful to the human body. Prolonged exposure to vibration can cause serious health impairment. There are two types of vibration in the industrial environment:

1. Segmental or localized vibration; and
2. Whole-body vibration.

Segmental vibration enters the hands and arms, mainly by the use of powered, hand tools (e.g. electric sanders, powered saws, pneumatic ranches, jack-hammers, and grinders). Whole-body vibration, which is of a lesser concern, is induced by operating heavy machineries, trucks, buses, etc. Its health effects have not found to be severe. However, it can inversely affect the person's productivity and concentration, and can expedite fatigue development. Nevertheless, industrial engineers and safety professionals should be concerned with safety hazard consequences due to the loss of control and concentration, and the induced fatigue.

Long-term exposure to segmental vibration can result in development of Raynaud's phenomenon or syndrome, which is also called vibration white finger (VWF). This disease is due to blood vessels constricting in the afflicted hand and causes tissue degeneration, symptomized by numbness, pain, stiffness, and blanching (paleness) in the fingers.

Like noise, vibration is described in terms of its frequency and intensity. The vibration frequency is measured in Hertz (Hz). Its intensity, especially for vertical vibration, is usually expressed by units acceleration (g). One g is the normal gravitational acceleration which is approximately 9.8 m/sec² (or 32.2 ft/sec²). The intensity of vibration by rotating or reciprocating motions is measured in terms of revolutions per minute (rpm) and the amplitude (magnitude) of displacement (in microns).

(a) Vibration controls

Vibration prolems are usually controlled by:

- Periodic maintenance and lubrication of machines and hand tools;
- Using support straps that absorb vibration energy before entering the hands and arms;
- Selecting equipment that generates low vibration;
- Using rubber or plastic mountings and couplings;
- Mounting vibrating equipment tightly on vibration isolators or dampers; and
- Minimizing vibration exposure by rotating the personnel.

16.5.4 Thermal factors

Heat stress due to exposure to a relatively high heat load places a physiological strain on the worker, resulting in potential health impairment. Although the human body possesses an excellent mechanism for maintaining an internal heat equilibrium with its surrounding, it has a limited capacity to adjust to extreme thermal conditions. When the limit is surpassed, heat cramps, exhaustion, collapse and heat stroke can happen. In addition, heat stress can impair the worker's mental functions and judgment of own condition, as a result of which serious accidents and injuries can occur. Cold stress, on the other hand, can also physiologically affect the human body, resulting in impairment of the individual's health as well as job performance. Exposure to severely cold stress can result in cold injuries and illness (e.g. frostbite, hypothermia when the body temperature falls below $35°C$).

In many occupations (e.g. mining operations, highway and building construction sites, glass and ceramics manufacturing plants), the workers are often exposed to high environmental heat loads. Heat stress, associated with such environments, is a health hazard and reduces workers' productivity. In some other occupations the workers are exposed to low environmental thermal conditions. Cold stress under such environmental conditions can present health hazards and deteriorate worker productivity.

The effects of work in hot environments on individuals have been studied for a long time. The results of these efforts have provided rational bases for the prevention of health impairment and physiological damage to workers due to heat stress (Stephenson *et al.*, 1974). Many industries, also, have become involved in heat stress investigations to provide safe work places for their workers and to comply with governmental standards for work in hot environments. Determination of heat effects on work quality and quantity has been another reason for industries to focus on heat stress problems.

(a) Physiological responses and heat exchange methods

Maintenance of life demands a constant flow of energy from the environment through organisms. The intake of energy is in the form of chemical potential energy of foodstuffs. The potential energy of foodstuffs is released in a form usable for organs. The released energy is finally returned to the environment in the forms of mechanical work and heat. The mechanical work may be external (e.g. physical activities) or internal (e.g. heart beats, respiration, digestion, and brain activities). The energy used in internal work, finally, leaves the body in form of heat. The heat produced by the body is called metabolic heat.

The body is continuously attempting to maintain a heat balance with its surrounding environment in spite of wide variations in the environmental condition. To prevent the internal heat build-up, the body has to dissipate some of its metabolic heat. The body attempts to achieve thermal equilibrium with its surrounding environment through the following heat exchange methods: metabolic heat, evaporation, convection, conduction, and radiation. The heat exchange follows the second law of thermodynamics, according to which heat from a substance of a higher temperature is transferred to another object with a lower temperature. The following are the body's methods of heat exchange:

- Metabolic heat is the body's internal heat generated through metabolism as it burns foodstuffs to release energy.
- Evaporation is the method of body heat transmission to the environment by evaporation of perspiration and respiration.
- Convection is the method of body heat exchange with air. In general, the convection heat transfer is made by a gas (air) or fluid through movement of its molecules. The gas/fluid molecules are warmed up touching (conducting) warmer objects in the environment; and colder objects in the environment, in turn, absorb heat from the warmer gas/fluid molecules. The circulation of the gas/fluid molecules are naturally caused by differences in density within the gas/fluid. This process is called *natural* or *free* convection. Sometimes the circulation is made by use of mechanical devices (e.g. pumps and blowers) which is referred to as *forced* convection. When the air temperature is lower than the skin temperature, the body transmits heat to the air (called heat loss). If air temperature is higher than skin temperature, the heat transmission is reversed (called heat gain).
- Conduction is the method of heat exchange by direct contact of body with other objects. In this method, heat is transferred from one substance to another, or among molecules within a substance, without physical movement of the substance itself or its molecules. Since clothing insulates the body, the amount of heat exchange by conduction may become negligible. Thus, in the absence of conductive cooling, it can be ignored.
- Radiation: Generally, any substance with a temperature above absolute zero emits heat radiation in the form of electromagnetic waves (similar

to light waves). Thermal radiation will travel through vacuum or transparent media. It can be reflected and its intensity will vary inversely with the square of the distance from the source, in the same way as for light waves and other electromagnetic waves. The amount of heat radiated depends on the temperature and area of the radiating object, regardless of its mass. The radiation heat exchange between the body and other objects (e.g., walls, the sun, or furnaces) takes place without direct contact.

The process of heat exchange between the body and its surrounding environment can be expressed by Gagge's equation (Robinson, 1949):

$$\pm S = M \pm CV \pm CD \pm R - E \qquad (16.8)$$

Where:

S = heat storage (positive sign indicates heat gain, while negative indicates heat loss. If the heat balance is achieved, S = 0);

M = metabolic heat (always positive);

CV = convective heat (positive sign indicates air temperature is higher than skin temperature, and negative indicates a reversed case);

CD = conductive heat (positive when the contacting objects are warmer than skin, and negative when the skin is warmer);

R = radiant heat (positive when surrounding objects are warmer than skin, and negative when the skin is warmer); and

E = evaporative heat (always negative).

To be able to work safely and effectively in hot environments, the worker must maintain a body thermal equilibrium. The maintenance of this equilibrium depends upon the body's ability to dissipate metabolic heat. Under conditions of high ambient temperature and/or high radiant temperature, heat gain in body tissues can result. The problem arises when environmental temperature approaches skin temperature, and heat loss by means of convection and radiation gradually ends. At temperatures beyond skin temperature convection and radiation reverse directions and increase the heat content of body, and the only remaining means of heat loss is evaporation. In hot environments with high humidity, heat loss by evaporation also comes to an end.

(b) Heat stress and heat strain

Heat stress may be defined as the total load of all the heat factors, both internal and external, upon the body. Internal factors include metabolic heat, degree of acclimatization and body temperature. External factors include ambient air temperature, radiant heat, air velocity and humidity.

Heat strain is the sum of all biochemical, physiological, and psychological adjustments in the person made in response to the heat stress. The severity of heat strain depends upon the level of heat stress, which can range from a feeling of discomfort to a serious heat disorder.

When an individual is exposed to a hot environment, he/she will experience:

1. A vasodilation (expansion of blood capillaries near the skin surface) in the skin which facilitates increased heat transfer from core to shell to be removed by evaporation, and the skin becomes redder; and
2. An activation of the sweat glands (in the subcutaneous layer under the skin) which causes evaporative heat loss (Åstrand and Rodahl, 1977).

These are done by the thermoregulatory system in order to restore a new thermal balance between the body and environment. Therefore, the steady-state core temperature will shift to a higher level as the environmental temperature increases. At some level of heat load the limit of heat dissipation will be reached. Beyond this physiological limit the establishment of a steady-state for the core temperature will become impossible and it may reach dangerous levels. Furthermore, the physiological limit is usually reached before the body temperature rises much above 40°C (Poulton, 1970).

Heat stress not only may deteriorate work performance, but also can lead to a heat disorder. At lower levels of heat stress there is no health damage risk, even though an individual may feel discomfort (Dukes-Dobos, 1981). However, when the heat stress exceeds the person's heat tolerance capacity, adverse health effects will occur.

The heat disorders and their causes have been well described in detail in the literature (e.g. Lee, 1964; Leithead and Lind, 1964; Martinson, 1977). The most critical heat disorders are heat stroke, heat exhaustion, heat cramps and prickly heat. An explanation of each of these disorders and methods of prevention and/or treatment of them can be found in many articles on the heat stress subject (e.g. American Red Cross, 1981).

(c) Cold stress

Cold stress is usually not as critical a problem as heat stress, except for the work performed in refrigerators/freezers or outdoor in cold winter climates. However, its effects on exposed personnel should not be underestimated. As Malley (1992) has stated, under severe cold conditions, a major proportion of the person's time and energy is used in self-preservation which reduces his/her efficiency. Winter clothing, which is rather bulky, decreases the person's efficiency. Wearing mittens or warm gloves will significantly reduce the sense of touch and the person's dexterity which may lead to accidents.

Cold exposure causes vasoconstriction (contraction of capillaries near the skin surface) to direct warm blood to the body core in order to prevent excessive heat loss through the skin and preserve the heat for maintaining vital body functions (Ramsey, 1985), and the skin becomes pale. Extreme cold exposure can cause shivering, low body temperature, numbness, weakness and drowsiness. The adverse effects of cold stress have been categorized by MacFarlane (1963) in three types of injuries: chilblains, wet cold syndrome (also known as immersion foot and trench-foot), and frostbite.

In a cold environment, metabolic heat generation must be increased by performing high intensity work to compensate heat loss, and warmer clothing can be worn to help preserve the metabolic heat in the micro-environment around the body. Shivering increases heat, produced by metabolism process, for warmth. Solar radiation also can offset some of the heat loss. Warm areas should be provided for personnel in extremely cold environments to escape and recover from continuous cold exposure.

Windchill factor is widely used as a thermal index for cold environments. Windchill factor combines the effects of temperature and air movement, and expressed as the number of kilogram-calories or kilocalories heat in an hour (Kcal/hr) that the atmosphere can remove from the body (Malley, 1992). It may also be measured in terms of kilocalories per hour per square meter of body surface area (Kcal/hr/m^2) absorbed from the body (Parker and West, 1964), or as the equivalent temperature sensed by exposed flesh (nude skin).

(d) Protection against cold stress

- The worker should wear warm clothing and, especially, protect his/her neck, chest, groin area, hands, ears, nose and feet.
- The worker should wear a waterproof, wind-resisting outer layer of clothing, preferably with a layer of wool inside.
- The worker should wear a warm hat because heat loss through the head takes place very rapidly.
- The worker should have spare dry clothing because the body loses heat more rapidly when it is wet.
- The worker should not work alone in cold environments.
- The worker should consume high-energy food and snacks that provide more calories to produce more body heat.
- The worker should be refrained from consuming alcoholic beverages to prevent dehydration.
- The employer should provide a warm escape place for workers who are exposed to cold stress.
- The employer should provide nonalcoholic liquids such as soda, milk, water and juice for the worker to maintain his/her body's fluid balance.
- Workers who take medications that make them more vulnerable to cold should not be exposed to cold.

As environmental conditions change to become cold or hot, body temperature tends to change, following the same direction; that is, it decreases or increases with the environmental temperature. Under such conditions, feelings of discomfort arise and, consequently, reduction in productivity may result.

The slow pace of work which is usually noticed in hot climates is not due to laziness; it is rather due to physiological causes (Edholm, 1967). Fanger (1972) concluded that work capacity and performance seem to be substantially impaired in hot environments. Ramsey (1983) pointed out that factors which affect the physical capacity to do work in hot environments include physical fitness, general health, age, sex, status of nutrition, and also the mental willingness to perform under physical workloads and heat. He also stated that physical strength is minimally affected by increasing levels of temperature, but localized or general feelings of fatigue due to exposure to high heat intensity and/or long exposure will affect performance.

The risk of heat collapse or heat stroke increases rapidly at deep body temperatures above 39°C (Van Graan, 1975). WHO (1969) and ACGIH (1991) recommended that workers should not be permitted to continue work when their deep body temperature exceeds 38°C (100.4°F). As environmental temperature and metabolic work load increase, the body is less able to maintain its thermal equilibrium with the surrounding environment and deep body temperature may rise beyond the acceptable level.

In many situations, if a worker is to maintain thermo-neutrality, minimize heat gain and the risk of heat collapse or heat stroke and/or, at least, be able to work for a reasonable time period before reaching an allowable level of heat storage, protective devices must be provided. Focusing on a system which can provide some complete or partial body cooling to workers in hostile environments seems to be a promising solution to the problem.

16.6 REFERENCES

ACGIH (1991) *1991–1992 Threshold Limit Values for Chemical Substances and Physical Agents and Biological Exposure Indices*, American Conference of Governmental Industrial Hygienists, Cincinnati, OH 45211.

Alexander, D.C. (1986) *The Practice and Management of Industrial Ergonomics*, Prentice-Hall, Inc., Englewood Cliffs, NJ.

American Red Cross (1981) *Multimedia Standard First Aid Student Workbook*, The American National Red Cross.

ANSI Standards RP-7-1979 (1979) *American National Standard Practice for Industrial Lighting*, Illuminating Engineering Society, New York, NY.

Åstrand, P.O. and Rodahl, K. (1977) *Textbook of Work Physiology*, (2nd edn) McGraw-Hill Book Company, New York, NY.

Bailey, R.W. (1982) *Human Performance Engineering: A Guide for System Designers*, Prentice-Hall, Englewood Cliffs, NJ.

Dukes-Dobos, F.N. (1981) *Scand. J. Work Environ. Health*, **7**, 73–83.

Edholm, O.G. (1967) '*The Biology of Work*,' *World University Library*, McGraw-Hill Book Company, New York, NY.

Fanger, P.O. (1972) *Biometeorology*, **5**(II): 31–41.

Grandjean, E. (1988) *Fitting the Task to the Man: A Textbook of Occupational Ergonomics* (4th edn), Taylor & Francis, London.

Jürgens, H.W., Aune, I.A. and Pieper, U. (1990) *International Data on Anthropometry*, Occupational Safety and Health Series, No. 65, International Labour Office, Geneva.

Kvålseth, T.O. (1985) *Industrial Ergonomics: A practitioner's guide*, (D.C. Alexander and B.M. Pulat, eds.), Norcross, GA: Industrial Engineering and Management Press, Institute of Industrial Engineers, pp. 57–76.

Lee, D.H.K. (1964) in *Handbook of Physiology*, (D.B. Dill, ed.), Ch. 35, American Physiology Society, Washington, DC.

Leithead, C.S. and Lind, A.R. (1964) *Heat Stress and Heat Disorders*, Cassell and Co., Ltd., London.

MacFarlane, W.V. (1963) in *Medical Biometeorology*, (S.W. Troup, ed.) pp. 372–417, Elsevier, New York, NY.

Malley, C.B. (1992) *Professional Safety*, **37**(1): 21–23.

Martinson, M.J. (1977) *Journal of Occupational Accidents*, **1**: 171–93.

NASA (1978) *Anthropometric Source Book, Volume II: A Handbook of Anthropometric Data*, NASA Reference Publication 1024, National Aeronautics and Space Administration, Scientific and Technical Information Office, Springfield, VA.

Nevins, J.L. and Whitney, D.E. (1991) in *Computer Integrated Design and Manufacturing*, (D.D. Bedworth, M.R. Henderson and P.M. Wolfe), Ch. 4, McGraw-Hill, Inc., New York, NY, 1991.

NIOSH (1973) *The Industrial Environment – its Evaluation and Control*, National Institute for Occupational Safety and Health (NIOSH), US Government Printing Office, Washington DC.

NIOSH (1981) *A Work Practice Guide for Manual Lifting*, Technical Report No. 81-122, National Institute for Occupational Safety and Health (NIOSH), US Department of Health and Human Services, Cincinnati, OH.

Parker, J.F. and West, V.R. (1964) *Bioastronautics Data Book*, US Government Printing Office, Washington, DC, p. 123.

Poulton, E.C. (1970) *Environment and Human Efficiency*, C.C. Thomas, Springfield, IL.

Ramsey, J.D. (1983) in *Stress and Fatigue in Human Performance*, (G.R.J. Hockey, ed.), John Wiley & Sons Ltd., New York, NY.

Ramsey, J.D. (1985) in *Industrial Ergonomics: A Practitioner's Guide*, (D.C. Alexander and B.M. Pulat, eds.), pp. 85–94, Industrial Engineering & Management Press, Institute of Industrial Engineers, Norcross, GA.

Robinson, S. (1949) in *Regulation and the Science of Clothing*, (L.H. Newburgh, ed.), W.B. Saunders Company, Philadelphia, PA.

Roebuck, J.A., Kroemer, K.H.E. and Thomas, W.G. (1975) *Engineering Anthropometry Methods*, John Wiley & Sons, New York, NY.

Stephenson, R.R., Colwell, M.O. and Dinman, B.D. (1974) *Journal of Occupational Medicine*, **16**(12): 792–5.

Van Cott, H.P. and Kinkade, R.G. (eds.) (1972) *Human Engineering Guide to Equipment Design*, (Revised Edition), US Government Printing Office, Washington, DC.

Van Graan, C. H. (1975) *Proceedings of International Mine Ventilation Congress*, Johannesburg, South Africa, September.

WHO (1969) *Health Factors Involved in Working Under Conditions of Heat Stress*, World Health Organization Technical Report Series No. 412, World Health Organization, Geneva.

Woodson, W.E. (1956) *Human Engineering Guide for Equipment Designers*, University of California Press, Berkeley, CA.

PART THREE

Cost Considerations in Concurrent Engineering

CHAPTER 17

Designing to cost

Mahendra S. Hundal

17.1 NOMENCLATURE

b	Width
c	Estimated cost
C	Cost
C_{LBD}	Direct labor cost
C_{MFT}	Total manufacturing cost
C_{MFV}	Variable manufacturing costs
C_{MTL}	Total material cost
C_{MTD}	Material direct cost
C_{MTO}	Material overhead cost
C_{PR}	Production cost
C_s	Cost per unit strength
C_{SET}	Set-up cost
C_v	Cost per unit volume
C_w	Cost per unit weight
d	Diameter
e	Relative error
E	Actual error
H_B	Brinnel hardness number
n	Lot size; number of items
S_u	Tensile strength
W	Weight
ρ	Density
ϕ	Proportionality parameter

17.2 INTRODUCTION

Product costs need to be recognized early, i.e. during designing, where they can be controlled the most. However, as a rule, for a new product the cost information is not available at this point. In fact, costs are influenced to the largest extent in the early stages of design, namely, the conceptual and embodiment stages. It has been estimated (Smith, 1988) that 70–80% of the product costs have been committed after only a small portion of the development resources have been expended in preliminary design. Figure 17.1 shows the data on costs incurred in individual activities and the potential for lowering product costs in each (Ehrlenspiel, 1985). For example, design and development account for only 7% of the product cost, but this phase can be responsible for 65% of the potential decrease in costs e.g., through value analysis.

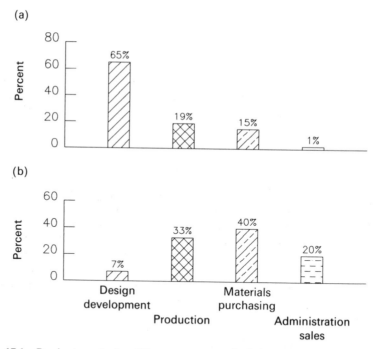

Fig. 17.1 Product costs in different stages and their reduction potential: (a) potential for reducing product costs; (b) costs in individual activities.

The designer often does not have sufficient knowledge of the manufacturing processes and the costs incurred therein. Thus a coordination between the design and production planning departments is needed. The crux of designing for cost is that the designer be able to calculate not only the physical properties, e.g., strength, wear, etc., but simultaneously, also the costs. Figure 17.2 illustrates this idea. In the traditional procedure, design

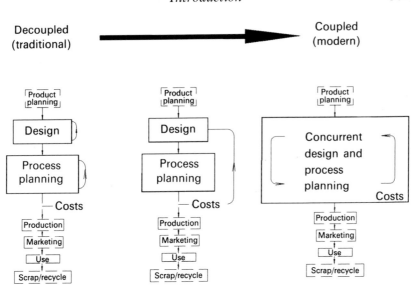

Decoupled (traditional) → Coupled (modern)

Fig. 17.2 Approaches to design and process planning.

and production planning are carried out sequentially and independently. The costs are calculated after the fact. In the ideal situation these steps are integrated; the design and production processes are finalized and the costs determined simultaneously.

17.2.1 Literature review

While large amounts of literature on cost estimating are available, the first definitive work relating to costs in mechanical design was published (in Germany) by Professor Ehrlenspiel in 1985. The book provides guidelines and rules for lowering product costs, and methods for estimating these during the design process. Extensive reference to German language literature on the subject is provided in the book. Other recent publications include the work by Ferreirinha (1990) who describes the operation of a PC-based program for calculating cost of machine parts. The interactive program requires as input:

1. Part features; number of primitives, dimensions with tolerances, surface quality, heat/surface treatment, quality features and joining methods;
2. Blank features; material, production process, shape and dimensions and pre-treatments; and
3. Production features; lot size, number of fixturings.

Ott and Hubka (1985) describe a method for calculating manufacturing cost of weldments based on weld dimensions. They calculate time requirements for each welding operation. A management perspective of designing to cost, oriented toward the defense industry is given in a recent book by Michaels and Wood (1989). Other noteworthy sources for cost models are Pugh (1977), Mahmoud and Pugh (1979), Dewhurst and Boothyroyd (1987), and Bevan, *et al.* (1989). A discussion of practices in manufacturing, construction, as well as chemical, electronic and mechanical industries is given by Sheldon, *et al.* (1991). Keys, *et al.* (1987) discuss electronic manufacturing process system cost modeling and simulation tools.

17.2.2 The design process

Before we consider means of reducing product costs during designing, we look first at the steps of the design process itself. Systematic design (Pahl and Beitz, 1988; Hundal, 1990) begins with presentation of the task, goes through the steps of conceptual, embodiment and detail design, and concludes when complete design documentation has been produced. The process is shown in Table 17.1. The first major document is the requirements (specifications) list which describes in precise terms what the product should be able to do, what properties it should and should not have, without pointing to a solution. The requirements pertain to the different life phases of the product, i.e. planning, design, manufacturing, marketing and use, and are of different types, i.e. technical, economic, ergonomic, legal, etc. Conceptual design begins with casting the problem statement in solution-neutral terms by removing all constraints which are not relevant to the product functions. This is the most abstract level of the design. Then the function

Table 17.1 Technical features and properties determined during designing

Step	Outputs (features; properties)	
Task clarification		
	Requirements list	
Concept		
	Functions	Function structures
	Physical effects	Physical structures
	Solution principles	Solution structures
Embodiment		
	Parts	
	Geometry	Shape structures
	Material	Assemblies
	Joining	Arrangement
	Surfaces	
	Standard parts	
	Part production	
	Assembly	
Detail		
	Final design, production, assembly, operation documents	

structure is derived, physics is considered and solution principles are ascertained, leading to the final concept. Development of the function structure is essentially a subdivision of the problem into subproblems (subfunctions). Embodiment design leads the process through a more concrete stage as the shapes, materials and motions (flows) are determined. During this stage it is no longer possible to change the functions and physical effects. Part shapes and sizes, fastening methods, materials and processes are in their final forms.

17.3 METHODS FOR REDUCING MANUFACTURING COSTS

While product costs are influenced at all stages of its life – from design order to sales – the most important factors are:

- The concept, including physical effects, material type, number and type of active surfaces;
- Size of the product, i.e. dimensions and amount of material; and
- Number of parts, including standard, similar and same parts.

These will be discussed next.

17.3.1 Influence at concept stage

During planning and conceptual design the methods for task clarification and recognizing the focal points for lowering product costs are the most important. These are market studies, economic analyses, ABC methods (Ehrlenspiel, 1985), and cost structures, some of which are explained later in the chapter. At the design order stage it is advantageous to keep the number of demands low and exclude any unnecessary demands in the requirements list. After this, it is the concept which has the single largest influence on costs. The concept determines the arrangement of elements through the function structure and thus the flows of energy, material and signals in the system. The physical effects determine how subfunctions are realized. The solution principles determine how complex the product will be. Major advances in technology are the results of new concepts rather than improvements in embodiment. Examples are: internal combustion engine replacing steam engine; ball point pen replacing fountain pen; fiber-optic cable replacing copper wire.

Methods for lowering product costs during concept development are:

- Set cost goal;
- Analyze similar products according to types of costs;
- Determine areas for influencing costs with cost structures, ABC analyses;
- Decide on what can or cannot be changed;
- Use fewer subfunctions and those easily realized;

- Use simple function structures, robust physical effects;
- Aim for function integration;
- Use same type of energy; and
- Use simple motions.

17.3.2 Influence at embodiment stage

Next to the concept, embodiment has the greatest influence on product costs. The thrust of embodiment (and detail design) is toward the selection of materials, shapes, manufacturing processes and systems. A number of individual topics are involved here: production quantities, sizes, layout, production process and assembly. Standardization affects product costs by use of standard parts, in the design of modular products and manufacture of a product in a series of size ranges.

Steps that can be taken for lowering product costs during embodiment are:

- Analyze similar products according to manufacturing costs, life cycle costs;
- Ascertain areas where costs may be reduced; what can or cannot be changed;
- Develop knowledge of costs (e.g. company data base);
- Consider alternative concepts for secondary functions;
- Consider alternative embodiment (shapes, materials, motions);
- Use fewer parts, similar parts, standard parts;
- Consider modular design, design in size ranges;
- Use standard material, more cost-effective material; and
- Consider production in own shop *v.* out-sourcing.

17.3.3 Influence of part size and production quantities

Models for calculating various types of costs and the influences of different parameters are available in the literature cited above. The components of product costs, as used in this chapter, are shown in Fig. 17.3. The total manufacturing costs can be classified in several ways. For example they may be divided into material and production costs, as shown. They may also be classified as direct costs, e.g. material, labor, which can be assigned to the product, and overhead costs such as heating and lighting of shops, which cannot be directly assigned. Further, the costs may be divided into variable costs which consist of direct costs and variable overhead costs, and fixed costs which remain constant over a period of time. It is the variable costs which can be influenced the most at the design stage. The variable manufacturing costs are given by

$$C_{MFV} = C_{MTD} + C_{LBD} \qquad (17.1)$$

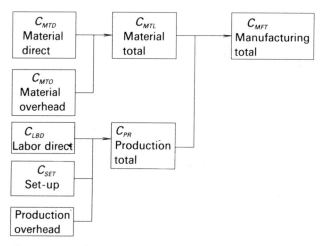

Fig. 17.3 Components of product costs.

Part size influences the set-up, material and production costs. Production in large quantities leads to reduction in unit cost. The fixed (initial) costs are divided up into two categories:

1. Start-up costs for design, development, prototype production, etc. (these relate to the total production quantity);
2. One-time costs arising for the production of a lot size n.

The basic equation for calculating the manufacturing costs per unit for lot size n, if the initial costs consist only of set-up costs, is

$$C_{MFT} = \frac{C_{SET}}{n} + C_{PR} + C_{MTL} \tag{17.2}$$

Part geometry affects the cost according to its linear dimension, surface area, volume and complexity. This is elaborated upon later by an example.

17.3.4 Reducing material costs

Material costs constitute 40–50% of the total costs in the general machine industry. They are higher for mass-produced items, where rationalization has led to the lowering of other costs and lower for one-of-a-kind products. Material costs include material direct costs and overhead costs. Direct costs are the product of material volume and the cost per unit volume. Methods are available for reducing each of these during embodiment.

Material volume can be reduced by applying the basic principle of form design: aim for uniform stress. Two corollaries of this are: avoid using material where it serves no purpose, and avoid bending and torsion. In the latter cases specially shaped parts are needed to achieve uniform stress. Figure 17.4 shows at (a) a conventional bell-crank lever in which the arms

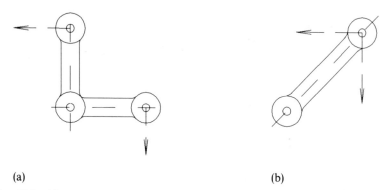

(a)   (b)

Fig. 17.4 Alternate designs for lever: (a) conventional design; (b) uniform stress.

are subjected to bending. At (b) the simpler design achieves uniform stress by causing axial force and thus using less material and/or a simpler cross section.

Other measures for reducing volume are:

- Use parallel paths for flow of energy (example: planetary and other gear trains);
- Use safety devices (examples: over-running clutch, relief valve) so the device does not have to be designed for high loading which occurs only occasionally;
- Use higher speeds for power transmission, thus reducing torque and consequently the amount of material required;
- Use higher strength materials (cost per unit volume must be kept in mind); and
- Reduce scrap generation.

Figure 17.5 shows single and double stage gear trains which provide the same overall speed reduction. The double stage trains are shown with one, two and three parallel paths. For larger powers transmitted, multiple path designs yield lower weight and manufacturing cost per unit torque (Ehrelspiel and Fisher, 1982/1983).

After material volume reduction the next steps to consider are those which reduce the cost per unit volume. This is achieved by using materials produced in large quantities, rather than speciality materials. A convenient means of comparing materials is to generate values of relative costs, using a common material such as mild steel as the basis. Special lists of this type can be useful, such as for simple cast parts of different materials.

Table 17.2 shows such a list for some materials if selection might be based on strength. Columns (1) and (2) show the costs per unit weight and volume, normalized to those of 1020 steel. The next four columns show properties: density, hardness, ultimate and yield strengths. Column (6) shows costs per unit tensile strength $\rho C_w / S_u$, which are normalized in column (7). In place of tensile strength, other properties could be used, e.g. yield or fatigue

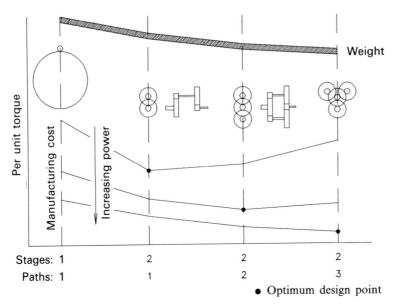

Fig. 17.5 Cost and weight of gear trains as a function of complexity.

strength. Dieter (1991) shows formulas for other loading conditions. (Cost figures are for illustration only. Companies should use their own procurement data.)

A further means of reducing material costs is to reduce the material overhead costs, which can amount to 5–20% of the material costs. The means of achieving this is to use fewer types of material and fewer different purchased parts. This reduces storage requirements, handling costs and administrative expenses. Figure 17.6 shows that a manufacturer was able to reduce the total number of rolled stock sections from 136 to 65 through such rationalization.

17.3.5 Improvement of a product by redesign

When a new product is put into service all of its parts are, as a rule, not optimally designed. They may need to be strengthened as failures occur. However, parts which may be too costly to begin with, due to over-design, do not show up in this manner. Figure 17.7 shows the progress of costs in a vacuum cleaner brush assembly as individual parts are redesigned. Beginning with the initial design, the motor exhibited failures and was replaced with a larger motor. This increased the total cost by 10%. Then the belt drive had to be redesigned, increasing the cost by another 10%. This 20% increase in the total cost could be lowered to 10% by reducing the size of other components, which were initially over-designed.

Table 17.2 Relative costs of materials

	C_w^* Cost per unit weight (1)	C_v^* Cost per unit volume (2)	ρ Specific gravity (3)	H_B Brinnel hardness (4)	S_u, kpsi Tensile strength (5)	C_s Cost per unit strength (6)	C_s^* Cost per unit strength (7)
Rolled sections							
1020 Steel	1	1	7.85	110	54	0.15	1
Cr-V Spring Steel	1.7	1.7	7.85	380	188	0.07	0.47
Al Alloy	7.9	2.8	2.8	110	62	0.35	2.43
Castings–simple shapes							
Cast Iron	1.5	1.4	7.25	185	26	0.42	2.85
Cast Steel	2	2	7.8	135	65	0.24	1.64
Al Alloy	2.7	2.25	2.65	55	25	0.28	1.95
Plastic sheet							
PVC	3.3	0.58	1.38	11.5	8	0.57	3.9
Polystyrene	4.3	0.57	1.05	15	11	0.42	2.86

* normalized to 1020 Steel.

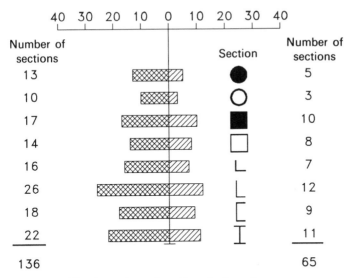

Before – standardization – after

Fig. 17.6 Reducing material overhead costs.

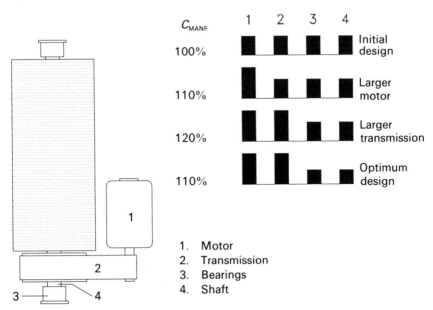

Fig. 17.7 Optimizing a design: vacuum cleaner power brush.

17.3.6 Choice of production process

The production process used for a part depends upon several parameters of embodiment design and is often the factor least understood by the designer. This is, of course, a vast subject and will be treated here only in brief. In a recent article Pighini, *et al.* (1989), describe a methodical procedure for design for manufacture. The choice of the production process is most closely related to the shape and the material(s). The requirements which are important in this regard are:

1. Function of the part;
2. Production quantities;
3. Physical properties;
4. Available facilities; and
5. Cost.

Processability of materials, i.e. forgeability, castability, machinability, weldability, etc., plays an essential role in production costs. Generally lower strength materials are easier to shape. For this purpose relative process-ability data must be used. It can be incorporated as an additional column in Table 17.2 (Pahl and Beitz, 1988). At the design stage it is essential to have cost figures for typical (simple) parts as functions of process, material, production quantity, etc. Often parts must possess mutually exclusive properties, e.g., toughness and wear resistance, strength and flexibility, etc. In such cases it can be more appropriate to subdivide the functions by making the part of separate materials, and/or geometry, or using specific processing techniques such as surface treatment.

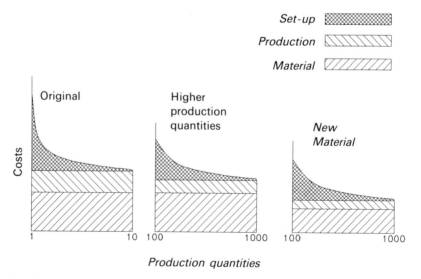

Fig. 17.8 Cost reduction by improved process and material.

For different materials and production processes there are limiting values of production quantities, above which costs per unit exhibit diminishing reduction. Figure 17.8 shows that, starting with the original material and process, an improved process lowers the costs for higher production rates. These can be further reduced by going to a more suitable material.

17.4 AIDS IN DESIGNING FOR COST

A cost goal for the product must be set by the company and provided to the design group. The designer then sets up cost structures according to parts or other criteria, based upon similar products. Only if cost calculation is carried out in parallel with design steps is there a reasonable chance to hold to the goal. This is illustrated later by an example. In case of products for which no cost guidelines are available a qualitative approach can be used in making decisions. By using relative costs for material, production process, assembly, etc., the optimum solution can be chosen.

17.4.1 Types of analyses

A cost structure shows the breakdown of the product cost according to one of several criteria: parts, type of cost, functions, production process, etc. Such a structure brings into focus the proportion of cost contribution from the various viewpoints. Thus a designer looking for reducing costs can address the areas which offer the greatest possibilities and not worry about minor points. One advantage of cost structures is that they remain fairly constant over time, unless a major design change takes place. As an example, the cost structures for a centrifugal pump are shown in Table 17.3, according to parts, types of cost and functions.

In the ABC analysis, as the term is defined by Ehrlenspiel (1985) (not to be confused with activity-based costing), parts of a product can be divided into three categories, according to some property, e.g. weight, cost. Parts in class A have the highest attributes, in class C the lowest; parts in class B lie in the middle. In Table 17.3 a classification according to manufacturing costs is indicated. ABC analysis highlights the more important aspects of the design according to the chosen category. The designer can ascertain where costs can be reduced the most, e.g., by lower material cost, combining functions, etc. Cost breakdown according to function is useful for several reasons:

- Comparing design variants according to cost per subfunction (more abstract than parts, possible application at concept stage);
- Comparing cost/benefit ratio of functions; and
- Combining subsolutions to achieve lower function costs.

The right-most column of the upper table in Table 17.3 shows how the functions of each part are divided. In the lower table costs allocated to each

Table 17.3 Cost structures for centrifugal pump

ABC class	Part	Manufacturing cost $	%	Types of cost (%) Material	Production	Assembly	Function allocation (%)
A	1. Housing	5,500	45	65	25	10	F1: 50%; F6: 50%
	2. Impeller	4,500	37	55	35	10	F2: 30%; F3: 70%
B	1. Shaft	850	7	45	45	10	F2: 60%; F6: 40%
	2. Bearings	600	5		Purchased		F6: 100%
	3. Seals	500	4		Purchased		F1: 100%
	4. Wear rings	180	2	35	45	20	F5: 100%
C	1. Key	15	<1	30	50	20	F2: 80%; F4: 20%
	2. Bolts	50	<1		Purchased		F4: 100%
	3. Gasket	10	<1		Purchased		F1: 100%
	4. Oiler	20	<1		Purchased		F5: 100%

Function	Cost contribution	Function cost $	%
F1: Contain liquid	$0.5 \times 5500 + 1.0 \times 500 + 1.0 \times 10$	3260	26.6
F2: Transfer energy	$0.3 \times 4500 + 0.6 \times 850 + 0.8 \times 15$	1872	15.3
F3: Convert energy	0.7×4500	3150	25.7
F4: Connect parts	$0.2 \times 15 + 1.0 \times 50$	53	0.4
F5: Increase life	$1.0 \times 180 + 1.0 \times 20$	200	1.6
F6: Support parts	$0.5 \times 5500 + 0.4 \times 850 + 1.0 \times 600$	3960	30.1

of the six functions are calculated. As shown by Fowler (1990) function analysis and cost allocation forms the basis of value analysis, and is used as an aid to increase the value (= 'worth'/cost) of a product. Extensive work in function costing has been reported by French (1990).

17.4.2 Aids in design selection

An important step in designing is the selection of one solution from a number of alternatives. This occurs both during concept development and embodiment. The selection takes place under technical as well as economic constraints.

The designer can call on several means to help in this decision-making. The most often used method is the use of so-called 'design rules'. These rules are based upon experience and generally not quantified or expressible in algorithmic form. Examples of design rules are: 'use few part types', 'avoid sudden cross section changes', in general, the 'dos' and 'don'ts' of design. The rules can sometimes contradict each other, in which case further analysis is required for decision. Rules relate certain parameters to properties of interest. They form the basis of expert systems. Use of design rules helps in decision-making, but can create a mind-set and preclude innovative solutions – just the opposite of what one strives for in design. A more definitive comparison of embodiments can be made on the basis of 'relative costs'. The simplest example is the relative cost of materials, given earlier. However, objects can also be compared on the basis of function (e.g. fastenings, couplings, bearings) or production (e.g. machined entities, production processes). Relative costs allow quick and rough calculation and do not change much over time (as opposed to absolute costs).

17.5 QUICK COST ESTIMATION

In obtaining estimates of costs concurrently with the design process there is a trade-off between the time involved in costing and the accuracy of the costs. Although in the earlier stages of design the cost estimation can only be approximate, decisions made in its absence can be costly. Designing for a cost goal is not possible unless such calculations are carried out simultaneously. In this respect the cost of class A parts needs to be most accurately estimated, and of class C the least. Quick cost estimation (QCE) refers to a simple method based on a defined cost entity which is at the designer's disposal at the given design stage. Among the many methods for cost estimating during designing the most noteworthy are those based on: weight, material, throughput parameters, physical relationships, regression analysis and similarity laws. Regression analysis can be used when, through statistical analysis, costs can be expressed as a function of the various parameters. Similarity laws, on the other hand are based upon physical relationships and expressed in the form of power laws. The exponents are derived from the equations for material costs, machining times, etc. The coefficients of the terms are found on the basis of some basic element, e.g. a part of nominal size, weight, etc. These methods are discussed below in brief.

17.5.1 QCE based on weight and material

The production costs per unit weight of similar products are proportional. At the low end are simple, heavy machines such as pumps and agricultural machines, at $5/lb. Complicated, high technology products can range upwards of $100/lb. This method allows cost estimation for products for which material costs predominate and only minor changes in the product are involved.

17.5.2 QCE based on physical relationships

Costs can be estimated, based upon physical laws applied in the design analysis of a product. For instance, stress equations can relate load for a structural member or pressure in a pressure vessel to the amount of material. Analyses can lead not only to costs but also their optimization. As a simple example, consider a pipe of constant thickness and of diameter d. Its annual cost is the sum of (a) operating (pumping) costs which are proportional to $1/d^4$, and (b) investment cost proportional to d. The total cost is

$$C = C_1/d^4 + C_2 d \qquad (17.3a)$$

which leads to an optimum diameter

$$d_{opt} = (4C_1/C_2)^{0.2} \qquad (17.3b)$$

This is shown in Figure 17.9.

17.5.3 QCE based on regression analysis

The dependence of costs on product characteristics such as throughput, sizes, weight, etc., can be found through statistical analyses. These analyses can be extensive and computationally intensive. Regression analysis shows this dependence in the form of coefficients and exponents in the regression equation. As an example, Pacyna, *et al.* (1982) have shown that for simple gray iron castings the cost per unit can be approximated by:

$$C_{MTD} = K_p K_c n^{-0.0782} v_s^{0.8179} r_e^{-0.1124} r_t^{0.1655} r_v^{0.1786} n_c^{0.0387} S_u^{0.2301} \qquad (17.4)$$

where,

K_p = a proportionality factor, $/unit
K_c = complexity factor (0.9 to 1.4)
v_s = volume of casting material

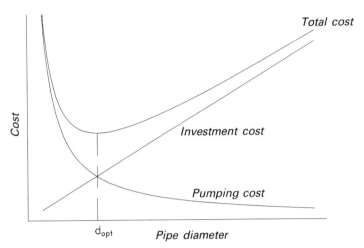

Fig. 17.9 Cost of pipe v. diameter.

r_e = elongation ratio = d_c/d_s
r_t = wall thinness ratio = s_s/f_c
r_v = volume ratio = v_c/v_s
n_c = number of cores

The terms needed for defining the ratios r_e, r_t and r_v are based upon considering a solid of similar overall shape and same volume as the net volume of the casting, as shown in Fig. 17.10.

17.5.4 QCE based on similarity principles

Similarity principles are used extensively in predicting costs of products made in size ranges. If a design has been built in one size, then the cost of designs in other sizes can be predicted by this method more accurately than using weight or material alone. The premise is that the linear dimension ratio of two products, $\phi_l = l_2/l_1$, is the primary parameter in relating their costs. For linear size ratio ϕ_l of two parts, their area ratio is ϕ_l^2 and the volume ratio, ϕ_l^3.

The product costs can be broken down into components which change with ϕ_l^i, $i = 0, 1, 2, 3$, i.e. fixed costs, those varying with linear size, with area and volume, respectively. Thus the manufacturing cost ratio, relating the cost of product 2 to the cost of product 1, is given by:

$$\frac{C_{\text{MFT2}}}{C_{\text{MFT1}}} = \phi_{\text{MFT}} = \sum_{i=0}^{3} a_i \phi_l^i \tag{17.5}$$

The parameters a_i can in general be calculated by using regression analysis. The exponents may have non-integer values as shown below.

In machining operations the similarity parameters for operation, idle and set-up times have been found to be of the form $\phi_t = \phi_l^n$. Set-up costs for

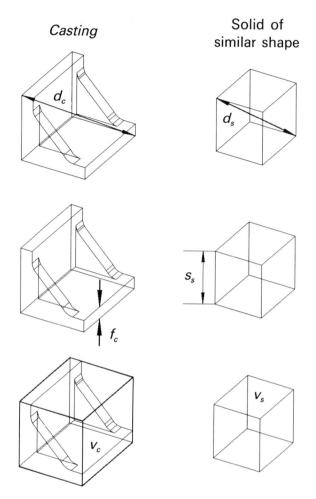

Fig. 17.10   Casting geometry.

machined parts have been found to be proportional to ϕ_i^s, where s (for gears) varies from 0.14 for 2–8 inch diameters, 0.56 for 8–40 inch diameters, to 1.8 for 40–60 inch diameters (Ehrlenspiel, 1985). Set-up cost per unit for product 2 in terms of cost for product 1 is given by:

$$C_{SET2} = C_{SET1}\, \phi_i^s \qquad (17.6)$$

The production costs for machined parts are affected by actual cutting time, idle time, etc., and are found to be proportional to the surface area (approximately to ϕ_i^p), i.e.:

$$C_{PR2} = C_{PR1}\, \phi_i^p \qquad (17.7)$$

where the exponent p varies from 1.8 to 2.2. Likewise, the material costs for similar parts are proportional to volume and have been found to vary as:

$$C_{MTL2} = C_{MTL1} \, \phi_l^m \tag{17.8}$$

where the exponent m varies from 2.4 for small gears to 3 for large sizes.

A combination of above equations yields the combined effects of product size and production rates. The manufacturing costs per unit for product 2 in terms of costs for product 1 is given by

$$C_{MFT2} = \frac{C_{SET1}}{n} \, \phi_l^s + C_{PR1} \, \phi_l^p + C_{MTL1} \, \phi_l^m \tag{17.9}$$

where the nominal values of the exponents are: $s = 0.5$, $p = 2$ and $m = 3$. This is a more generalized and useful form of equation (17.2).

(a) Example

As an example, consider a machined part such as a gear or gear coupling. A company has produced the following size part:

Diameter, $d_1 = 8$ in
Weight, $W_1 = 40$ lb
Manufacturing cost per unit, $C_{MFT1} = \$210$, made up of
    Set-up cost, $C_{SET1} = \$110$,
    Production cost per unit, $C_{PR1} = \$70$, and
    Material cost per unit, $C_{MTL1} = \$30$.

It is required to estimate the costs for a similar part of diameter, $d_2 = 32$ in. We find that the linear size ratio, $\phi_l = d_2/d_1 = 4$. Therefore, $\phi_l^{0.5} = 2$, $\phi_l^2 = 16$, and $\phi_l^3 = 64$.
For the new part:

Weight, W_2	$W_1 \phi_l^3$	40×64	2560 lb
Set-up cost, C_{SET2}	$(C_{SET1})(\phi_l^{0.5})$	110×2	\$220
Production cost, C_{PR2}	$(C_{PR1})(\phi_l^2)$	70×16	\$1,120
Material cost, C_{MTL2}	$(C_{MTL1})(\phi_l^3)$	30×64	\$1,920

Manufacturing costs per unit for the new part are $220 + 1,120 + 1,920 = \$3,260$, if produced in single quantities. Thus for the larger part, material costs dominate.

For cylindrical objects in general, e.g. gears, drums, bearings, etc., the ratio of diameter d and width b affect the costs. For spur gears the torque ratio ϕ_t is related to the diameter ratio ϕ_d and the width ratio ϕ_b, and thus to the linear size ratio ϕ_l by

$$\phi_t = \phi_d^2 \, \phi_b = \phi_l^3 \tag{17.10}$$

17.5.5 Accuracy and errors in QCE

The total estimated costs of a unit (part, assembly, etc.) arise by summation of many separate items: costs of individual parts, production steps, material costs, etc. If the individual estimated costs c_i have errors E_i which are uniformly distributed about the true cost C, the total estimated cost $c_{tot} = \sum_{i=1}^{n} c_i$ will exhibit a smaller total error E_{tot} than do the individual costs c_i. This comes about due to the partial cancellation of positive and negative errors.

Kiewert (1982) has shown that the total relative error for a unit is given by:

$$e_{tot} = \left[\sum_{i=1}^{n} (c_i e_i^2)/c_{tot}^2 \right]^{1/2} \tag{17.11}$$

where $e_i = E_i/c_i$ is the relative error in c_i. If the individual costs are equal: $c_1 = c_2 = \cdots = c_n = c$, and have the same relative error: $e_1 = e_2 = \cdots = e_n = e$, the equation (17.11) simplifies to:

$$e_{tot} = e/n^{1/2} \tag{17.12}$$

i.e., the total error is $1/n^{1/2}$ of the individual error.

In order to determine the accuracy e_{al_i} to which the individual cost c_i must be estimated such that the total cost c_{tot} has the error e_{tot}, Kiewert (1982) gives the equation (all quantities are percentages)

$$e_{al\_i} = 10e_{tot}/c_i^{1/2} \tag{17.13}$$

as shown in Fig. 17.11. Thus if the allowable error in the total cost is to be $e_{tot} = 10\%$, a cost which is $c_i = 25\%$ of the total cost, it must be estimated to

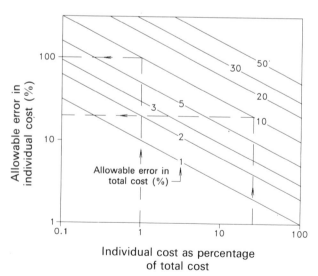

Fig. 17.11   Allowable errors in cost estimation.

within $e_{al\ i} = \pm 20\%$. For a cost which is 1% of the total, the allowable error for it is $\pm 100\%$.

17.6 DESIGNING TO A COST GOAL

As an example of designing to a preset cost goal we consider a centrifugal pump similar to the one cited above. The total manufacturing cost is to be $10,000, within $\pm 5\%$. The first step is to set the cost goals of parts by using cost structures of similar pumps. The calculations are shown in Table 17.4 where the part numbers and classification of Table 17.3 have been used. All parts in category C have been lumped together. On the basis of the percentage of total cost of each part (column 3), the maximum allowable deviation is found from Fig. 17.11 and shown in column 4. Column 5 shows the maximum cost of each part. By using cost models and methods described previously, the cost of each part in the initial design is calculated, shown in column 6. At this stage the cost of more expensive parts (category A) needs to be calculated more accurately, whereas category C parts require only a rough estimation. Comparing the figures in columns 5 and 6 we see that parts A2 and B1 exceed the limits and must therefore be redesigned. After redesign and calculation the final cost estimates are shown in column 8. The total cost of the redesigned pump is $10,070 which is close to the preset goal.

17.7 CONCLUSION

The chapter has addressed the following points:

- Importance of calculating product costs during the design process;
- Cost models and methods of estimating costs; and
- Methods for reducing manufacturing costs.

In conclusion, the following basic rules can be given for reducing product costs:

- Produce in large quantities – this results not only in an economy of scale, but also that more optimum manufacturing processes can be used;
- Reduce complexity – use fewer parts and production operations;
- Reduce size – reduces material volume; and
- Ask for only the minimum accuracy and tolerances.

The systematic design method can enable the designer to recognize product costs early and to realize the design which meets a preset cost goal for quotation, prototype development or production. A coordination between the design and manufacturing departments is needed to bring this about. The conceptual and embodiment design are the most important

Table 17.4 Calculations for designing to a preset cost goal

Cost goal: $10,000(c_{tot}) \pm 5\%(e_{tot})$; Maximum cost: $10,500

Part (1)	Part cost goal, $ (2)	% total cost c_i (3)	% allowable deviation e_{al-i} (4)	Maximum cost, $ (5)	1st design cost, $ (6)	Decision (7)	Redisigned cost, $ (8)
A1	4,500	45	7	4,835	4,400	acceptable	4,400
A2	3,500	35	8	3,796	4,000	redesign	3,600
B1	700	7	19	832	900	redesign	750
B2	600	6	20	722	600	acceptable	600
B3	400	4	25	500	400	acceptable	400
B4	200	2	35	271	220	acceptable	220
ΣC	100	1	50	150	100	acceptable	100
Total	10,000	100			10,630		10,070

phases in which the product costs can be influenced. Information provided here should aid in the development of rules and knowledge bases of an intelligent CAD system which is able to integrate design, manufacturability and cost estimating, and is thus able to produce low cost designs.

17.8 REFERENCES

Bevan, N., Jebb, A. and Wynn, H.P. (1989) *Proc. ICED 89*, 415–24, London: Inst. Mech. Engrs.

Dewhurst, P. and Boothroyd, G. (1987) *J. Manufacturing Systems*, **7**(3): 183–91.

Dieter, G.E. (1991) *Engineering Design* (2nd end), New York: McGraw-Hill.

Ehrlenspiel, K. and Fischer, D. (1982 and 1983) 'Relative costs of spur gears in single and short production runs,' Final report, parts I and II, Frankfurt: FVA.

Ehrlenspiel, K. (1985) *Design for Cost*, Berlin, New York: Springer Verlag (in German: *Kostengünstig Konstruieren*).

Ferreirinha, P. (1990) *Proc. ICED 90*, **3**: 1346–53, Zurich: Heurista.

Fowler, T.C. (1990) *Value Analysis in Design*, New York: Van Nostrand Reinhold.

French, M. (1990) *J. of Engineering Design*, **1**(1): 47–53.

Hundal, M.S. (1990) *Mechanisms and Machine Theory*, **25**(3): 243–56.

Keys, L., Balmar, J.L. and Creswell, R.A. (1987) *IEEE Trans. Component Hybrids and Manufacturing Technology*, **CHMT-10**(3): 401–10.

Kiewert, A. (1982) *VDI-Z*, **124**: 443–6.

Mahmoud, M. and Pugh, S. (1979) *Proc. Information for Designers Conf.*: 37–42, Southampton: University Press.

Michaels, J.V., and Wood, W.P. (1989) *Design to Cost*, New York: J. Wiley & Sons.

Ott, H.H. and Hubka, V. (1985) *Proc. ICED 85*, **1**: 478–87, Zurich: Heurista.

Pacyna, H., Hillebrand, A. and Rutz, A. (1982) 'Early cost estimation for castings,' VDI Berichte Nr. 457, *Designers Lower Manufacturing Costs*, Dusseldorf: VDI Verlag.

Pahl, G., and Beitz, W. (1988) *Engineering design – a systematic approach*, Berlin, New York: Springer Verlag.

Pighini, U., Long, W. and Todaro, F. (1989) *Proc. ICED 89*: 399–414, London: Inst. Mech. Engrs.

Pugh, S. (1977) 'Manufacturing cost data for the designer,' Proc. 1st National Design Conf., London.

Sheldon, D.F., Huang, G.Q. and Perks, R. (1991) *J. of Engineering Design*, **2**(2): 127–39.

Smith, M. (1988) *Mechanical Engineering*, **110**(9): 130.

CHAPTER 18

Economic design in concurrent engineering

James S. Noble

18.1 INTRODUCTION

The integrated economic design of products and associated manufacturing processes requires a thorough understanding of the breadth and relevance of the problem domain. This section discusses the scope of the concurrent engineering problem as it relates to design and economic issues.

18.1.1 Concurrent engineering

Concurrent engineering is typically defined as the integration of both the product and the manufacturing design processes. The goal of this integration is to reduce the product development time, to reduce the cost, and to provide a product that better meets the customer's expectations. The motivation for concurrent engineering is primarily an economic one. As depicted in Fig. 18.1, the best time to impact the economics of the product design is in the conceptual/design phase. Therefore, it is critical that tools be developed to integrate economic issues within a concurrent engineering approach.

Design is a complex activity. When designing a product or a system, it is important to remember that the design should be considered as a whole and therefore all related components and information should be considered interactively. The typical definition of concurrent engineering tends to limit the concurrent aspects of the design process to the engineering functions, but if economics are to be considered then the scope must be broadened to include all functions within the organization. Figure 18.2 illustrates the necessary integration of management, marketing, finance, purchasing, accounting, product design, process design, manufacturing and suppliers to meet the product expectations of customers. From an economic viewpoint, a more restricted perspective of the players in the design

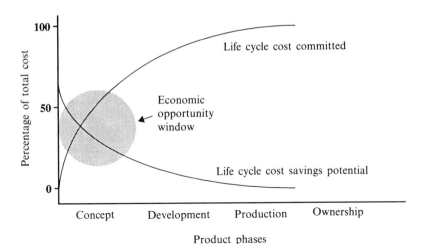

Fig. 18.1 Window of economic opportunity for concurrent engineering.

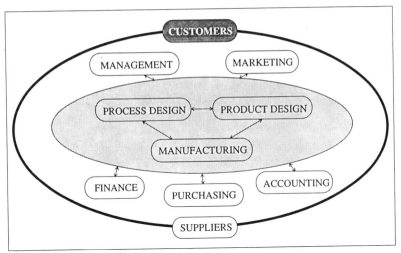

Fig. 18.2 Sphere of economic players in concurrent engineering.

problem would result in only a partial solution. However, much of the research in economic design has isolated the different functions of the organization and then solved the problems that are immediately concerned with them. This approach fails to address the total design problem. When viewed in light of the inherent complexity of a product/system this design approach is not necessarily incorrect. However, since it does not take into consideration the interaction of other functions, it can result in a suboptimal design. A primary goal of an integrated economic design approach should be to incorporate all design function perspectives so that decisions reflect overall organization economic criteria.

To better understand the importance of considering each function of the organization within an economic design approach, consider the example of the ramifications of material selection. The material selected affects the types of processes that can be used, which ultimately impacts the capital budget of the organization. The type of material also affects the process of procurement, which impacts the operating budget. The type of material also affects the performance of the product which in obvious and subtle ways, impacts the cost of the final product and its marketability. The list of interactions is endless, but is clear that it is important to consider the whole organization in an approach to economic design within the concurrent engineering context.

The next section will review the work that has been done in economic design in both product design and manufacturing system design.

18.2 ECONOMIC DESIGN APPROACHES

The primary difference between economic justification and economic design is that in economic jusification the goal is to determine if a product or system meets the economic criteria for acceptance, hence justification. In economic design the goal is to consider economic issues as part of the design decision process. Therefore, a different approach must be taken with each to include economics in the design process. However, the results of economic justification and economic design are similar in that both provide a basis for determining when economic criteria have been satisfied. (Note: for literature related to economic justification approaches the reader is referred to literature reviews by Canada (1986) and Wallace and Thuesen (1987).) The following two sections will review the economic design work that has been done in product design and manufacturing system design, which are the two fundamental issues in a concurrent engineering approach.

18.2.1 Economics in product design

Traditionally economic issues were included implicitly in the product design process rather than explicitly. This entailed the use of design principles that were based on economic considerations for selecting a certain type of material or process. Therefore, minimization in the allocation of resources was incorporated in the design, but an actual economic evaluation was not conducted until the design was completed.

One of the first researchers to address providing economic information directly to the product designer was Pugh (1974). He established several criteria for consideration in the design of a product cost estimating system to ensure that the designer has sufficient cost information. This early work focused on the development of target costs for the product and utilized relative, parametric, or standard costing approaches as the basis for developing them. Later, cost equations based on more detailed design parameters (hybrid parametric/detailed cost models) were developed for a variety of design problems (cost of turned components during design, Mahmoud and Pugh, 1979).

Two cost model based approaches have been proposed for economic design: parametric and detailed. Fad and Summers (1988) suggested that parametric cost models are particularly applicable in the early phases of a design project when the details associated with the design are not clearly formulated. Daschbach and Apgar (1988) went on to say that parametric cost models, when properly applied, provide better estimates of actual cost than traditional 'bottom-up' approaches to cost estimation, particularly in large product design (i.e. aerospace). Detailed cost models have been developed for a variety of applications. Dewhurst and Boothroyd (1988) developed detailed cost models of both the machining and injection molding processes so that a designer can explore the trade-offs between machining or casting a major component of a product. Boothroyd and Reynolds (1989) derived a detailed cost model for typical turned parts to assist designers in the early phases of product design.

The integration of both the economic and design functions into one environment was first proposed by Westney (1983). He reaffirmed the need for cost information in both the conceptual and detailed design stages of the product design process. He also noted that the difficulty that typically exists in integrating the two functions was primarily due to a lack of an appropriate mechanism for transferring information. He went on to propose that the functions of the design engineer and cost engineer be combined within a computer aided design (CAD) system. An integrated CAD system would contain both design and cost models, and display the design both graphically and numerically. Ferreirinha developed a CAD based product cost estimating program called HKB as was reported by Flemming (1987). The HKB approach is a CAD based estimating system which uses 'form-systems' to estimate the manufacturing cost of machine parts and suggests how to alter the design to reduce costs.

Integrating CAD systems with economic design also enables the product design and process planning functions to be linked in order to explore economic design trade-offs. Fleischer and Khoshnevis (1986) proposed a general knowledge based design system that links design, process planning and cost analysis within a CAD system. Later, Knoshnevis and Park (1989) implemented an integrated system that provides real time process planning information on each design feature so that the designer can see the cost consequences of design decisions.

In a review of economic product design work by Ehrlenspiel (1987) it was noted that design organization greatly affects the cost of a product. Ehrlenspiel categorized the methods by which product and process cost information have been brought to the designer into four areas:

1. Use of cost structures (i.e. percentage breakdown);
2. Relative costs (in reference to a standard material);
3. Rapid costing (parametric techniques); and
4. CAD systems that utilize product geometry to develop material and process costs.

Wierda (1988) also reviewed a variety of cost estimation methods that can be used to assist in product design (i.e. weight-method, dimensioning-method, cost functions, function costs, relative costs, part classification) He stated that the current trend is to link cost models with CAD systems, using CAD geometrical data as the primary input, to enable the designer to do cost estimation.

The approaches discussed are representative of economic design approaches in product cost estimating that provide economic information that could be used in concurrent engineering.

18.2.2 Economics in manufacturing system design

The work done on economic design of manufacturing systems can be classified as either 'bottom-up' or a 'top-down'. A 'bottom-up' approach combines performance and economic factors through a search strategy that focus on the analysis and evaluation phases of design. Bottom-up design results in solutions that focus on the satisfaction of economic and performance requirements. A 'top-down' approach utilizes a mathematical formulation of the problem to consider performance and economic issues in either the synthesis or analysis phases. Top-down design seeks an optimal solution to the problem as represented in the mathematical formulation.

(a) Bottom-up design approaches

There are a variety of bottom-up design justification approaches which combine design performance and economics. The bottom-up design approaches can be classified as: cost model based, queueing network based, simulation based, multiattribute based, or expert system based.

Cost model based approaches to economic design utilize specific economic and performance relationships to guide the design search strategy. Cost models are often applied during the design synthesis phase, particularly when the design lends itself to an incremental approach. In other instances, cost models are primarily used to evaluate a design that has already been generated. Lynch (1976) proposed a cost model based approach in his work on the design of assembly machines. A price-time relationship was developed based on cost models of manual, transfer and programmable assembly processes to select the type of assembly system, the number of assembly stations and a reasonable system configuration efficiency. Scott *et al.* (1983) developed a combined performance and cost model design methodology for the design of unit load conveyors. A cost estimate is used to search for a design solution, as well as to aid in the development of operating conditions such as motor size, belt tension, and gear/sprocket ratios. Blank and Carrasco (1985) developed a framework for a manufacturing system life-cycle costing model and its utilization throughout the design process. System models are developed by selecting equipment alternatives from which system cash flows are generated for economic analysis. As the design develops, an economic model is generated from which life-cycle cost issues are analyzed (Carrasco and Blank 1987). Alberti *et al.* (1989) developed a cost efficiency index (CEI) consisting of the ratio of the minimum theoretical cost to actual cost where the goal of each design iteration is to raise the CEI to 1.0. Michaels and Wood (1989) focused on designing within budgetary limits and proposed what is termed, 'design-to-cost'. The concept proposes the use of budgetary constraints as design limits to highlight areas of economic improvement which then can be used to direct the initial design and redesign process.

Queueing network models have been used to generate economic data within bottom-up approaches to economic design. Leimkuhler (1981) used CAN-Q (a queueing model developed by Solberg (1977)) to analyze and generate manufacturing system design and operation information that is then combined with the related economic factors. The combination of design and economic information allowed for marginal analysis to be conducted with respect to different design criteria. Co and Liu (1984) added a simulation of product inputs to a CAN-Q based interactive design and justification procedure. Model output is analyzed from an integrated performance and economic perspective to determine system design modifications.

Simulation is another modeling technique used to combine performance and economic factors in the design of manufacturing systems. Simulation has been used to generate economic data so that more information is available for the justification process. Haider and Blank (1983) proposed that simulation be used to estimate product and revenue costs for system performance measures. Similarly, Falkner and Garlid (1986) developed a simulation model that calculates the material, cell operating, rework, inventory and maintenance costs and the annual revenue, and then combines revenue and cost factors to get annual cash flows. Suresh (1990) presented an integrated evaluation structure that is predominantly a bottom-up approach using simulation to provide performance data. The structure maintains an overall view of the manufacturing organization, such that strategic objectives and various measures of system flexibility are considered. He proposed that a mixed integer programming model be used to reduce the solution space that is evaluated using simulation. A breakdown of the relationships between design inputs and physical and financial variables is developed to enable simultaneous consideration of performance and economic factors.

Muliattribute analysis has been used to assist in economic design by considering economics as one component of a subjective preference structure. Gabbert and Brown (1987) proposed a knowledge based multiattribute design approach for specifying the types of material handling equipment to use in a given manufacturing scenario. The decision makers attribute preferences are reflected in the updating of attribute weights. Liang *et al.* (1989) developed a multiattribute approach for the specification of material handling equipment that utilized the analytic hierarchy process combined with a database. Troxler (1989) developed a system value model that allows for evaluation of the broad attribute categories of suitability, capability, performance and producibility at lower levels of detail.

Expert systems have been proposed for combining performance and economic factors of a design. Expert systems offer advantages over other methodologies as they allow a variety of design analyses that can be combined in a knowledge guided manner to incorporate economic issues within the design process at appropriate times. Fisher and Nof (1987) developed an expert system that combines an assembly technology selection and cost model with a replacement analysis model to provide the designer further economic insight into design decisions. Mellichamp and Wahab (1987) incorporated economics in an expert system for FMS design. The system seeks to achieve the design goal of minimizing total capital investment and cost per part. When the goal is not satisfied, an attempt is made to substitute equipment with the maximum cost differential, while still maintaining production requirements.

(b) Top-down design approaches

Top-down economic design approaches typically contain a mathematical formulation of the design problem that characteristically minimize total capital and operating cost, so that machine utilization, production requirement, and budget constraints are satisfied. This approach integrates performance and economic factors explicitly by forcing the design to meet economic constraints or objectives. The solution to the design problem formulation is guaranteed to meet both the stated performance and economic objectives of the design to the extent to which they are able to be captured in the mathematical formulation. The following will cover a range of manufacturing system design problems that deal with economics in this manner.

Miller and Davis (1978) formulated a mixed integer programming model for the machine requirements problem that minimizes machine investment cost, overtime operating expenses, excess capacity opportunity costs, and machine disposal cost, such that the budget, floor space and capacity constraints are met. Bard and Feo (1991) formulated a nonlinear cost minimization model to determine the type, number, and utilization of machines required for each operation, while also accounting for machine flexibility. The problem is solved with a depth-first branch and bound routine that utilizes a greedy set covering heuristic to find feasible solutions.

Tanchoco *et al.* (1979) formulated the belt conveyor specification problem to minimize total capital cost, such that the operating constraints of conveyor speed, belt tension, and required horsepower are met. A decomposition and implicit enumeration procedure is given to obtain solutions. Ziai and Sule (1989) developed a material handling equipment selection methodology that uses a mixed integer program to obtain an initial solution. The initial solution is then used as the basis of a cost driven improvement heuristic to improve the solutions equipment utilization. Sharp and Liu (1990) used a combined top-down/bottom-up design justification approach to configure fixed guidepath material handling systems. First, a cost per time unit versus total vehicle arrival rate graph was generated from queueing network and simulation analyses. Second, the initial data are transformed into a network flow problem that has as its objective to minimize the sum of the vehicle travel/waiting, and guidepath construction costs. Webster and Reed (1971) addressed the complete material handling selection problem. The problem of minimizing capital, operating, and unit load changing cost with respect to material flow requirements and transporter capacity, was modeled as an integer program and solved using a heuristic solution procedure. Hassan *et al.* (1985) reformulated Webster and Reed's model without the unit load selection problem and developed a construction algorithm that utilizes the formulation's greedy structure to find solutions. Egbelu (1990) proposed a design methodology that combines the selection of machining parameters and the size of unit loads for two different kinds of material handling systems by minimizing the sum of the labor/machining, tool, overhead, and material handling costs. The solution procedure used a combined unidimensional search technique and a dynamic programming algorithm to find the minimum cost machining and material flow system.

The most extensive approach to the integration of economics in design has been developed in connection with the Charles Stark Draper Laboratory for the design of assembly systems (Nevins and Whitney, 1989). Initial work focused on mathematical formulations of the assembly design problem. Graves and Whitney (1979), Graves and Lamar (1983) and Graves and Redfield (1988) all formulated variations of the selection of assembly station and operation assignment problem such that the total capital and operating cost is minimized, station capacity not exceeded and the annual production requirement for each task satisfied. Solution procedures were proposed that utilized a Lagrangian relaxation, a combined pure linear relaxation and mixed integer–linear relaxation, and a graphical shortest-path approach, respectively.

The equipment selection/task assignment problem approaches tend to result in solutions that are operationally inefficient, therefore, Gustavson

(1983) approached the problem through heuristics based on a general unit cost model of a manufacturing system. The general unit cost model included the fixed unit cost of hardware, software and installation costs and the variable unit costs of labor, maintenance, and operating costs. Based on the unit cost relationship, unit cost/production quantity curves are constructed that show which system would be lowest cost for different production quantities. The unit cost model was then incorporated into a dynamic programming type heuristic that synthesizes cost effective assembly systems for a given production quantity (Gustavson, 1984, 1988). The heuristic utilizes aspects of dynamic programming by initially exploring the versatility of resources. It then allocates resources based on a quality rating of the resources for a given task until all tasks are assigned. The heuristic does not ensure optimal results, and as a consequence it is recommended that systems be analyzed further for performance and economical viability. The output of the design procedure is a listing and graph of the unit cost of all usable systems and the availability for a specified production rate. Further output can be obtained for a specific system including a listing of resource availability and related data, a schematic layout of the synthesized system, and a graph of unit cost versus batch size versus utilization. Gustavson also proposed a procedure for creating cost effective systems that combine his system synthesis algorithm with traditional economic justification to ensure that economic criteria are satisfied. The procedure iterates between the synthesis and justification procedures until the assembly system design is acceptable. Figure 18.3 depicts the logic of the economic design approach for assembly systems.

18.3 ECONOMIC DESIGN ISSUES IN CONCURRENT ENGINEERING

The economic design problem in a concurrent engineering context addresses the same issues that are dealt with in product or manufacturing system design, however, it is further complicated by the integration of more design, people, and information issues. The issues that must be resolved can be categorized as either information based or analysis based.

Concurrent engineering by definition integrates the functions of an enterprise. This integration of functional tasks creates the fundamental problem in concurrent engineering: communication of information. If there is to be a truly concurrent design effort then the proper type of information, the correct amount of information, and the availability/accuracy of information must all be addressed. The following will discuss each of the concurrent engineering information issues as they relate to economic design.

There are four types of information required for an economic design effort. The first type of information includes the development of accurate product material and manufacturing process cost models (cost estimating

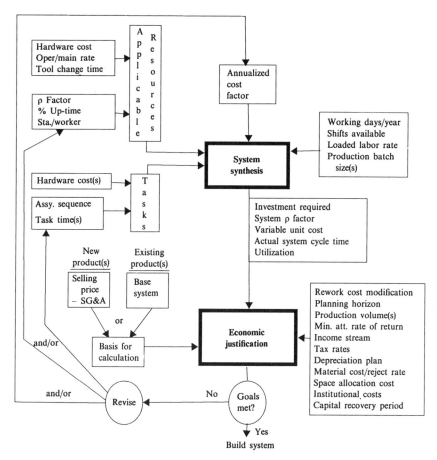

Fig. 18.3 Logic diagram for the creation of cost effective systems. (Reproduced by permission: Nevins and Whitney, 1989.)

systems specifically for concurrent engineering have been developed by Wong *et al.* 1991). The second type of information includes the development of cost models of the marketing function and the broader competitive environment. The third type of information required includes capturing a products function, manufacturability, and marketing forecast information so that an accurate and representative view of the underlying physical realities is obtained. The fourth type of information needed within an economic design effort is a measure of how design decisions support the organizations business strategy from an economic perspective.

The issue of obtaining the correct amount of information is paramount so that design decisions are based on sufficient data and in an economic manner. The problem in determining the correct amount of information that typically arises is that nobody has a comprehensive view of the design problem. Therefore, each function obtains all the information needed for them to make decisions and then passes it all to the usual integrating function, i.e. management. The result is that the design process becomes over burdened with unnecessary information. The difficulty then in determining the correct amount of information is balancing the need to have sufficient, but not extraneous information from which to base decisions.

The information issues of availability and accuracy require that not only the correct type, but accurate information be available when critical economic decisions must be made. If correct information of the type and accuracy needed is not available when major design decisions are made, then the value of the economic information is greatly diminished. The accuracy and integrity of information is especially a critical issue in concurrent engineering since a broad range of design functions are integrated and must work together. Maintaining up-to-date and consistent economic information is required if meaningful decisions are to be made. Accounting for all of the mentioned issues in a large-scale concurrent engineering effort is not a small task, however, research is being conducted on the types of database and information architectures that will be required to support this type of integrated design process.

The analysis-based issues of economic design are a consequence of the informational requirements of concurrent engineering. When the proper information is available for an economic design process, then the task becomes how to conduct the product and process performance/cost trade-off analysis. The trade-off analysis is further complicated in the concurrent engineering context since decisions are based on a greater number of decision criteria than has been traditionally considered. One method for considering multiple decision criteria is to use multiattribute decision analysis techniques (i.e. weighted evaluation method, utility functions, analytic hierarchy process, goal programming). Another method for exploring trade-offs between multiple criteria is to use an integrated marginal analysis. This approach will be presented in more detail in Section 18.4.

18.4 AN INTEGRATED ECONOMIC PRODUCT AND PROCESS DESIGN APPROACH

The economic design approach, termed *design justification*, provides a framework for integrating both the product and process design from a combined economic and performance perspective. In the design justification approach the economic ramifications of design decisions are considered concurrently with design development and are used to guide the design process (Noble and Tanchoco, 1989, 1990; Noble 1991). Figure 18.4 illustrates the basic concept of the design justification approach. This section describes the design justification approach, presents examples of its application to both product and process design, and discusses a comprehensive

System design goals and objectives

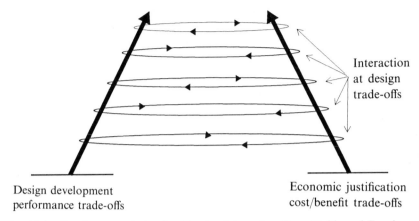

Interaction
at design
trade-offs

Design development
performance trade-offs

Economic justification
cost/benefit trade-offs

Fig. 18.4 Simultaneous design justification interaction (from Noble and Tanchoco, 1990).

economic design framework for concurrent engineering based on the design justification approach.

18.4.1 Economic design approach of design justification

The design justification approach is based on four foundational concepts. The first and second foundations address how information is processed concurrently in design justification. The first foundation is that evaluation of design performance and economics occurs concurrently. A concurrent consideration of performance and economic issues provides trade-off information that is utilized at critical design decision points. Potential design trade-offs are analyzed so that the designer can make decisions that support the design objectives. The concurrent trade-off analysis enables the design to be justified simultaneously as it is developed. The second foundation is that information generated within the evaluation process in design justification is used in a design modification strategy. The modification strategy assists the designer in knowing where to alter the design to reach the design objectives. The goals/objectives of the design are obtained through a branching and bounding procedure that directs the designer throughout the design process. Design branching provides for the exploration of various design alternatives and design bounding provides the designer with a measure of how good the design is in terms of cost and performance. Bounding is accomplished using a marginal analysis of cost versus performance factors. This procedure allows for the objectives/goals to be considered within the branching and bounding and to modify them if it is deemed necessary.

The third and fourth foundations address how information is used to guide the design process in design justification. The third foundation is that a knowledge-guided design synthesis is used that combines top-down and bottom-up design approaches. A combined top-down/bottom-up knowledge-guided design synthesis provides a more robust approach to design synthesis that focuses on directing the design process to satisfy design objectives. In a knowledge-guided design procedure a critical decision point is the redesign or modification of the current design. A modification strategy which guides the design efficiently and enables the designer to determine which modifications to pursue is provided by the knowledge component of the design approach. The fourth foundation is that information is processed and presented to provide the designer with the ability to make good decisions, hence a decision-maker centered approach. A decision-maker centered approach maintains the designer as the key controller in the design development and analysis, so that designs reflect the range of objectives/constraints which are inherent to complex design probems and so that final decisions incorporate designer experience.

The following two sections will provide implementation examples of the design justification approach. Within the examples the application of the four design justification foundational concepts will be noted (i.e. DJ foundation number).

18.4.2 Economic design of products in design justification

The design justification concept has been applied to the design of an EMI (electromagnetic interference)/RFI (radio frequency interference) shield (Noble and Tanchoco, 1990). The major objective of these shields is to reduce EMI/RFI in the metering circuitry through the development of a physical barrier to shield components. The design of an EMI/RFI shield includes the selection of design criteria, structural material, forming processes, shielding material, and shielding application process, subject to considering shield performance and other marketing factors. All of the shield design decisions have direct cost implications, therefore two economic models were developed to explore design trade-offs. The first economic model considered the cost implications of a design decision from an incremental perspective. An incremental cost model is constructed during the design process so that the economic impact of each design decision on the overall cost can be explored. During the development of a completed alternative, multiple design combinations are explored considering the engineering and economic trade-offs (DJ foundation 1). The second economic model, an incremental payback analysis, was applied when the designer desired to see a fixed/variable cost comparison of two different design alternatives.

The EMI/RFI design justification implementation included a knowledge base consisting of information on plastic materials, plastic forming processes, conductive coating materials, coating application processes, and conductive fill materials. This information included the current material and equipment costs, and provided the ability to explore the feasible technical path resulting from the selection of a given material or manufacturing technology (DJ foundation 3). Information was presented to the decision-maker in several forms. The decision-maker determined the initial design criteria which were posted on a bulletin board. The bulletin board posted and updated design criteria, and posted the decision-maker's decisions with their associated economic and performance implications. The bulletin board also allowed the decision-maker to see the economic/performance ramifications of a decision and to explore other possible design alternatives. Information was also presented graphically to the decision-maker. Graphs showing design criteria versus design criteria (i.e. surface resistivity versus attenuation), design criteria versus cost (i.e. coating thickness versus coating cost), and incremental payback period versus production level were simultaneously presented to the designer for analysis (DJ foundations 2 and 4).

18.4.3 Economic design of manufacturing systems in design justification

The design justification concept has also been applied to the selection and specification of material handling systems (Noble, 1991). The implemented material handling system (MHS) design justification framework has six major components: system designer, design interface, design inference model, model generator, rule base, and database. The following will describe the three major phases within the material handling system design justification procedure:

1. The development of manufacturing system data;
2. The development of MHS alternatives; and
3. the analysis and modification of system alternatives, resulting in a final selection and specification of a MHS.

The first phase of the design justification procedure consists of the importation and preparation of maufacturing system data. The MHS design justification procedure requires an instance of the product description, process plan, physical layout, and unit load description. The data for the physical layout are developed using CAD software.

The second phase of the design justification procedure consists of a two-step procedure to develop MHS alternatives. This stage can be considered to be a solution branching phase where the MHS alternatives that merit analysis are generated (DJ foundation 3). The two steps work together with the first step reducing the number of equipment options for each material flow so that the second step can configure a complete system efficiently. The first step of the alternative generation process utilizes a knowledge based ranking of equipment alternatives for each material flow path in the system. The results of the equipment ranking are passed to the equipment selection/system generator step of the solution branching stage.

The second step of the alternative generation process utilizes an integer programming formulation of the equipment selection problem that is solved using a modification of the construction algorithm developed by Hassan *et al.* (1985).

The third phase of the design justification procedure addresses the crux of the framework: the analysis and modification of a design. The analysis and modification phase is comprised of four functions: system classification, system performance modeling, system cost and flexibility modeling, and combined performance and economic analysis. The following addresses each of these four functions and discusses how they are implemented within the framework.

The first function in the analysis phase of a MHS is to classify systems based on the equipment composition. A MHS classification scheme was developed that is based on each equipment's share of the total system material flow.

The MHS classification scheme provides the basis for the second function of modeling MHS performance at different levels of model abstraction (DJ foundation 3). Three levels of modeling abstraction were implemented: deterministic capacity analysis, queueing analysis (CANQ, Solberg, 1977), and simulation (SLAM-MH, Pritsker, 1986). In each abstraction level an automatic model generator was developed.

The third function measures a MHS's flexibility and cost characteristics. A MHS flexibility metric was developed to measure a MHS's combined ability to reconfigure, so as to handle new and additional material flow. An extensive cost modeling structure for each type of material handling equipment was developed to measure the economic impact of design decisions. The cost models developed capture capital, installation, operating, and maintenance cost.

The fourth function of the analysis and modification phase is a combination of system sensitivity analysis and an interactive marginal analysis directed modification strategy (DJ foundations 1 and 2). Sensitivity analysis is supported in two main areas of the framework: capacity and design objective. These two types of analysis are facilitated through a variety of decision graphics that are provided to the designer. The modification of a system design to satisfy the design objectives is accomplished through an integrated performance and economic trade-off analysis. The trade-off analysis is based on a marginal analysis network that shows the marginal or relative incremental effect of and absolute or incremental percentage alteration to the system on the performance measures of throughput, total MHS cost, material handling unit cost, MHS flexibility, and unit flexibility cost. The marginal analysis networks are developed using incremental calculus concepts of Eilon (1984). The marginal analysis provides the designer with the ability to explore the effect of changing individual parameters on the overall peformance measures (DJ foundation 4).

18.4.4 Design justification in concurrent engineering

The design justification approach as illustrated in the EMI/RFI shield and the material handling system implementations provide a basis for economic design in a concurrent engineering context. Though in both cases there are further design issues and parameters that must be addressed if they are fully to support a concurrent engineering effort, the integration of design parameters provides the necessary broader decision perspective. This section describes how the concepts illustrated in the design justification implementations could be combined into a fully integrated concurrent engineering approach. Figure 18.5 provides the conceptual

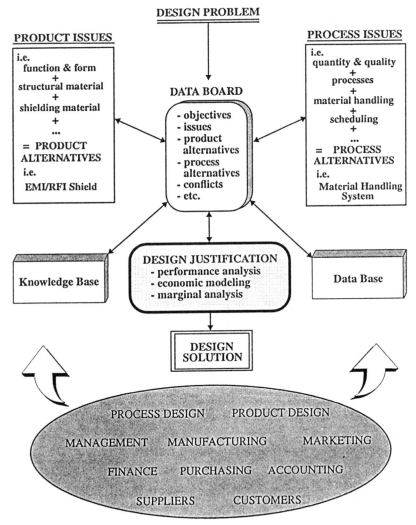

Fig. 18.5   Integrated economic design framework for concurrent engineering.

framework for this application of the design justification concept in concurrent engineering.

An integrated economic design approach for concurrent engineering must have as its central focus the combining of product and process design issues in the context of the overall organization. The combination of such a broad range of design issues and perspectives requires that the overall design approach incorporate both the communication of information within the organization and the integration of design analysis results.

In the proposed framework, information issues such as design objectives, product and process parameters, design conflicts, etc. are integrated and communicated through the use of a data board. The data board serves to maintain consistency and integrity of information since information is required from a variety of design analyses and perspectives at different points in time. Information is obtained and posted concurrently from both the product and process design analyses as they proceed. There are instances when the value of a decision variable is required that is yet undetermined. In the preliminary design phase it may be sufficient to obtain a 'temporary' value from an overall organization design knowledge base or database. These temporary decision values tend to be information that reflects the overall organization perspective and are fairly constant across design projects. As more detailed design decisions are required, the appropriate decision function would be prompted and decisions that reflect a dynamic overall organization perspective would be made and integrated through the data board.

The integration of design analyses is handled in the proposed framework through both the combined performance and economic modeling, and the marginal analysis trade-off approaches that are the central issues in design justification. Performance models that represent the breadth of decision issues that need to be addressed are used. Models are selected/ developed so that they can share design parameters through the data board. Detailed economic models are developed that capture capital, operating and other life-cycle cost factors that are based on the performance measure modeled. The performance and economic models are combined in a marginal analysis framework that enables the impact of different product, process and organizational factors to be explored. Exploring the design trade-offs from an overall organizational perspective allows for strategic and tactical issues to be considered at the detailed design level.

The application of the design justification concept to concurrent engineering is a logical and natural extension to the individual product and system implementations that have been developed. The concept is intuitively appealing and the technology is readily available to make the framework a usable design tool.

18.5 CONCLUSION

This chapter has addressed the need for, and difficulty associated with economic design approaches in concurrent engineering. A review of the literature related to both the economic design of products and for manufacturing system design was presented to gain an understanding of the issues related to the problem of economic design.

An economic design framework, termed design justification, was presented that forms the basis for an integrated economic design approach for concurrent engineering. Examples illustrating the application of the design justification concept were given, followed by an overall conceptual framework for economic design in concurrent engineering. The application of the design justification concept within concurrent engineering is one approach to integrating the range of design issues and creating a usable concurrent engineering design tool.

18.6 REFERENCES

Alberti, N., La Commare, U., Noto La Diega, S. (1989) in *Proceedings of the 3rd ORSA/TIMS Conference on Flexible Manufacturing Systems*, (eds. K. Stecke and R. Suri), Elsevier Science Publishers D.V., Amsterdam, pp. 67–72.

Bard, J.F. and Feo, T.A. (1991) *IIE Transactions*, **23**(1): 83–92.

Blank, L.T. and Carrasco, H. (1985) in *Proceedings of the 1985 International Industrial Engineering Conference*, Los Angeles, CA, May, pp. 161–7.

Boothroyd, G. and Reynolds, C. (1989) *Journal of Manufacturing Systems*, **8**(3), 185–93.

Canada, J.R. (1986) *The Engineering Economist*, **31**(2), 137–50.

Carrasco, H. and Blank, L.T. (1987) in *Proceedings of the 1987 International Industrial Engineering Conference*, Washington, D.C., May, pp. 211–16.

Co, H.C. and Liu, J. (1984) in *Proceedings, 1984 Winter Simulation Conference*, (eds S. Sheppard, U.W. Pooch, C.D. Pegden), Dallas, TX, November 28–30, pp. 406–12.

Daschbach, J.M. and Apgar, H. (1988) *Engineering Costs and Production Economics*, **14**(2), 87–93.

Dewhurst, P. and Boothroyd, G. (1988) *Journal of Manufacturing Systems*, **7**(3), 183–91.

Egbelu, P.J. (1990) *International Journal of Production Research*, **28**(2), 353–68.

Ehrlenspiel, R. (1987) in *Proceedings of the 1987 Intl. Conference on Engineering Design*, (ed. W.E. Eder), ASME, New York, pp. 796–806.

Eilon, S. (1984) *The Art of Reckoning – Analysis of Performance Criteria*, Academic Press, London.

Fad, B.E. and Summers, R.M. (1988) *Engineering Costs and Production Economics*, **14**, 165–76.

Falkner, C.H. and Garlid, S. (1986) in *Proceedings 1986 Fall Industrial Engineering Conference*, Boston, MA, December, pp. 99–108.

Fisher, E.L. and Nof, S.Y. (1987) *Journal of Manufacturing Systems*, **6**(2), 137–50.

Fleischer, G.A. and Khoshnevis, B. (1986) in *Proceedings of the 1986 International Industrial Engineering Conference*, Dallas, TX, pp. 163–74.

Flemming, M. (1987) in *Proceedings of the 1987 International Conference on Engineering Design*, (ed. W.E. Eder), ASME, New York, pp. 379–93.

Gabbert, P.S. and Brown, D.E. (1987) in *Proceedings of the 1987 International Industrial Engineering Conference*, Washington, D.C., pp. 445–51.

Graves, S.G. and Lamar, B.W. (1983) *Operations Research*, **31**(3), 522–45.

Graves, S.G. and Redfield, C.H. (1988) *International Journal of Flexible Manufacturing Systems*, **1**(1), 31–50.

Graves, S.G. and Whitney, D.E. (1979) in *Proceedings of the 1979 IEEE Decision and Control Conference*, Ft. Lauderdale, FL, pp. 531–6.

Gustavson, R.E. (1983) in *Proceedings of the 13th International Symposium on Industrial Robots and Robots 7*, **1**, Society of Manufacturing Engineers, Chicago, IL, pp. 4.85–4.104.

Gustavson, R.E. (1984) in *Proceedings of the 1984 International Symposium on Industrial Robots*, (ed. N. Martensson), Elsevier Science, Gothenburg, Sweden, pp. 95–106.

Gustavson, R.E. (1988) *SME Technical Paper # AD88–250*, Society of Manufacturing Engineers, Dearborn, MI.

Haider, S.W. and Blank, L.T. (1093) in *Proceedings of the 1983 Winter Simulation Conference*, (ed. S. Roberts, J. Banks, B. Schmeiser), Arlington, VA, pp. 199–206.

Hassan, M.M.D., Hogg, G.L. and Smith, D.R. (1985) *International Journal of Production Research*, **23**(2), 381–92.

Knoshnevis, B. and Park, J. (1989) in *Proceedings of the 1989 International Industrial Engineering Conference*, Toronto, pp. 135–41.

Leimkuhler, F.F. (1981) *Report 21, NSF Grant APR74 15256, Optimal Planning of Computerized Manufacturing Systems*, Purdue University, West Lafayette, IN.

Liang, M., Dutta, S.P. and Abdou, G. (1989) in *Proceedings of the 1989 International Industrial Engineering Conference*, Toronto, Canada, pp. 255–30.

Lynch, P.M. (1976) Economic-Technological Modeling and Design Criteria for Programmable Assembly Machines, Unpublished Ph.D. Thesis, Massachusetts Institute of Technology, Cambridge, MA.

Mahmoud, M.A.M. and Pugh, S. (1979) in *Proceedings of the Information Systems for Designers*, Southampton, pp. 37–42.

Mellichamp, J.M. and Wahab, A.F.A. (1987) *Simulation*, **48**(5), 201–8.

Michaels, J.V. and Wood, W.P. (1989) *Design to Cost*, John Wiley and Sons, New York.

Miller, D.M. and Davis, R.P. (1978) *AIIE Transactions*, **10**(3), 237–43.

Nevins, J.L. and Whitney, D.E. (eds.) (1989) *Concurrent Design of Products and Processes*, McGraw-Hill, New York.

Noble, J.S. and Tanchoco, J.M.A. (1989) in *Cost Analysis Applications of Economics and Operations Research, Lecture Notes in Economics and Mathematical Systems*, (ed. T.R. Gulledge, Jr., L.A. Litteral), Springer-Verlag, New York, pp. 197–213.

Noble, J.S. and Tanchoco, J.M.A. (1990) *International Journal of Production Research*, **28**(7), 1225–38.

Noble, J.S. (1991) A Framework for the Design Justification of Material Handling Systems, Unpublished Ph.D. Dissertation, Purdue University, West Lafayette, IN.

Pugh, S. (1974) in *Proceedings of the 2nd International Symposium on Informations Systems for Designers*, University of Southampton, UK, pp. 12.1–12.8.

Pritsker, A.A.B. (1986) *Introduction to Simulation and SLAM II*, (3rd edn), Halsted Press, John Wiley and Sons, New York.

Scott, H.A., Tanchoco, J.M.A. and Agee, M.H. (1983) *International Journal of Production Research*, **21**(1), 95–107.

Sharp, G.P. and Liu, F-H F. (1990) *International Journal of Production Research*, **28**(4), 757–83.

Solberg, J.J. (1977) in *Proceedings of the 4th International Conference on Production Research*, Tokyo, pp. 22–30.

Suresh, N.C. (1990) *International Journal of Production Research*, **28**(9), 1657–72.

Tanchoco, J.M.A., Davis, R.P., Wysk, R.A. and Agee, M.H. (1979) in *Proceedings of the 1979 Spring Industrial Engineering Conference*, San Francisco, CA, pp. 112–19.

Troxler, J.W. (1989) *Journal of Manufacturing Systems*, **8**(3), 175–83.

Wabalickis, R.N. (1988) *Journal of Manufacturing Systems*, **7**(3), 175–82.

Wallace, W. and Thuesen, G. (1987) *The Engineering Economist*, **32**(3), 247–57.

Webster, D.B. and Reed, Jr., R. (1971) *AIIE Transactions*, **3**(1), 13–21.

Westney, R.E. (1983) *Engineering Costs and Production Economics*, **7**(3), 205–12.

Wierda, L.S. (1988) *Engineering Costs and Production Economics*, **13**(3), 189–98.

Wong, J.P., Parsaei, H.R. and Liles, D.H. (1991) in *Factory Automation and Information Management*, (eds. M.M. Ahmad and W.G. Sullivan), CRC Press, Roca Raton, FL, pp. 600–7.

Ziai, M.R. and Sule, D.R. (1989) *Computers and Industrial Engineering*, **17**(1–4), 55–60.

PART FOUR

Artificial Intelligence in Concurrent Engineering

CHAPTER 19

Application of expert systems to engineering design

Gary P. Moynihan

19.1 INTRODUCTION

There exists a spectrum of applications software that ranges from the traditional data processing systems to advanced artificial intelligence applications (see Fig. 19.1). The sophistication and complexity of these applications softwares is steadily increasing. While conventional systems, such as data processing and management information systems, still exist and are quite appropriate for some manufacturing applications, there is expanding interest in artificial intelligence (Savino, 1990).

	Level of sophistication/complexity			
System type	Data processing	Management information systems	Decision support systems	Artificial intelligence
Processing method	Simple arithmetic	Algorithms	Complex algorithms; operations research models	Heuristics
Example	System to report statistics regarding engineering bill of material	Computer-aided process planning	System to optimize specific design criteria	Design expert system

Fig. 19.1 Spectrum of information system applications.

Artificial intelligence (AI) can be considered to be the function of making machines achieve human-like capabilities (e.g. seeing, hearing or thinking). Expert systems are a subset of AI that attempts to produce expert levels of performance in solving problems within a very specific area (Leonard-Barton and Sviokla, 1988). As with most valuable commodities, human expertise may be both scarce and costly. Automation of this expertise, i.e. an expert system, is a logical alternative.

Unlike conventional software systems which are based on algorithms, expert systems are founded on heuristic processing. These heuristics, essentially problem-specific rules of thumb, are obtained from the human expert through a process known as knowledge engineering. This process also translates the knowledge into a structure that the computer can utilize. In an expert system, it is stored in a knowledge base for a specified problem-solving domain (e.g. the construction of high-rise buildings). The knowledge base is a primary component of the expert system, and contains the facts, beliefs and associations unique to the domain expert (see Fig. 19.2).

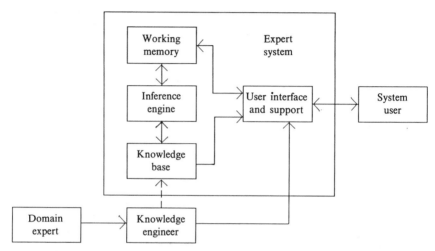

Fig. 19.2 Generalized expert system architecture.

The second key ES component is the inference engine. Essentially, the inference engine may be compared to a control mechanism that organizes the correct sequence of heuristic activation. Its function is to access and manipulate the knowledge base according to the specific problem data contained in the working memory. The inference engine incorporates reasoning methods to solve the stated problem and to provide an explanation for the solution.

Currently, the field of expert systems (ES) is a very active area for both research and actual implementation. The rationale for this rapid growth has been well documented in the technical literature. It relates to ES providing the relevant capabilities of preservation of knowledge; wide dissemination of scarce expertise; consistency of approach and response; perpetual accessibility; solution of problems with incomplete or uncertain data; and the explanation of logic used to arrive at a conclusion (Hayes-Roth *et al.*, 1983; Parsaye and Chignell, 1988; Ignizio, 1991). Because of these advantages, expert systems have been constructed and implemented in a variety of successful applications, including engineering design (see Table 19.1).

Table 19.1 Expert system applications by function

Function	Problem Addressed	Examples
Prediction	Inferring probable consequences of specific situations	Projection of shop floor status
Design	Configure specifications to meet given objectives and constraints	Building design; electronic circuit layout
Planning	Formulating a specific method of action	Capital equipment requirements planning and scheduling
Monitoring	Comparing actual data to planned specifications	Nuclear power plant alarm analysis
Interpretation	Analyzing situation based on observed data	Diagnosis of quality control problems and resulting recommendations for solution
Diagnosis	Inferring the cause of system problems based on observed data	
Prescription	Suggest remedies for inferred or observed system malfunction	
Maintenance	Develop and execute plan to remedy diagnosed problem cause	Automated computer repair systems
Instruction	Provide adaptive feedback of information to study users	Computer-aided instruction
Control	Adaptively govern system behavior	Equipment operator aids

19.2 THE RELEVANCE OF ES TO ENGINEERING DESIGN

Engineering design is the process by which high quality and innovative products are conceived and specified for fabrication. Judgment and previous experience are applied in concert with technical knowledge to select the correct values of the design parameters. Since engineering design is an open-ended problem domain, a given design may have a near infinite number of possible alternatives. Conventional computer programs, even those utilizing mathematically sophisticated approaches, are not successful under these circumstances. Design problems exhibit combinatorial explosiveness, i.e. 'the time required to solve such problems increases exponentially with problem size; and at some point it is simply unrealistic to consider solving such problems by exact, analytical approaches – even on the most powerful existing computer' (Ignizio, 1991). Selection of an optimum (or even satisficing) design becomes extremely challenging. The application of expert systems to this problem domain appears to be a reasonable approach to limit and analyze the feasible set of alternatives.

Brown and Chandrasekaran (1989) identify the following taxonomy of engineering design:

1. *Creative design* as characterized by true innovation;
2. *Variant design* which entails the modification or replacement of whole components (e.g. frequently automotive design); and
3. *Routine design* which tends to be limited variations on existing frameworks with known constraints.

As subsequently discussed by Dym and Levitt (1991), routine design requires 'significant amounts of design domain knowledge because of complex interaction between subgoals and between components .... Thus even in this 'simple' class of routine design, there is more than ample scope for deploying design knowledge'. Given this classification scheme, expert system applications in engineering design have focused primarily on routine design problems, and to a much lesser extent on variant design.

19.3 KNOWLEDGE REPRESENTATION PARADIGMS

Expert systems tend to represent knowledge via two primary means: rules and frames. Rule-based systems are the most popular mode of knowledge representation, due to ease of understanding for both developer and user; ease of modification; and widespread availability of rule-based expert system development packages and shells. Rules are essentially sophisticated IF–THEN computer statements, frequently incorporating logical ANDs, ORs, negation and confidence factors. The primary drawback of a rule-based expert system is, that for some situations, rules may be an inefficient means of representing knowledge. This is particularly pertinent to combinatorially explosive domains such as design. In such cases, the number of rules required to solve the problem in a comprehensive manner would be prohibitive.

Frames, or object-oriented representations, provide a more efficient means of structuring complex knowledge. A frame is essentially a concept of a specific object. Each frame contains a series of 'slots' for the attributes and attribute values of the object. In addition to storing these parameters, slots may contain default values, sets of rules or procedures that may be activated, or linkages to other frames. This last consideration provides the key to the power of using frames. By defining one frame as a subset of another, the first object inherits the properties of the second. This inheritance capability provides considerable value in reducing computer memory storage requirements. In Fig. 19.3, for example, the frame PART#0011 would inherit all of the attributes associated with CAM.

FRAME = CAM
MATERIAL: STEEL
FABRICATION METHOD: MACHINING
MACHINE TO BE USED: MILLING MACHINE #3
TOLERANCES: +/- .001 IN

FRAME = PART #0011
IS-A: CAM
NEXT ASSEMBLY: PART #1054

Fig. 19.3 Example of inheritance using frames.

19.4 EXPERT SYSTEM PROCESSING

The framework of heuristics, linking initial data with possible solutions within an expert system, is referred to as an inference net. A variety of control strategies are utilized to select the next heuristic for activation within an inference net. The most basic strategy is referred to as generate-and-test. In this strategy, all possible alternatives are generated and evaluation heuristics are sequentially activated to test if the alternative leads to the goal. These evaluation heuristics prune the set of feasible solutions down to a number that can be further analyzed by the system. Pure generate-and-test is effective for domains with a relatively small number of alternatives, and is thus inappropriate for combinatorial explosion.

Expert systems utilize two primary strategies to sequence the activation of individual heuristics. This type of activation mechanism is called 'chaining'. In forward chaining, the characteristics of the input data trigger the initial activation. Forward chainers are utilized for applications where the number of possible goals, or conclusions, are more numerous than the number of inputs. Design, production planning and scheduling expert systems normally use this method.

In the other strategy, referred to as backward chaining, the activation proceeds from a limited number of conclusions to the more numerous input data. Essentially, backward chainers assume a given final conclusion, then activate a series of heuristics to test for its validity. Frequently, this control strategy is used when all of the input data may not be relevant to the problem. Selectivity in the prompts for data might be necessary. Backward chaining would be appropriate for a quality control diagnostic expert system.

There are also applications designed to utilize a bidirectional search, i.e. simultaneous forward and backward chaining. Theoretically, this would be extremely efficient as a chain of reasoning would be established when the two searches intersect. Controlling the search paths so that they do intersect, however, is very difficult.

There are two primary methods of exhaustive search used within these chaining strategies: breadth-first and depth-first. Breadth-first search investigates the inference net by evaluating every heuristic node at a specific level before advancing to the next level in the net. This search strategy is most effectively used when the inference net is relatively shallow (a small number of heuristic levels linking the data and goal states).

Depth-first search may be considered to be an intensive burrowing into the inference net. By convention, this search mechanism selects a heuristic node at the initial level of the network, then descends to one of the nodes branching from it to the next level. This process continues until a terminal node is reached. The primary advantage of depth-first search is its smaller computer memory requirements than breadth-first search. Conversely, its principal problem is that depth-first search may continue indefinitely if there are no natural termination points. Modifications to this strategy have been suggested to address this problem (Stefik, 1990). Figure 19.4 provides a representation of these control strategies.

Variations on these fundamental control strategies frequently occur, but are normally targeted to specific problem domains. The strategy of constraint satisfaction is particularly germane to engineering design. Constraint satisfaction is a variant of generate-and-test. The key to this methodology is 'an emphasis on the representation and manipulation of the constraints' (Dym and Levitt, 1991). In other words, domain knowledge is incorporated into the generator portion in order to reduce the number of alternatives to be evaluated. This adddresses the primary disadvantage of the pure generate-and-test method mentioned previously. The constraints themselves can be organized as part of the inference net, whose framework can assist in guiding the process of selecting interacting variables (Stefik, 1990).

Stopping criteria for any of these control mechanisms are based on finding a satisfactory solution to the problem. By definition, expert systems are satisficers rather than optimizers. They will focus on obtaining an acceptable solution rather than a single 'best' solution. Optimization is normally the function of algorithmic-based decision support systems.

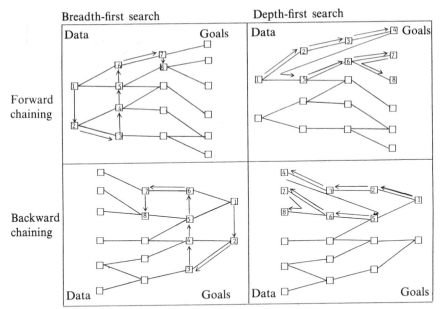

Fig. 19.4 Common expert system control strategies.

Hybrid ES/DSS systems have been created where a mixture of analytical tools are used to address the specific nature of the problem. Normally, the expert system component is utilized heuristically to determine a limited number of feasible solutions, then an optimal solution is derived via the DSS component.

19.5 SPATIAL REASONING

The human ability to do a variety of mundane tasks (such as reasoning about the positions and volumes of objects) is often taken for granted. These spatial reasoning tasks are fundamental for an engineering design problem, yet are very difficult to model and program.

The chief concept relating to spatial reasoning for design is shape representation, i.e. the shapes of objects and the spaces between them. McDermott (1987) identifies three primary approaches: part-whole, surface and volume representations.

Part-whole representation describes objects in terms of the components from which they are made. The components are arranged in networks which, depending upon the intricacies of the problem domain, range from simple rule-based inference nets to more complicated structures involving frames and inheritance capabilities. From a design engineering perspective, a part-whole description would also include considerations of functionality. The XCON system, described later in this chapter, utilizes part-whole representation to configure VAX minicomputers. It considers both spatial and functional requirements, and satisfies constraints regarding cable lengths, parts fitting into limited spaces, etc.

The second approach to shape description is surface representation. This method portrays objects based on the shape of their boundaries. This approach works well with design engineering projects involving numerically-based data in graphic design systems. The SightPlan expert system, also described later in this chapter, incorporates this general approach.

The final method is volume representation. Objects are described in terms of their volumes, which are normally accessed from a library of primitives. The object volumes are then assembled by specifying locations, dimensions and orientations. As with part-whole representation, volume description may incorporate function properties. Cunningham and Dixon (1988) have applied this approach to the design of manufactured parts.

19.6 INTEGRATION WITH CAD DATABASES

Expert systems frequently utilize internal data sources. As identified by Ben-Arieh and Moodie (1987), the capability of an expert system is increased with access to external factory data sources. A particularly effective ES architecture would provide a linkage of facts (available from a computer-aided design (CAD) database) and heuristics (from a design expert system).

A CAD database may contain several types of descriptive information that can be used by a design expert system. These include component geometry (volume, surface area), functional properties, technical specifications (for assembly, test of operation) and administrative information (e.g., vendor, contract). Topological information, provided by a purely graphic CAD system, can be interpreted and converted to a limited degree. (Dym and Levitt, 1991) Several software packages are available that can access CAD data and translate it into a frame representation. Each component is then represented as a single frame at a level of detail appropriate to the problem area. Data would then be automatically available for expert system processing for routine and variant design problems.

19.7 EXPERT SYSTEM APPLICATIONS IN DESIGN

A variety of expert systems have been constructed in order to support the function of engineering design. A representative listing is discussed in this section. Among the first identified in the literature is ALADIN (ALuminum Alloy Design INventor). Its primary function is to suggest an alloy composition that would satisfy an input set of criteria, such as tensile strength (Rychener *et al.*, 1979). The system used a line of reasoning similar to the generate-and-test model. The generator selected a known material that had properties similar to the input criteria. This known material provides a baseline for progressive iterations regarding the projected effects of changes to its composition and/or processing methods. The effects of these changes on the alloy properties were noted, and provided the next baseline. This iterative approach continued until the modified composition would match the input constraints.

Perhaps the best documented design expert system is XCON. XCON (initially called R1) was developed to support the configuration of minicomputers at the Digital Equipment Corporation (DEC). The configuration problem resulted from DEC's marketing strategy of offering system customization to customer requirements. All possible component combinations did not necessarily work together. This problem was obviously combinatorially complex and knowledge intensive, thus a series of heuristics were applied to create an acceptable configuration.

The XCON/R1 development project began in 1978, and a prototype system was demonstrated in 1979. It was implemented in 1981, and since has been used on a routine basis successfully to configure 97% of all VAX orders (Dym and Levitt, 1991). The knowledge base contains properties of various components, configuration rules and general sequencing rules. Processing utilizes a hierarchical generate-and-test approach to intermingle spatial and functional reasoning using part-whole representation.

XCON has been continually updated and its functions expanded to include production planning, delivery scheduling and interactive sales assistance. The latest report in the literature identified XCON as incorporating more than 8,000 rules executing under the OPS5 expert system environment (Ignizio, 1991). OPS5 is a LISP-based software tool that normally executes in a forward-chaining mode. The rapid expansion of its rule base has caused some maintenance problems.

The success of XCON has resulted in the development of a series of other configuration-support expert systems. One example, referred to as PRIDE, addresses the design of mechanical paper-handling systems inside of photocopiers. This design problem involves the consideration of geometry, timing, forces and spatial layout. PRIDE utilizes an extensive knowledge base that includes specific design steps, dependencies between these steps, how to execute them, relevant constraints, and heuristics for testing and modifying the solution (Mital *et al.*, 1986). It uses a frame-based inference mechanism, in conjunction with a constraint satisfaction approach, to achieve a set of input design goals. PRIDE was implemented in the LOOPS expert system environment, with supporting LISP code, on a XEROX 1100-series LISP processor.

IBDS (Intelligent Boiler Design System) is a configuration-support ES that utilizes a different approach. Rather than utilizing hierarchical generate-and-test, IBDS assembles solutions from basic components. This is a feasible approach for both the preliminary and detailed designs of semi-custom-built products. IBDS is utilized by Tempella Power Industries of Finland to design boilers and other power plant components. Specifically, it aids in the design of boiler piping layout. Like PRIDE, IBDS is frame-based, however it has been integrated with a ComputerVision CAD system and an Oracle relational database to extend its capabilities (Dym and Levitt, 1991).

Expert systems have also been developed to support civil engineering tasks. HI-RISE was created for the preliminary design of large buildings. The system utilizes a hierarchical generate-and-test approach. It attempts to generate all possible design solutions at a very summary level, then heuristically reject most of these from further consideration (Maher, 1984). Remaining solutions are expanded in detail, then evaluated again. This series of iterations continues to the lowest level of design detail, where all remaining solutions are ranked according to a set of evaluation criteria. Another expert system, called SightPlan, was developed to assist in the location of temporary facilities at a construction site. It utilizes constraint-based spatial reasoning to locate facilities in a two-dimensional plane, by considering facility dimensions; worker travel time between facilities; access for large equipment; safety and other concerns (Tommelein, 1989).

19.8 ES AND CONCURRENT ENGINEERING

There is accelerating interest in design expert systems as their benefits are increasingly acknowledged in the literature. For example, Dym and Levitt (1991) identified that the design of circulation pipes using manual methods frequently took two months, while utilization of the integrated IBDS system could reduce this task time to two days.

The application of expert systems to support concurrent engineering is an area of particular interest. Concurrent design, for routine types of engineering tasks, often involves parallel work by disparate specialists with periodic coordination for decision-making (as in a design review). An alternative type of expert system architecture has been developed which provides an initial approximation of the concurrent engineering process. This architecture utilizes multiple knowledge bases which interact though what is referred to as a 'blackboard'. The blackboard is essentially a type of working memory shared by all of the knowledge bases, allowing each to concentrate on one aspect of the problem domain. The blackboard records intermediate states and decisions by the knowledge bases, and is supported by a scheduler that determines the sequence of knowledge base access. This blackboard architecture provides an incremental approach to problem-solving by multiple expert sources, with a strict control mechanism. Several design expert systems have been developed utilizing this approach. The previously-mentioned SightPlan provides one example.

The concept of Cooperative Distributed Problem Solving (CDPS) may provide a better model of concurrent engineering. CDPS is an extension of the blackboard architecture which utilizes a network of discrete expert systems. Each ES node then works in cooperation with the others to solve a complex problem. Computer hardware technology has evolved to a point where the implementation of such networks is now both technically and economically feasible. CDPS is a recent area of research. This decade will be marked by active research and application development for concurrent engineering problems using both blackboard architectures and CDPS.

19.9 CONCLUSIONS

The function of design is one of the most challenging responsibilities of the engineer. Design encompasses considerable domain-specific heuristics. As such, the development of expert systems to support this function is a logical approach. Because true creativity is difficult to model, existing design expert systems primarily support the routine aspects of component selection, component parameter design, configuration selection and configuration design. Design expert systems represent an emerging technology, one that will have an increasing value to the engineering community.

19.10 REFERENCES

Ben-Arieh, D. and Moodie, C. (1987) *Journal of Manufacturing Systems*, **6**(4), 287–97.

Brown, D. and Chandrasekaran, B. (1989) *Design Problem Solving*, Pitman, London and Morgan Kaufman, Los Altos, CA.

Cunningham, J. and Dixon, J. (1988) *MDA Technical Report No. 3-88*, Department of Mechanical Engineering, University of Massachusetts, Amherst, MA.

Dym, C. and Levitt, C. (1991) *Knowledge-based Systems in Engineering*, McGraw-Hill, New York.

Hayes-Roth, F., Waterman, D. and Lenat, D. (eds.) (1983) *Building Expert Systems*, Addison-Wesley Publishing Co., Reading. Ma.

Ignizio, J. (1991) *Introduction to Expert Systems*, McGraw-Hill, New York.

Leonard-Barton, D. and Sviokla, J. (1988) *Harvard Business Review*, **88**(2), 91–8.

Maher, M. (1984) HIRISE: a knowledge-based expert system for the preliminary design of high rise buildings, Ph.D. Dissertation, Department of Civil Engineering, Carnegie-Mellon University, Pittsburgh.

McDermott, D. (1987) in *Encyclopedia of Artificial Intelligence* (ed. S. Shapiro), John Wiley & Sons, New York, pp. 863–70.

Mital, S., Dym, C. and Morjaria, M. (1986) *IEEE Computer*, **19**(7), 102–14.

Moynihan, G. (1992) Expert systems and their application to the foundry, to be published in *Industrial Engineering in the Foundary* (ed. J. Matson), American Foundrymen's Society, Desplaines, Illinois (in press).

Parsaye, K. and Chignell, M. (1988) *Expert Systems for Experts*, John Wiley & Sons, New York.

Rychener, M., Farinacci, M., Hulthage, I. and Fox, M. (1979) in *Proceedings of the Fifth National Conference on Artificial Intelligence*, Philadelphia, pp. 878–82.

Savino, J. (1990) *Chilton's I&CS*, **63**(1), 90–1.

Stefik, M. (1990) *Introduction to Knowledge Systems*, Xerox Palo Alto Research Center, Palo Alto, CA.

Tommelein, I. (1989) SightPlan: an expert system that models and augments human decision-making for designing construction site layouts, Ph.D. Dissertation, Department of Civil Engineering, Stanford University, Stanford, CA.

CHAPTER 20

A knowledge-based approach to design for manufacture using features

Eoin Molloy and J. Browne

20.1 INTRODUCTION

To start with a quotation from Canty (1987)

> CE (Concurrent Engineering) is both a philosophy and an environment. As a philosophy, CE is based on each individual's recognition of his/her own responsibility for quality of the product. As an environment it is based on the parallel design of the product and the processes that affect it throughout its life-cycle.

It is the provision to the individual of tools to aid their contribution to product quality, as well as the facilitation of the parallel development of product and process design through new methodologies as well as tools which must be primary goals of current research work in CE and design for manufacture (DFM).

At its most basic level CE involves a team approach, where people from design, manufacturing, marketing, maintenance, etc., work together to define the product and the processes that affect it during its lifetime. Most authors agree that it is essential to form multidisciplinary design teams at the start of the design process (Corbett *et al.*, 1991). Thus from a manufacturing viewpoint CE requires that the individual functions leading to the finished product are coordinated to work together from the concept stage to arrive at a product that is easy to make and test. This should lead to a shortening of the product life-cycle, and stricter control of all phases of product development. To accomplish this, manufacturing engineers must be provided with an input into the design process.

Improved communication procedures and computer-based tools must be provided to enable CE to take place (see for example the US Department of Defence DARPA initiative in CE – acronym DICE – whose mission is to create an open computer-assisted CE environment (GE Aircraft Engines, 1989).

Indeed, without bringing about this team approach to product development, the long-term benefits of introducing specific design or manufacturing DFM tools are dubious. Real DFM tools can only be developed through design and manufacturing deciding how they want to work together, and jointly designing the tools which will help them achieve integration.

Stoll (1988) presents the most common DFM methodologies. Most of the currently available DFM/A techniques are spreadsheets, requiring the designer to answer questions relating to the product and its components, their form and functionality, and how they interact. Examples of these are the Lucas DFA technique (Lucas Engineering Systems, 1988), the Hitachi Assemblability Evaluation Method (Miyakawa and Ohashi, 1986), and the Design for Assembly Spreadsheet (Poli, 1984). The following issues are still to be overcome, however:

- It is desirable that DFM systems should analyze the design directly, not rely on the designer correctly to reply to questions concerning the design and its components;
- The actual analysis done on a product design should reflect the actual manufacturing concerns of the user;
- The results of the analysis should be *quantitative*, offering design recommendations/alternatives; and
- There should be a mechanism for incremental capture of manufacturing rules and decisions.

Most methodologies evaluate a design for criteria which may not be relevant to the specific company's processes, thus achieving a very low level of integration into the design and manufacturing functions of the company.

Current research in DFM is focusing on a knowledge-based approach to the design of DFM expert systems (or computer aided DFM–CADFM). These would consist of a knowledge base and an inference mechanism for applying the knowledge in the knowledge base to product designs. An example is the prototype artificial intelligence (AI) based design system CADEMA (O'Grady *et al.*, 1988). This system analyses design from a manufacturing and assembly viewpoint and recommends changes to improve the functional and manufacturing aspects of the design (Yap, 1991).

Given that there is a need for DFM, the following questions should be answered before an implementation can be considered:

1. What/who are the DFM knowledge sources in the company? How can the knowledge be organized in a form accessible either by design/manufacturing personnel, or by some form of DFM expert system?
2. Which products and processes will be affected by DFM?

3. Is the final solution capable of being used at all stages in the product design, or is a finished design required? Obviously, the earlier in the design process the DFM solution can be applied, the greater the savings in time and money.
4. What training will be required to integrate DFM techniques into the product life-cycle process, and what will be the cost in terms of additional time and effort in using it?

It is plain that there are many issues to be addressed in the implementation of DFM, not least of which is that the designer must have information on the implications of the possible choices of manufacturing, assembly, or joining which are implicitly determined by the designer. The designer can only take into consideration these choices, and make the optimum decision, if he has a proper source of manufacturing information, presented in a format which he can relate to his design.

From reviewing the current literature on DFM techniques, it is evident that there are few generally applicable DFM solutions; instead various DFM systems are evolving to be applied to specific types of manufacturing. In spite of this specialization, there are common areas and trends in current research into DFM, such as:

• The means by which the DFM knowledge is acquired;
• The use of expert systems techniques, most recently with object-oriented programming methods; and
• Interfacing of DFM with CAPP (Computer Aided Process Planning) and the interpretation of design data by DFM and CAPP systems. Feature-based CAD systems may help provide a solution.

20.1.1 Expert systems in DFM

An expert system can be described as any computer system that exhibits expert performance in a given domain (Yap, 1991). It offers the opportunity to organize human expertise and experience into a form that the computer can manipulate (Buchanan and Shortliffe, 1985). This build up of expert knowledge allows users to access information without the need of an expert.

Professor Edward Feigenbaum of Stanford University defined an expert system as:

> An intelligent computer program that uses knowledge and inference procedures to solve problems that are difficult enough to require significant human expertise for their solution. Knowledge necessary to perform at such a level, the inference procedures used, can be thought of as a model of the expertise of the best practitioners in the field. (Harmon and King, 1985)

In general, expert systems consist of three main parts: a knowledge base; an inference engine; and a user interface.

The knowledge base is the part of the expert system containing application specific reasoning knowledge that the inference engine uses in the course of reasoning. Hence the knowledge supplied by experts must be transformed into a formalized structure and a set of operations that expresses the description, relationship and procedure of the body of knowledge (Yap, 1991).

The contents of the knowledge base are used by the inference engine, which applies them to particular situations (in our case a design). The inference mechanism in an expert system controls the sequence and application of the rules. It is the inference engine which poses the biggest problem in defining generic DFM systems, as it is here that the actual analysis of the design takes place.

Last but not least, the means by which the users access the knowledge storage and analytical subsystems of the DFM system are controlled by the user interface, which must always be programmed with the user and working environment in mind.

20.1.2 Knowledge acquisition

A great deal of time and effort is involved in collecting and categorizing sufficient manufacturing knowledge to populate a comprehensive DFM knowledge base. As well as identifying the design factors influencing the manufacturing process, the impact of these factors must be quantified in some way (e.g. in terms of criteria such as assembly time, cost, addition of process steps). Also links between alternative design choices may be established. A study of the literature of artificial intelligence shows that there are no clearly established techniques for the task of knowledge acquisition, most attention being given to the structuring of this knowledge (for example in the form of rules) once it has been acquired. As stated by Bryant (1988): 'Knowledge acquisition, the bottleneck of developing applications, is not a clearly defined science. At best it is an art practised by the few.' One issue which must be resolved in the design of the knowledge base is the conflict between the way in which the expert expresses his knowledge and the way it is stored in the knowledge base.

The DFM knowledge base will act as a manufacturing tool and reference database, as well as input to a DFA expert system used primarily by designers. It should therefore be open to access from other tools that may be developed in the future.

Swift (1987) suggests some ways of aiding the task of acquiring knowledge from experts, such as intelligent interfaces that can interview the expert and formulate the rules directly, or a learning system that can be used to induce rules from examples or by reading textbooks and papers. In his expert system, Swift used case studies and discussions to acquire knowledge, thus using only the third suggestion. In this chapter we look at a means of acquiring DFA knowledge in a structured fashion.

20.1.3 Design data exchange standards

Central to the analysis of design information by a DFM system is the interface with the design tool, normally a CAD system. In the end the basic problem is that of feature recognition. As it involves recognition of 'high level' features from 'lower level' geometric entities, the problem varies enormously depending on the geometric representation used for the object containing the features in the CAD database. Many different methods of feature recognition have been investigated, but none is perfect. Joshi and Chang (1990) present a good up-to-date survey of the different recognition methods.

Feature recognition is a complicated issue, with as yet no one method significantly outperforming all the others. Feature-based design seems to offer the way forward to the problem of feature recognition for process planning purposes, but feature recognition will still probably be necessary in the future due to such problems as recognition of intersections between geometric elements, conversion of features from one set of processes to another. Also feature-based design is restrictive, and it would not be possible to create a big enough library of features for all future designers' needs. Even if it were, the task of searching for the correct feature would be prohibitively slow.

However, in the absence of suitable feature recognition techniques, we have opted for the use of a feature-based approach in our research, using the Pro/Engineer CAD system for feature-based design.

(a) Standards and data exchange

Access to the design data can be either directly to the CAD model using the proprietary software of the CAD system, or using a neutral interface. The latter method which we have adopted has the disadvantage that it is not real-time, but the advantage that it allows the easy adaptation of the DFM system to many CAD systems.

In the late 1970s, in answer to the need for data transfer between CAD systems as well as between CAD and CAM systems, the IGES (initial graphical exchange standard) was jointly developed by Boeing, General Electric and the USAF (CASA/SME Technical Council, 1989). IGES functions well for 2D drawing applications, but is not designed for solid modeling and assembly applications. Nowadays, the Autocad standard DXF is also widely accepted and supported for 2D applications. Information such as features are not represented. This led to the setting up of the PDES (product data exchange specification) initiative in 1984, with the objective of defining all the information needed to design, manufacture and support a product. This initiative is supported by the CALS (computer-aided acquisition and logistics support) initiative of the US Department of Defence (Jones, 1990). PDES has since joined forces with the international

(ISO) initiative, STEP (standard for the exchange of product model data), which is to supersede IGES. The first concrete results to emerge from STEP have been in the Express data exchange and modeling language. This is currently being used in numerous EC projects as the basis for CAD data exchange, and in the system presented in this paper.

Current EC research
The exchange of CAD geometric entities was a main objective of the ESPRIT project 322, CAD*I, which ran from 1984 to 1989. It specified and developed a neutral file format and the associated pre- and post-processors. This project was the basis for the geometrical data exchange aspect of the ISO initiative STEP. The STEP program is currently supported by projects such as the ESPRIT projects CADEX, NEUTRABAS and IMPPACT.

CAD geometry data exchange (CADEX)
This project is mostly concerned with data exchange. The aim of the CADEX project (Heinrichs and Helpenstein, 1990) is to develop processors to perform data exchange using neutral files defined by STEP. The specifications are developed by an ISO committee, who provided the Integrated Product Information Model (IPIM). This contains an EXPRESS description of entities. CADEX uses a reduced IPIM geometry set, and specifies application protocols which prescribe which entities and which attributes can be exchanged. The CADEX project contributes to STEP by its feedback of requirements to ISO.

Further information on the actual structures and entities handled by CADEX, and the internal structure of STEP is not yet available to the author at this time.

Neutral product definition database for large multifunctional systems (NEUTRABAS)
The objective of this project (Wilckens and Nizery, 1990) is to develop standardized methods for the storage and exchange of data defining shipbuilding and other large multifunctional systems. It is, in effect, destined to define a neutral format for product definition for marine structures. As an example of a project aiming to specify the product defining data for a large complex structure like a ship, the progress of this project should be followed. It is intended to define all the data required to define the ship over its life-cycle, with strong links to the STEP neutral file.

Integrated modeling of products and processes using advanced computer technologies (IMPPACT)
Esprit II project 2165 (1991) is designed to develop and demonstrate new computer integrated modeling systems for integrating product design,

process and operation planning, and the generation of machine tool data. This project, initiated in 1987 and started in 1989, aims to develop a product database to support the integrated product design and process planning environment.

The project is divided into five main work packages:

WP1. Reference Model and Interface;
WP2. Part modeling (this is focused on the development of improved design software and feature modeling);
WP3. Process modeling;
WP4. Product database (the development of an integrated database which serves as the main integration factor between the software modules developed in WP2 and WP3); and
WP5. Complex shaped parts application and sheet metal parts application.

The part modeler and product database from the core of the product modeler, with the feature modeling acting as a link between different areas of product and process modeling. Interfaces between the CIM modules will be based on the integrated information reference model developed in the language Express.

While this type of representation is useful for machining, it is not possible to create such links between design and assembly via features. Up to now, IMPPACT have implemented and integrated their product and process models using only NIAM (Nijjsens Information Analysis Method) representation techniques, which they intend to translate to tangible Express representation.

20.2 A GENERIC APPROACH TO DFM SYSTEMS

The BRITE DEFMAT project is an EC project of three years' duration (1991–4) whose main aim is to develop a generic DFM systems architecture. The current, first phase of the project is concerned with the development of several research prototype DFM systems, which will contribute to the research and specification of a generic architecture, and eventual industrial prototype. Currently four research prototypes are being developed in Ireland (PCB assembly), England (miniature electromechanical assembly), Germany (electromechanical assembly), and Belgium (metal working). The results of these research prototypes will be used as the basis for the development of a generic architecture and methodologies for DFM.

The DFM system presented here is being developed with Digital Equipment Corporation, Galway, Ireland, and is addressing the surface mount technology PCB assembly process. Specifically, it is aimed at the analysis of designs for Galway-specific design/manufacturing conflicts.

20.2.1 The manufacturing problems

We concern ourselves here uniquely with PCB assembly, as this is the work of the Digital, Galway manufacturing plant. The following process steps take place:

- Bar coding;
- Etch cutter;
- Screen printing;
- Epoxy dispenser;
- Pick and place;
- Infrared prebake;
- Vapour phase reflow;
- Aqueous wash;
- Inspection and rework; and
- Automatic testing.

(a) Main design problems to be addressed

The manufacturing engineers have found that the main design problems may be classified as follows:

1. Proliferation of part numbers;
2. Proliferation of process steps caused by technology mix;
3. Repair of non-standard material;
4. Thermal management of circuitry;
5. Poor component clearance; and
6. Insufficient reference points on component and boards for recognition purposes.

Table 20.1 summarizes the DFM goals generated by the preceding manufacturing concerns.

A large number of rules dealing with particular component and process combinations have been generated to deal with these issues.

20.2.2 The design/manufacturing link

In Digital, module design is performed using PCB layout systems such as VLS. Designs at any stage of development can be converted into the Digital proprietary standard for CAD data, the ADS file. This is the format in which Digital design data are made available to this first prototype (see Fig. 20.1). In order to demonstrate the use of a commercial 3-D feature-based design system with the DFM system, the Pro/Engineer CAD system is also interfaced using the STEP/Express standard. This gives the ability to solve placement problems involving 3-D component geometry, though initially the prototype system is concentrated on simple 2-D problems.

Table 20.1   DFM goals generated by manufacturing concerns

Measure	Action
Reduce feeding cost and time	Fewer components, no new components
Reduce assembly cost	Avoid components needing hand-assembly
Reduce assembly time	Avoid components that need baking
Reduce number of processes	Minimize through-hole components
Reduce number of defects likely	Avoid submitting active components to wave-soldering
Increase process reliability	Observe board thickness limits
Decreased inspection and touch-up	Avoid heat sinks on 25ml components
Ease materials handling	Observe component marking standards
Observe machine capability	Observe circuit size limits
Ease testing	Ensure correct component clearances and pad sizes and clearances

20.3 SYSTEM DESCRIPTION

We have based our approach, in so far as possible, on the use of a common language of features linking the design data, rule acquisition, and the process model. The overall approach is that the process engineer can specify the manufacturing problems he identifies in terms of those product features/parameters involved and the process features/parameters which are involved. Obviously, the domain is different to machining where product design could in fact be done using machining operations as the building blocks. Instead, we have tried to create a system which allows the process engineer to identify the relationships between certain process and product features using DFM rules.

20.3.1 Overall system requirements

The two main aspects to the requirements of the system are the designs which must be analyzed and their representation, and the actual manufacturing knowledge used to analyze them. The outputs are the results of the design analysis and the associated design advice.

The DFM knowledge is available in the form of corporate and Galway-specific rules/guidelines. Some means must be found to store these rules, and also to provide process engineers with new rule generating capability.

20.3.2 Specification of major DFM system components

In this section we describe the functionalities of the five major components of the proposed DFM system (knowledge system, product model, process model, inference engine, user interfaces).

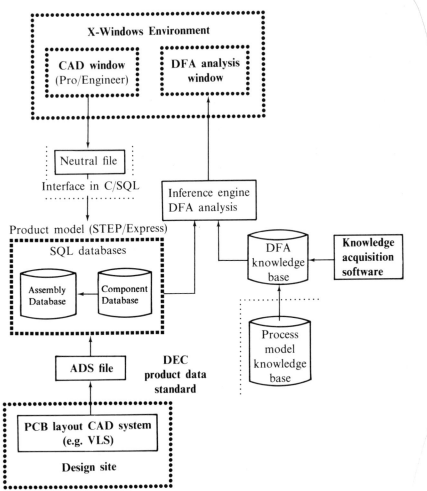

Fig. 20.1 Overall DFM system architecture.

(a) The knowledge acquisition system and knowledge base

In the architecture described here (see Fig. 20.2), the principles of feature-based design and DFM are contained: the manufacturing engineer is to input new rules described in terms of features used by the designer, thus creating a common linking language. Here we have retained the principle of basing the rules on known features and component parameters. This is to be achieved by only presenting to the manufacturing engineer a set of possible features and component/product parameters which he may use in composing a new rule. As the user interface to the knowledge acquisition system is menu-driven (written in X-windows), it is possible to offer the engineer a list of these features and parameters which he may then select and automatically place in the rule.

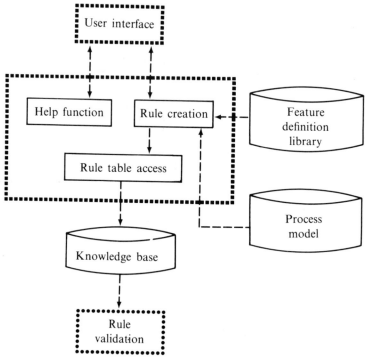

Fig. 20.2 Knowledge acquisition system architecture.

It was decided to implement the rules in an IF/THEN type format, i.e.:

IF There is evidence that A and B are true

THEN Conclude there is evidence that C is true

One side of this rule would then be evaluated with reference to the product model database, and if this succeeds, the action specified on the other side is performed. The action to be performed will be either to execute another rule, or to signal the rule violation to the user.

We decided to define the rules in terms of features (product and component features). These features may be either explicitly defined in the product model, or they may have to be derived from it by the inference engine software. Therefore an expanding knowledge base necessitates an expanding 'intelligence' on the part of the inference engine.

Link between process and product DFM

An important feature of this system is that it must allow manufacturing engineers to create new rules. The rules must reflect the current manufacturing capabilities, however. Therefore the critical values in the rules must be linked to the process model. A system designed to achieve this is described in Fig. 20.3. When creating a new rule, lists of product features are

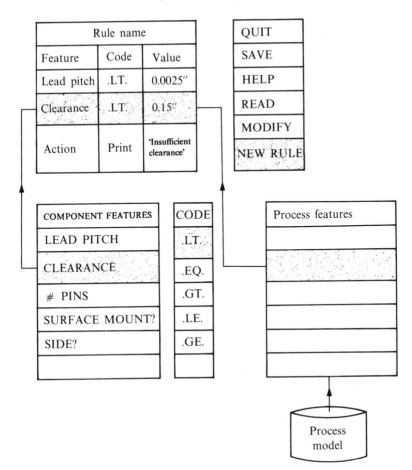

Fig. 20.3 User interface to knowledge acquisition system.

presented to the user in menu form. When a feature is selected from the lists, it is automatically placed in the rule. The critical value associated to that product feature which will trigger the rule is a function of some process feature/parameter, the current values of which are retrieved from the process model and made available. Any number of product and process features may be referenced by the rule in this way.

Once the rule has been fully defined, it is stored in the relational database using the same structure. The user interface also allows the user to search through the rule base using the SQL languages functionality.

(b) Product model

Different types of product and component information must be available for implementation of the rules. In general, geometric knowledge will only be needed to calculate certain derived features, such as component clearance.

The product model is comprised of product-specific or assembly-specific information stored in one database, and component information stored separately. Therefore, for each different product, only the information in the product database should change. The CAD/ADS interfaces and product model have been implemented using the STEP/Express standard (ISO, 1990a, 1990b, 1990c).

The number of different part numbers used in any design is very great. However the number of different package types used is much less. Therefore, any component specific information which is also package specific (for example the number of pins) is not stored separately for each component, but is written into the entry for each package type. This package information is stored in the component database. The structure of the product and component data tables are shown in Figs 20.4 and 20.5. Rather than choose a set structure for the tables, with a column per feature, features can be written in at random, being identified by their name. While this requires slightly more interpretation of the tables to find the value of a given feature, it allows for easier expansion of the tables. Also, as derived features (such as component clearance) are calculated, they can be simply written into the database.

		Assembly table			
Part No.	Ref	Posx	Posy	Rot	Side
10240011	R1	10.1	25.0	180	2
11356600	E40	100	133	0	1
11356600	E41	205	230	0	1
10234545	R2	14.1	25.0	180	2
11356600	E42	109	133	0	1
11356600	E43	115	133	0	1
10240011	R3	18.1	25.0	180	2
11356600	E44	130	133	0	1
11356600	E45	201	133	0	1

Fig. 20.4   Assembly database table.

(c) Process model

The system must include some modeling of the assembly process. This is because the actual values of parameters specified in the DFM rules are normally a function of the actual process or the equipment being used. For example, the amount of package clearance required will depend on the actual width of the placement tool being used. Therefore it is preferable to have the rule in the knowledge base actually depend on an up-to-date process model, rather than have such values hard-coded in the rules themselves.

The processes and equipment used in them are defined in terms of sets of process parameters. Each process is defined by a separate set of parameters,

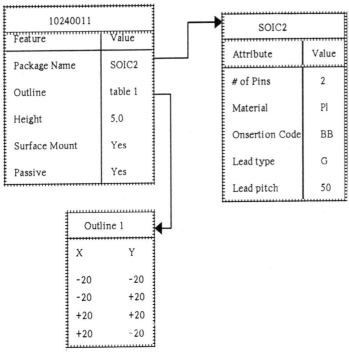

Fig. 20.5 Component database tables.

contained in a data file or database. For each process one machine may be replaced by another. Therefore it is necessary that the process data format is not specific to particular makes of machine. These parameters are then available to the process engineers for the creation of DFM guidelines using the process parameters and product features to define the guidelines.

(d) Inference engine

The function of the inference engine is to analyze the product data for the PCB assembly using the available process information. This is achieved by comparing the values of the product and component features with those specified in the rules. It functions by checking each component against the contents of the knowledge base, and contains the necessary analytical functions to derive those features/parameters not explicitly defined in the product model. By having a common set of features/parameters between the product model and the knowledge base we minimize the intelligence necessary in the inference engine. The output from the analysis is written to the user interface window as the analysis is being performed.

20.4 CONCLUSIONS

In this chapter we have presented the current state of a developing DFM expert system. The main preoccupations in developing this system have been to avoid depending on particular software tools, and to provide links to a dynamic model of the manufacturing capability. As much as possible, this architecture is based on providing manufacturing engineers a simple means of storing DFM knowledge in a usable form; and on creating a comprehensive design database accessible by the DFM systems inference engine. The use of feature, as opposed to purely geometrical, information as the currency of information exchange between the design and manufacturing databases is seen as primary in this approach.

Through this prototype, and the others which are also being developed currently in the BRITE DEFMAT project, it is hoped to synthesize a more generic approach to DFM systems design.

20.5 REFERENCES

Bryant N. (1985) *Managing Expert Systems*, J. Wiley & Sons, Chichester.
Buchanan, B.G. and Shortliffe, H. (1985) *Rule Based Expert Systems*, Addison-Wesley, Wokingham.
Canty, E.J. (1987) *Simultaneous Engineering: Expanding scope of quality responsibility*, Digital Equipment Corporation White Paper.
CASA/SME Technical Council (1989) *PDES: The Enterprise Data Standard*, Society of Manufacturing Engineers, Dearborn, Michigan.
Corbett, J. *et al.* (1991) *Design for Manufacture: Strategies, principles and techniques*, Addison-Wesley Publishing Company, Wokingham.
Harmon, P. and King, D. (1985) *Expert Systems – Artificial Intelligence in Business*, John Wiley and Sons, Inc., New York.
Heinrichs, H. and Helpenstein, H. (1990) *CADEX Ways to STEP Data Exchange Processors*, Project No. 2195, Esprit, CIM, Results and Progress of Selected Projects in 1990, November, pp. 25–41, XIII/388/90.
GE Aircraft Engines (1989) *DARPA Initiative in Concurrent Engineering (DICE)*, paper published by Concurrent Engineering Programs, Cincinnati, Ohio 45215-6301.
IMPPACT Esprit II No. 2165 (1991) Proceedings of the Workshop, IWF Berlin 26–27 February.
ISO TC184/SC4/WG6 (1990a) *STEP Part 1: Overview and Fundamental Principles*. Contact: Howard Mason, British Aerospace, CTU, Dept 357-Q191, P.O. Box 87, Farnborough GU 14 6YU, UK.
ISO TC184/SC4/WG5 (1990b) *EXPRESS language reference manual*. Contact: Philip Spilby, CADDETC, 171 Woodhouse Lane, Leeds LS2 3AR, UK.
ISO TC184/SC4/N64 (1990c) *ISO CD 20303-11. Exchange of Product Model Data – Part 11: The EXPRESS Language*. Secretariat: NIST, A127 Building 220, Gaithersburg, MD 20899, USA.
Jones, P.F. (1990) *Computer-Aided Design*, **22**(6) July/August, 388–92.
Joshi, S. and Chang, T-C. (1990) *Journal of Intelligent Manufacturing*, **1**, 1–15.
Lucas Engineering Systems (1990) *Design For Assembly*, version 7.1.

Miyakawa, S. and Ohashi, T. (1986) *The Hitachi Assemblability Evaluation Method,* Proceedings of First International Conference in Product Design for Assembly, April, 1–13.

Poli, C. (1984) *A Design for Assembly Spreadsheet,* Proceedings of the International Symposium on Design and Synthesis, Tokyo, July.

O'Grady, P., Ramers, D. and Bowen, J. (1986) *Computer-Integrated Manufacturing Systems,* **1**(4), 204–9.

Stoll, H.W. (1988) *Manufacturing Engineering,* **January**, 66–73.

Swift, K.G. (1987) *Knowledge-Based Design for Manufacture,* Kogan Page Ltd., London.

Wilckens, H. and Nizery, B. (1990) *Neutral product definition database for multi-functional systems – NEUTRABAS,* Project No. 2010, Esprit, CIM, Results and Progress of Selected Projects in 1990, November, 25–41, XIII/388/90.

Yap, B.J. (1991) *Development of Computer Aided Tools To Support Strategic Planning,* CIMRU, UCG, Galway.

CHAPTER 21

Concurrent accumulation of knowledge: a view of concurrent engineering

Robert E. Douglas Jr. and David C. Brown

21.1. INTRODUCTION

21.1.1 Concurrent engineering

Concurrent engineering (CE) has been identified as a vital ingredient in America's attempts to modernize its design and manufacturing practices. In this paper we present a discussion of the ingredients of a CE system, an overview of some of the current CE research, and an alternative view of CE.

21.1.2 Ingredients of CE

Based on our research, we have identified twelve major issues in developing tools to support CE. These are the ingredients which have been found to be present in some form or other in many, if not all, of the systems we have studied.

1. Design agents are the principle actors in developing a design. They make decisions, monitor progress, or analyze parts of the design. These agents can be humans, expert systems, or other computer tools. They interact in order to develop a design.
2. Multidisciplinary goal and specification representation is a means of expressing goals and design specifications in terms used by different disciplines, even though they use different terminologies. It is an attempt to unify the terminologies used by the design agents within the scope of a particular product design.

3. A catalog contains descriptions of parts that can be used in the design. These parts can be components or partial designs. They are a means of relating past experience into the current design, and speeding the design process.
4. A design database contains the current representation of the object being designed. This is necessary to keep current information to be used by the agents when designing.
5. An accumulated database contains knowledge about downstream aspects accumulated along with the design. This accumulated knowledge is discussed in Section 21.3 below.
6. Shareability of design information allows all design agents to use all design information, as they see fit. No information should remain the sole property of any agents, as there may be knowledge about a different aspect hidden in that information.
7. Communication allows the design agents to communicate despite their different backgrounds. Support for this may come in translators, discussed below in Section 21.4, as well as through the development of effective communication channels.
8. A manager schedules agents, oversees negotiation, keeps track of resources used, and monitors and evaluates the progress of the design.
9. A design history captures knowledge about alternatives and decisions made, and the rationale behind those decisions. This is useful in determining if the best design was chosen, for credit/blame assignment, for learning, and for use in restoring the design if a design decision has to be retracted.
10. A checklist indicates important decisions that still must be made.
11. A standard interface to all tools keeps the user from having to learn too many different interfaces. This lends a feeling of working in a single environment, and the addition of new tools into the system does not require a long learning period.
12. Virtual collocation of people makes all the agents appear to be in the same room. This promotes unity among the team members and encourages cooperation in determining the next course of action. It also encourages communication.

This list of issues was developed primarily from Bedworth *et al.* (1991), Cunningham and Dixon (1988), Jagannathan *et al.* (1991), Kott *et al.* (1991), Kroll *et al.* (1988), Lemke and Fischer (1990), Lemon *et al.* (1990), Londoño *et al.* (1989), Stoll (1986) and Subramanian *et al.* (1990).

21.2 OTHER RESEARCH

There are several research efforts into computer support for CE currently underway. The best known of these is the DARPA Initiative in Concurrent

Engineering (DICE) at the Concurrent Engineering Research Center (CERC) in West Virginia. There are also the CAD Framework Initiative (CFI), the Open Systems Architecture for Computer-Integrated Manufacturing (CIM-OSA), and the Engineering Information System (EIS). These four are reviewed in Jagannathan *et al.* (1991).

Jagannathan *et al.* identify three areas in which CE research is progressing: management processes, technical practices and information technology. They then focus on the information technology issue, which the four previously mentioned systems support. Within this category, they identify five areas in which computers can be used to support CE: sharing information, collocating people and programs, integrating tools and services with frameworks, coordinating the team, and capturing corporate history.

Sharing information is necessary to promote cooperation among the members of multidisciplinary design teams. The models each of the members uses in the design process may be quite different, however, and the system should be able to translate models done by one designer into the format used by another.

Collocating people and programs is achievable by networking. The key to maintaining this virtual collocation is in making the access to programs, people, and data across the network transparent to the user.

Integrating tools and services with frameworks is a means of allowing designers to use different tools with ease. The idea is to support a single means of user interaction with all of the tools, to make them more uniform. This would lead to less training time on new tools and a greater chance that designers would use all of the tools available to them.

Coordinating the team is a means of keeping all members of the design team apprised of the current state of the design. Whenever a design decision is proposed, all members of the team are informed.

Capturing corporate history is a means of keeping track of design decisions and the reasons for them. It uses an electronic design notebook, or some other means of recording decisions.

Londoño *et al.* (1989) are doing research to build a system to support CE for DICE. This system is designed to help designers participate in cooperative design. They have chosen to use a blackboard for communicating and for control of information flow. All data about parts is held in the product, process and organization (PPO) database. This database is accessible by all of the product's designers. Individual designers use local object workspaces (LOWs), in which they can modify current designs and test hypotheses before committing everyone on the design team to a decision. The whole design process is overseen by the project lead (PL). The PL must be able to keep track of new tasks and follow the progress of the tasks, as well as generate new tasks, in order to change the focus of the design, and assign them to designers.

Representations of the designed products are stored in frames and pointers to AutoCAD files. This information is stored in the PPO database.

Designers must have access to all the information in the PPO database. The database should support user friendly searches for parts on which a designer wishes to work. Though the system defaults to helping the designer in a bottom-up design fashion, there exists the option to do all or part of the design top-down. Initialization of the blackboard includes adding design specifications, determining dependencies, deciding on initial tasks, and including known heuristics.

Global, local, and implicit constraints are supported by the system. Global constraints refer to overall specifications that apply to the design as a whole. Local constraints refer only to individual parts. Implicit constraints are constraints which need to be inferred from the constraints on subparts. Reasoning with constraints can be done to allow the system to spawn new tasks to be done. Dependencies are kept by the system to indicate the other aspects of the design that may be affected by a design decision about a part. Those on the dependency list are notified when a change occurs. Maintenance of versions has not yet been addressed in the system.

Heuristics are another form of design knowledge used in the system. They can be used to help schedule tasks and develop uniform plans of action during a design. Negotiation is an important consideration involving multiple designers. The designers are allowed to vote on accepting a change to the design.

The function advisor (Kott *et al*, 1991) is another CE support system being developed. Their main concerns are those research issues which can be used in allowing 'a computer-based advisory system to support the cooperation between multiple engineering agents'. These include: planning and management of the activities of multiple agents; representing and modeling the product; managing multiple representations and versions of the product; developing engineering databases capable of supporting concurrent engineering; managing and propagating constraints and avoiding inconsistencies in the current state of the product description; sharing the information between the multiple agents without creating an 'information blizzard'.

The objective of the system they are building is to help a group of knowledge workers (designers) design an engine and all of its component parts. They categorize the functional objectives of the function advisor they are developing into four groups: advising the human organizer of tasks to be performed; retrieving, organizing, and conveying pre-stored information relevant to the current needs of the organizer; inferring information that is not explicitly stored; and monitoring the design for consistency, completeness, and correctness.

The more specific functional goals for Kott's system are given below:

- To provide an inference mechanism to reason about the design organization;
- To alert human workers of abnormalities or inconsistencies between the design and goals;

- To suggest a plan of design activities to human designers;
- To advise the human designer on dependencies between parts of the product;
- To find parts, constraints, or other design knowledge that is useful to a particular designer;
- To prevent attempts to finalize a design before designs which affect it are finalized;
- To collect and store all product documentation;
- To detect those parts which are dependent on other design decisions;
- To identify the need for special purpose design aids; and
- To guard against recurring design errors.

Subramanian *et al.* (1990) are developing a design support environment for CE. When different groups work on a design from different aspects, the design with regards to those aspects is usually done coincidentally. They used their system as a testbed for their theories about which kinds of information are used during group design. They discuss the idea that a computer system designed to help groups from the various aspects of product design should work similarly to the way that these groups do. The groups share computations, figures, and any other data they feel necessary. This means having mechanisms to represent designs in different ways and to maintain copies of all the current design data. It also requires a means of scheduling the use of the data.

Knowledge-based systems which communicate by means of a blackboard are used in a project called DICE (Distributed and Integrated environment for Computer-aided Engineering) (Sriram *et al.*, 1989). DICE is made up of a control mechanism to coordinate the activities of its modules and users, a blackboard to post the current state of the design, and knowledge modules to represent various aspects of the design. The system uses an object oriented approach to design. It allows for negotiation between agents. It provides support for the databases that are needed for this kind of distributed design. The knowledge modules have various roles: Strategy (helps the control mechanism by determining the next course of action); Specialist (an aggregation of expert systems to help make decisions); Critics (keep track of consistency in the design); and Quantitative (algorithmic evaluation or CAD tool). The Blackboard is divided into three partitions: Coordination (keeps bookkeeping information); Solution (contains the object hierarchy and current design); and Negotiation (contains constraints on the design and a trace of the negotiation process). All of this allows the user to interact with various autonomous agents when designing part of a building.

Possible tools to be included in a CE system are mentioned in Boothroyd and Dewhurst (1988) and Lemon *et al.* (1990). Both sets of tools are concerned with cost estimation over the entire product's life cycle during the design phase. Incorporating these tools and their methods allows designers access to knowledge about another aspect of the product design and the processes to be used, since cost is affected by the processes chosen in manufacturing, assembly and disposability as well as material selection and other, more concrete costs.

The issues concerned with version control are covered in Katz (1990). In his work, Katz attempts to unify terms used in version management and propose a scheme for developing a complete version management system. The details are quite extensive. What he proposes is an organization in the database which connects versions of different parts together to form a design. But it also connects those versions to their parents and offspring. So the entire database will contain links to different versions of the parts, and these can be combined in different ways to obtain different current designs. These links can be used to search for alternatives and to propagate constraints to all versions of a part.

There are several different thrusts in the area of computer support for CE. CERC and the DICE initiative concentrate on the availability of information and information sharing. Their main goal is to provide as many people as possible with as much useful data as possible so as to make decision-making easier. This is also the thrust of the CAD framework initiative (CFI), the open systems architecture for computer-integrated manufacturing (CIM-OSA), and the engineering information system (EIS). Londoño *et al.* are also working to support this view. They are also concerned with supporting negotiation.

Subramanian *et al.* are also concerned with sharing information among the members of the design team and supporting this kind of inter-change. Sriram *et al.* are concerned with communication between intelligent agents involved in the design. They are also concerned with the computer's handling of negotiation in the event that some decision is not amenable to all parties. Kott *et al.* concentrate on using the computer to keep track of the design, maintain the focus, and monitor the consistency of the design.

In this section we have attempted to present a summary of some of the current CE research, and have tried to reflect what their main concerns are. We have also attemped to characterize (perhaps wrongly) what they consider as the key aspects to the problem of producing CE systems. In the next section we will discuss a different view of CE.

21.3 CE AS KNOWLEDGE ACCUMULATION

Clearly, as CE systems require many ingredients, many views of what is 'key' are possible. Most of these views concentrate on the needs of the agents in the system (human or computerized) to communicate, or to have the right information available. We would like to propose an alternative view.

We suggest that the key issue in CE is the accumulation of knowledge. Thus the focus of this view is on 'what' is being decided or learned, as opposed to 'who' is deciding or 'how' it is being done. The primary purpose of CE is to produce a design. That is clearly a process of knowledge accumulation – from requirements to a design.

In addition, a CE system should produce descriptions of all other aspects – of the manufacturing process, the assembly process, the design for packaging, etc. Thus the goal is to accumulate knowledge about all of the aspects of the life cycle. Consequently, this view of CE as knowledge accumulation can be seen to underlie all CE activities, and is independent of the processes used to generate the knowledge, and the strategies for controlling them.

Figure 21.1 presents a diagram that characterizes this knowledge accumulation process. Knowledge is accumulated during the design process about all relevant aspects of the life-cycle. In general, over time this knowledge moves from abstract to concrete, although in fact this transformation is more complex (Brown, 1993). The design is complete when all relevant knowledge has been accumulated, and not merely when the component or product design is complete.

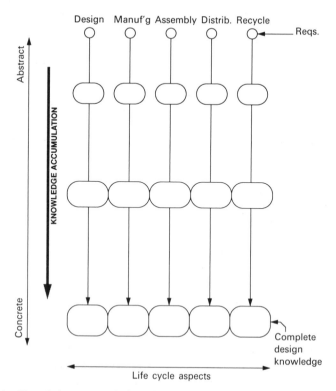

Fig. 21.1 Knowledge accumulation along multiple paths.

21.4 CONSEQUENCES OF THE ACCUMULATION VIEW

One consequence of this view is that it separates what is to be decided from how it is to be decided. In particular, it allows for a variety of knowledge accumulation strategies.

In many discussions of CE, the strategy for including inputs from agents about all downstream aspects of the life-cycle is not discussed. Some CE articles assume that all the effects on the product design from all aspects of the life-cycle somehow act at the same time, without describing the control strategy to accomplish that. This is somewhat acceptable when describing CE systems that consist of teams of people, but is not adequate when describing automating CE using computers. The strategy is often described as negotiation.

However, in many cases it is assumed that the goal of completing the product design drives the whole process, and that, apart from some early feedback about downstream concerns, highly detailed consideration of other aspects besides the design are addressed in their normal life-cycle sequence.

Of course, other strategies are possible. In Altamuro (1991) 'Strategic Product Design' is described as a form of concurrent engineering system where the dominant strategy is working backwards (p. 44). Thus, disposal and recycling could be used to drive maintenance, and so on back to the product design itself. An extension of this idea is the possibility of using the accumulation of knowledge about any of the aspects as the driving force for the whole design process. Thus, for example, the whole design process could be driven by packaging design. Decisions made about packaging would have ramifications for other aspects, which would lead to knowledge being accumulated about all aspects.

The knowledge accumulation view of CE produces an emphasis on learning as much as possible about all aspects during the design and not just focusing on the design. For convenience, let's refer to vertical lines in Fig. 21.1 that represent accumulation as *paths*. The knowledge accumulated down each path could be from an expert or experts. This would occur when a decision is made, or information is retrieved. For example, on the packaging path it could be decided to use bent cardboard inserts to protect the component in its box, while on the maufacturing path, database retrieval of company design policy about objects of a certain size could fix the manufacturing processes to be used.

Alternatively, a decision on path A may be seen to produce knowledge relevant to path B. This may require some inference. For example, the design path decision to use a certain material could affect the cost, manufacturing, and recycling paths. A similar example is shown in Fig. 21.2. Clearly some form of 'translator' is required to propagate this information from one path to another. These are shown in the figure by arcs from one path to another.

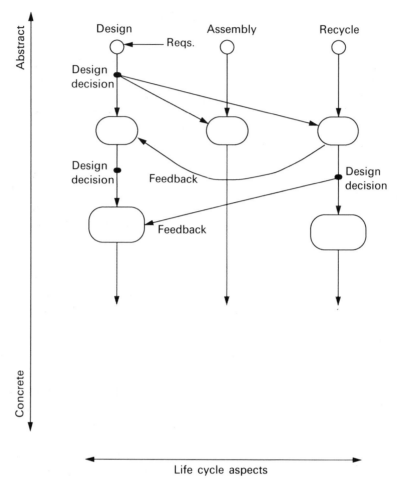

Fig. 21.2 Translators and agents.

The translators are data analysis expert systems whose function is to translate knowledge accumulated along one path into data usable by a different aspect. Data to be propagated to another aspect must go through a specialized translator first. This helps solve some of the communication problems between the different aspects in CE systems. Only with these translators will one aspect be able to communicate data to another.

In some cases it may be appropriate for decisions to be made about some aspect other than the design, such as recycling (Fig. 21.2), and for the consequences of that to be fed back to the design, or to any other 'earlier' path.

21.5 CONCLUSION

The view we have presented requires representatives from several different aspects of the design. There is no requirement that these 'agents' be human. Several different types of artificially intelligent agents can take on the responsibilities of accumulating the knowledge and making design decisions. The upper feedback arrow in Fig. 21.2 shows the action of an agent. We are currently developing a system in which all of the agents, except for the user, are expert systems. The knowledge accumulation view is still quite valid.

In our research, we have identified six different types of intelligent design agents (Brown, 1992). There may be others. These six types of agents have different tasks and are not tied to a particular aspect. There could potentially exist at least one type of agent for each aspect. However, some of the agent types may counter the need for others. The six agent types are as follows:

1. A designer advisor is able to make abstract to concrete design recommendations. A design advisor is an expert on some aspect of the overall design. It would know how and when to use the analysts (see below).
2. A critic compares the design to certain standards, or the design specifications, and decides what is missing from the design, or how it holds up against those standards.
3. A suggestor takes the criticism of a critic and offers suggestions for satisfying the criticism. These suggestions may not take into account other aspects of the design. That issue is for the negotiator.
4. A negotiator uses knowledge about the design specifications to help resolve a negotiation situation, such as conflicting suggestions. It also keeps track of constraints across multiple aspects in order to avoid violations.
5. An analyst offers numerical analysis of strengths, capacities, or time to do something. For example, it could evaluate a standard formula. Another agent must activate it.
6. An evaluator evaluates the design with respect to some quality issue. It takes into account the whole design and not just parts of it (as the critic does). There may be several aspects that can be evaluated separately.

These agents have different responsibilities and can act collectively to provide a variety of different analysis and decision-making tools.

In summary, we have tried to present in this paper an alternative view of CE as the concurrent accumulation of knowledge. This view does not exclude other views of CE, but should provide a new perspective from which to look at the problem of supporting CE.

21.6 ACKNOWLEDGMENT

This work was partially supported by a grant from the Design For Manufacturing Institute, Digital Equipment Corporation, Northboro, MA. We would like to acknowledge the WPI AI in Design Group, and Prof. David Zenger.

21.7 REFERENCES

Altamuro, V.M. (1991) *Concurrent Engineering*, **1**(2), 39–45.

Bedworth, D.D., Henderson, M.R. and Wolfe, P.M. (1991) *Computer-Integrated Design and Manufacturing*, McGraw-Hill, New York, pp. 134–76.

Boothroyd, G. and Dewhurst, P. (1991) In *Design for Manufacture*, (Eds. J. Corbett, M. Dooner, J. Meleka and C. Pym), Addison-Wesley, Reading, MA, pp. 165–73.

Brown, D.C. (1992) *Encyclopedia of AI*, (2nd edn.), (Ed. S. Shapiro), J. Wiley, New York.

Brown, D.C. (1993) Routineness revisited, *Mechanical Design: Theory and Methodology*, (Eds. M. Waldron and K. Waldron), Springer-Verlag, New York.

Cunningham, J.J. and Dixon, J.R. (1988) *Computers in Engineering*, ASME, pp. 237–43.

Jagannathan, V., Cleetus, K.J., Kannan, R., Matsumoto, A.S. and Lewis, J.W. *Concurrent Engineering*, **1**(5), 14–30.

Katz, R.H. (1990) *ACM Computing Surveys*, **22**(4), 375–408.

Kott, A., Kollar, C. and Cederquist, A. (1991) The role of product modeling in concurrent engineering environments, submitted to the *Journal of Systems Automation*.

Kroll, E., Lenz, E. and Wolberg, J.R. (1991) In *Design for Manufacture*, (Eds. J. Corbett, M. Dooner, J. Meleka and C. Pym), Addison-Wesley, Reading, MA, pp. 203–14.

Lemke, A.C. and Fischer, G. (1990) *Proc. Nat. AI Conf.*, *AAAI-90*, pp. 479–84.

Lemon, J.R., Dacey, W.E. and Carl, E.J. (1990) Concurrent product/process development, Technical Report, International TechneGroup Incorporated.

Londoño, F., Cleetus, K.J. and Reddy, Y.V. (1991) In *Computer-Aided Cooperative Product Development: Proceedings of MIT-JSME Workshop on Cooperative Product Development*, (Eds. D. Sriram, R. Logcher and S. Fukuda), Springer-Verlag, New York, pp. 26–50.

Sriram, D., Logcher, R.D., Groleau, N. and Cherneff, J. (1989) DICE: An object oriented programming environment for cooperative engineering design, Technical Report No: IESL-89-03, Mass. Institute of Technology, Cambridge, MA.

Stoll, H.W. (1991) In *Design for Manufacture*, (Eds. J. Corbett, M. Dooner, J. Meleka and C. Pym), Addison-Wesley, Reading, MA, pp. 107–29.

Subramanian, E., Podnar, G. and Westerberg, A. (1990) A shared computational environment for concurrent engineering, Engineering Design Research Center Technical Report, Carnegie Mellon University, Pittsburgh, PA.

CHAPTER 22

Integrated knowledge systems for adaptive, concurrent design

Steven H. Kim

22.1 INTRODUCTION

Every organization faces a world in flux. The dynamism arises primarily from internal developments such as retiring experts, or exogenous factors such as advancing technologies, shifting demographics, and rising consumer expectations.

Even when the environment is relatively tranquil, an organization faces the threat of competition from other extant players or potential entrants. It must therefore continually strive to improve itself against past performance. The enhanced ability of a system to fulfill its objectives has been called learning (Kim, 1990a). A key example lies in the need to traverse quickly down the learning curve of production costs. A doubling of cumulative production often results in a decrease in production by a factor of 3%, 10%, or some other number depending on the industry. The decrease arises from the simplification of designs, the substitution of cheaper materials, or other learned schemes.

The learning process must be actively managed, and will not occur of its own accord. A complacent organization which fails to learn is one that comes to play a diminished role in the market place, and all too often disappears altogether.

The capability of an organization to serve its societal role is reflected in its ability to market its products, whether in the form of goods or services. For this reason, the design of the product in conjunction with the design of the production, distribution, marketing, and administrative strategies embody the mission and competence of the company.

22.2 PERFORMANCE FACTORS AND KNOWLEDGE BASES

The factors of corporate performance may be broadly categorized into internal and external groups as indicated in Fig. 22.1. These factors

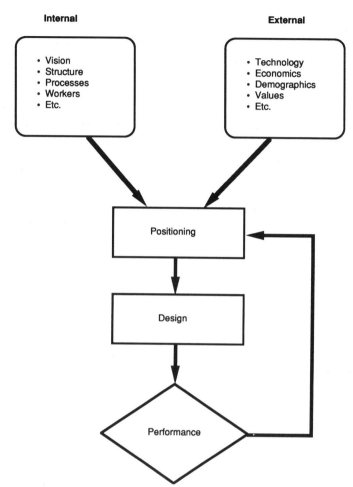

Internal

- Vision
- Structure
- Processes
- Workers
- Etc.

External

- Technology
- Economics
- Demographics
- Values
- Etc.

Positioning

Design

Performance

Fig. 22.1   Sources of corporate performance.

collectively determine an appropriate role for the organization, which can be translated into its positioning in the competitive arena. The design of products, processes, and systems follows, leading finally to the organizational performance.

The tasks of positioning and design are continual activities, since the level of performance in the market has implications for the positioning strategy. Moreover, both the internal and external environments are in constant flux, thereby necessitating an iterative process of learning and adjustment.

An individual organization has limited ability to affect the external environment. From a macrolevel perspective, however, the collective performance of all organizations is itself a critical factor in the environment. A classic example is the galloping pace of computer technology; increasing benefit/cost ratios lead to the infusion of intelligence in various activities, from reducing materials costs in production to conserving energy in operations across diverse industries.

The heterogeneity of the sources of performance underscores the need for a rational means for their comprehension and management. This objective can be supported by an integrated knowledge system to assist in strategic decision-making.

A number of critical components within such a Strategy Support System (S^3) is depicted in Table 22.1. The components relate to external factors such as international trade flows and industry structure as well as internal factors such as specialized resources (Kim, 1990a, Ch. 12; Kim, 1991, Ch. 17; Knight and Kim, 1991; Porter, 1985, 1990; Roush, 1991; Scheel, 1991; Wiseman, 1988).

Table 22.1   Components of a knowledge base for competitiveness

- International competitiveness
  - Trade patterns
  - Worldwide coordination
  - Efficient processes
  - Other systemic advantages
- National competitiveness
  - Primary resources
  - Infrastructure
  - Technology concentration
  - Availability of supplies
  - Labor pool
  - Demand factors
- Industry attractiveness
  - Competitors
  - Buyers
  - Suppliers
  - Substitutes
  - Potential entrants
  - Entry barriers
- Positioning factors
  - Differentiation features
  - Cost issues
  - Production capabilities
  - Specialized resources
  - Reputation
- Design advantage
  - Products
  - Processes
  - Systems

22.3 SYSTEM ARCHITECTURE FOR STRATEGY SUPPORT

To be effective, a company must draw on the entire gamut of knowledge available within the organization and in the external environment. This can be achieved by the use of integrated knowledge systems which encode reasoning as well as serve as a substrate for diverse databases.

A network configuration for such a strategy support system is illustrated in Fig. 22.2. The network encompasses knowledge bases internal to the company, and can access external databases as needed for timely decision making. The reasoning modulus of the S^3 can be adapted from a number of systems developed in the past. Examples of such relevant systems are described below.

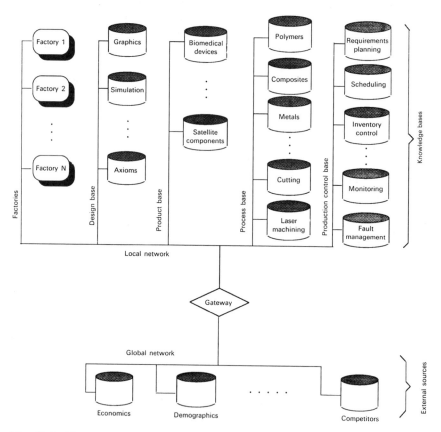

Fig. 22.2 Network configuration for a strategy support system (see Kim, 1990a, p. 153).

22.3.1 Creativity support system

Creativity has been defined as the product of resolving difficult problems; it springs from the deliberate selection of options to satisfy a challenging objective. By encoding the decision rules of creative thinkers in specific domains, computer systems can assist in generating solutions to difficult problems (Kim, 1990b).

The creativity support system (CSS) represents a general purpose architecture to assist in addressing difficult tasks. This general structure has been tailored to a package called the concurrent design advisor (CDA), a system for simultaneous synthesis and analysis of products. The CDA focuses the human expertise of a product design group toward creative solutions for satisfying customer needs.

The issues involved in computer-assisted group decision making have a long history. One topic concerns resource allocation and control for project management (Malcolm *et al.*, 1959). However, aside from tools such as Gantt charts and project network diagrams, very few group projects are managed using such computer support systems (Liberatore and Titus, 1983). The management of product development requires the coordination of many different areas of expertise, functional departments, and evolving specifications, and hence has proven too complex to be supported by a simple computer-based approach.

The integration of cooperative knowledge bases, each possessing a unique perspective on the product development task, has proved successful for tackling problems which are beyond the scope of individual experts. This approach recognizes the fact that, to be effective, an organization must make full use of all the knowledge at its disposal (Lander and Lesser, 1989). The research in distributed artificial intelligence has led to approaches such as a constraint-directed negotiation technique for coordinating distributed knowledge sources. Such systems provide advantages in speed and reliability, in addition to a conceptually clear, modular structure.

The creativity support system and its domain-dependent modules have been organized using the result-sharing form of distributed cooperation. That is, its independent knowledge bases assist each other by sharing results as they become available. The computation performed by each module depends on the current structure of the solution as it evolves as a result of the actions of the other modules.

The creativity support system (Kim, 1990b, p. 111) consists of two primary modules, the domain-independent and the domain-dependent modules. The former component contains knowledge independent of the problem domain, while the latter contains information specific to the task at hand.

The Domain-dependent module contains a repertoire of the uses to which the CSS may be put. The system also has access to diverse fields or disciplines that may pertain to the problem, such as knowledge of statistical methods, financial accounting, experimental design, or others. Some of this knowledge may reside within the system, while other portions may be available in databases accessed through telecommunication networks. These knowledge bases reflect the need for diversity in addressing difficult problems.

The case base contains knowledge of a given problem, including specification of the task and the solution as it develops. For the example of product design, the problem definition is stated in terms of the customer needs, while the design concepts which satisfy this problem definition are stored in the case base.

The domain-independent module contains generic capabilities such as general-purpose strategies and representation techniques. Knowledge is represented and utilized through a set of linguistic and sensory vehicles. Declarative knowledge may be represented through mechanisms such as facts, frames or rules. These serve as the foundation for linguistic constructs or images such as visual or auditory messages.

These structures may be manipulated through operators such as inference schema for transforming knowledge from one stage to another. The use of specific operators throughout the reasoning process is directed by metalevel strategies.

New knowledge is examined and encoded through the acquisition module, which serves as an interface between the environment and the internal base of domain-specific knowledge. The need for externalization is served by two modules. First is the user interface, which communicates with the human decision-maker through audiovisual input and output. The second vehicle for externalization is the effector, which provides for a durable embodiment of the solution. The 'hardcopy' may take the form of printed prose, drawings, mechanical prototypes, or other modes of externalization.

22.3.2 Concurrent design advisor

The concurrent design advisor, as encoded in conjunction with the creativity support system, possesses knowledge of a general framework for the entire product development process. The process, which has been called quality function deployment (QFD), encompasses all elements of product design and production after a target market has been identified. Developed in Japan in the late 1970s, this approach to product development has proved effective in reducing lead time and delivering high-quality, low-cost products. The techniques in QFD channel the expertise of marketing, design, and production personnel toward the ultimate goal of customer satisfaction. As such, QFD offers a good framework for helping human experts to reach creative solutions for complex products.

Each module of the concurrent design advisor corresponds to one key step of QFD. The module is responsible for guiding the user through that process.

The concurrent design advisor can be included in a network configuration of knowledge bases for both design and manufacturing (Kim, 1990a, p. 153). The advisor interfaces with these knowledge bases to obtain the information required to design and produce the product. A complete system for a product line would possess knowledge of all resources and constraints (such as part families or available production equipment) involved with the overall development of the product.

The main thrust of the concurrent design advisor, however, is the knowledge of the framework for product development rather than information on task-specific resources which constrain that development. In this way, the advisor can rely on the domain expertise of human experts in addition to encoded knowledge bases.

22.3.3 Product design process

The driving force behind QFD takes the form of multifunctional teams during all phases of product development. By combining the expertise of each field – marketing, design, production, and service – the process addresses all relevant issues simultaneously. In this way, the process of 'throwing it over the wall' from design to production is avoided.

The concurrent design advisor is based on this principle of simultaneous engineering. Its main purpose is to help combine the expertise of various individuals working as a team. To this end, it relies on the following techniques: house of quality, Pugh's selection and synthesis matrix, competitive benchmarking, and the Taguchi method.

The concurrent design advisor employs the following procedure for product design:

1. Product specification;
2. idea generation;
3. evaluation; and
4. enhancement
   (a) comparison,
   (b) consolidation, and
   (c) optimization.

These steps are repeated four times: once for generating the product concept and subsequently for subsystem, part parameter, and process parameter designs. The particular techniques incorporated into the advisor for performing these tasks are described more fully below.

22.3.4 Product specification

The first stage, product specification, refers to the identification of the functional requirements and constraints on a design. The technique used by the CDA is the *house of quality*, a planning table for organizing and targeting the needs of the customer (Clausing and Hauser, 1988). A house of quality illustrating the design of an electromechanical product is shown in Fig. 22.3.

The rows of the house, called customer needs, are taken directly from customer responses. Effort is taken to avoid 'interpreting' these

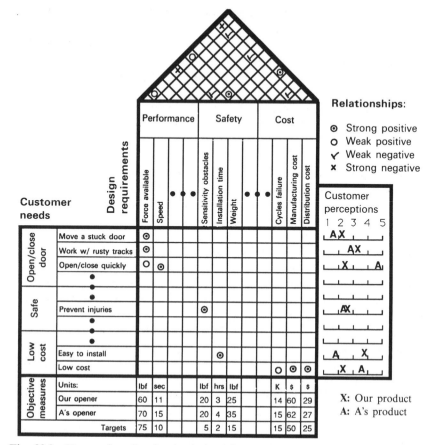

Fig. 22.3 House of quality for the design of a sample product. Three categories of customer needs have been translated into design requirements, and competitive benchmarking has been used to set design targets. Areas delineated by thick borders define different types of knowledge (Knight and Kim, 1991).

requests before recording them, but are rather transcribed using the original wording.

The columns of the house, called design requirements, are specific engineering measurements that are expected to be achieved in the new product. These parameters simply restate the customer requirements using the engineering expertise of the design team. They serve to guide subsequent design activity in terms of more objective or quantitative terms. During these steps the design team begins to interpret the customer needs; by discussing, debating, and clarifying the functional requirements, the team establishes a common understanding of the goals of the project.

The relationship matrix indicates each area where the design team feels that a relationship exists between the given row (customer need) and column (design requirement). That is, an entry indicates the degree to which achieving the design requirement (column) will have a tendency to satisfy the customer requirement (row).

Customer perceptions, to the right of the house, graphically display how customers feel about the ability of various products to meet each of their needs. This information is useful for identifying aspects of existing designs which must be improved.

Objective measures, located below the house, facilitate competitive benchmarking for each of the design requirements. The performance of competitive products is recorded in this area, allowing for technical comparisons of different products.

The correlation matrix, the roof of the house, shows the relationships between pairs of engineering characteristics. Any negative correlation between columns signals the need for special attention to an appropriate trade-off between the associated pair of requirements.

The house of quality is helpful for design tasks of varying complexity. As described above, the initial design work with the house will help translate customer needs into design requirements for the overall product. But subsequent steps will be required to translate these design requirements into the proper product, the correct process plan for its production, and the right process parameters to realize the product.

To determine these specifications, the targets of the first house are inserted on the rows of a second house for parts deployment. The team then works to create specifications for the components or subsystems which will address the requirements of the overall product. This process of translating the 'targets' of the previous house into the 'needs' of the subsequent house continues for a third, process planning house, and a fourth, production planning house. These houses specify the key process operations and the production control parameters, respectively.

The four linked houses of quality represent the format for 'deploying' the customer needs from planning to production, and provide the structure for quality function deployment. Design teams working with the concurrent design advisor create these four linked houses of quality for each level of specification complexity. After each house is created, the team proceeds to generate ideas for evaluation and enhancement. After the appropriate concept has been enhanced to the point where it meets the targets in the overall house of quality, the team proceeds to the houses for the subsystems, then process planning, and finally production planning.

22.3.5 Idea generation

The stage of idea generation refers to the creation of candidate solutions to the problem at hand. In this phase, directed brainstorming can produce a variety of possible solutions, and the generic creativity enhancement tools encoded in the creativity support system are employed to expand and enhance the list of alternatives.

The first phase of group ideation involves brainstorming. Members suggest possible solutions, or components of possible designs, as they are conceptualized. These items are quickly recorded into the case base of the CDA in the form of a few descriptive words or a simple sketch.

The ideation stage terminates at the end of the designated time period. More alternatives are sought by producing words, phrases, or pictures that help spark thoughts. A number of creativity enhancing utilities encoded in the creativity support system are used to create these key words or pictures. For instance, assumptions are identified to highlight the design constraints which the team members may have been using, whether consciously or unconsciously. The team is asked to list these assumptions, and then speculate on the benefits which would result if each were violated or eliminated. The team can then speculate on methods for removing these constraints.

Resources are considered in order to identify people, equipment, or techniques to aid in the design process. This list includes people who might benefit from the solution, and what they might do to help.

Such tools produce a variety of creativity triggers which can generate more design alternatives. Additional tools within the creativity support system assist in the deleting, editing and combining which may be necessary to construct the preliminary list of designs considered 'feasible'.

22.3.6 Evaluation

The next step of the process involves the comparative evaluation of the multitude of ideas proposed at the previous stage. The technique used in the CDA is Pugh's selection and synthesis matrix (Pugh, 1981). This matrix is a tool for evaluating as many as several dozen alternative concepts against a set of criteria, and allowing for the synthesis of additional candidate concepts. Based on the evaluation of the old and new concepts, the 'best' idea can be selected.

This process offers all the design team members a better understanding of the requirements of the product and the candidate solutions, and provides a stimulus to produce other concepts by combining the strong points of older concepts. At the conclusion of the process, the team selects the best concept for meeting the customers needs as listed in the house of quality.

22.3.7 Enhancement

Once the concept has been selected, it must be detailed or fleshed out to ensure that it consistently meets the targets specified in the columns of the house of quality. The enhancement phase refers to the elaboration of the basic concept to incorporate effective features.

Competitive benchmarking can be used to establish standards for the subsystems of a product once a general concept has emerged from the selection phase. The design team evaluates competing products and strives either to surpass these designs or use them in the product under development.

The results of competitive benchmarking are useful in value analysis/value engineering (VA/VE). This procedure translates functional requirements of the product into hardware designed to meet these requirements. By placing a monetary value on satisfying functional requirements, the design team is able to specify the worth of the corresponding hardware, and thereby begin to specify part parameters.

After focusing on preliminary hardware configurations, it is possible to direct the efforts of the design team toward specific design issues which might cause the product to fail to meet the customer needs. Fault means – effects and analysis (FMEA) helps the team to identify the possible failures of the product, and then determine which product parameters would cause such failures. For example, a garage door opener could 'fail' to open if the lifting force were insufficient, which would result if the motor power were deficient. After recognizing this, the design team can concentrate on optimizing the motor power to prevent this failure. The lowest level in the hierarchy of other possible failures leads the design team to the critical parameters for the design.

At this stage in the design, the basic structure of all the subsystems have been selected, and critical parameters have been identified. All that remains is to optimize these parameters at the component level. A popular tool for efficiently optimizing these parameters has been developed by Taguchi (Phadke, 1989; Taguchi, 1977, 1978). The concurrent design advisor uses knowledge of this procedure to design tests which can be performed on a limited number of prototypes, so that the best values for part parameters can be chosen. The tests are used to determine the signal-to-noise (SN) ratio for each prototype parameter; this SN ratio is an indication of how uniformly a product can perform in the face of manufacturing and environmental variations.

It is especially important during this phase to ensure that the design is 'robust'. This means that the design will be able to perform well – i.e. meet the customer's requirements – over the full range of expected operating conditions. By optimizing Taguchi's signal-to-noise ratio, the design team minimizes the variation in the performance of the product, thereby ensuring a robust design.

Taguchi's methods of optimization can also be employed to design the required production processes. Again, the concurrent design advisor incorporates the knowledge to assist in designing tests to ensure a reliable process in the face of production variances.

22.4 INTEGRATION FOR STRATEGIC DESIGN

The creativity support system and the concurrent design advisor are examples of packages which can be integrated into a comprehensive strategy support system. Further, the overall system can evolve by capturing the knowledge of product development programs. The encoded knowledge can take various forms of multimedia, including text, figures, voice and video.

By searching the case base of past experience, the system can automate many functions such as the retrieval of benchmark products or processes to pace the design task at hand. The same type of performance enhancement can occur in the context of the external environment. For instance, the implications of the pricing strategies of competitors can be recorded in the knowledge base of cases for future reference. In this way the system can learn to be more effective with experience.

Another extension lies in the capability of the strategy support system to better assist teams separated by geography. Improvements in computer-based conferencing systems, coordination methodology, and other approaches in distributed artificial intelligence suggest the means for enhancing the integration of human and computer-based expertise.

The effective coordination of decision-makers requires the provision of common data and knowledge bases. This coordination is crucial for the successful use of the strategy support system. While communication systems have been designed which allow for real-time conferencing systems, the development of multiuser applications has been hampered by the high costs of custom programming. More experience is required before common functions can be incorporated into a portable conferencing toolkit (Sarin and Grief, 1985). The system can act as a bulletin board for displaying collective decisions; until real-time conferencing techniques become prevalent, however, the system will depend primarily on traditional group interaction to integrate human expertise.

These issues apply to the coordination of encoded expertise as well. Successful conflict negotiation among distributed knowledge bases will permit the differing goals and perspectives of each processing node to contribute to the final solution. Innovations in this area will be needed to make full use of the knowledge networking scheme shown in Fig. 22.2. In this way, a comprehensive understanding of coordination principles and methods will allow the strategy support system to enhance the creative work of interacting decision-makers.

22.5 CLOSURE

The design of a product embodies the mission and competence of an organization. As the environment changes, both in terms of technological enablers as well as consumer requirements, so must the product.

The design of new models is therefore a perennial task, both to adapt to the changing environment as well as to incorporate internal learning such as reducing production cost. To this end, a strategy support system for product design must be able to accommodate the knowledge of the external environment as well as the full spectrum of internal organizational knowledge. Moreover, the system must be capable of learning new knowledge to reflect these dynamic environments.

22.6 REFERENCES

Clausing, D.P. and Hauser, J.R. (1988) *Harvard Business Review* **66**(3), May-June, pp. 63–73.

Kelly, J.E. Jr. and Walker, M.R. (1959) *Proceedings of Eastern Joint Computer Conference*, pp. 160–173.

Khan, M. and Smith, D.G. (1989) *Proceedings of International Conference on Engineering Design*, Harrogate, UK, pp. 605–19.

Kim, S.H. (1990a) *Designing Intelligence*, Oxford University Press, New York.

Kim, S.H. (1990b) *Essence of Creativity*, Oxford University Press, New York.

Kim, S.H. (1991) *Knowledge Systems through Prolog*, Oxford University Press, New York.

Knight, T.P. and Kim, S.H. (1991) *Journal of Intelligent Manufacturing*, **2**, pp. 17–25.

Lander, S. and Lesser, V.R. (1989) in *IEEE International Symposium on Intelligent Control*, (eds A.C. Sanderson, A.A. Desrochers and K. Valavanis), Albany, New York, September, pp. 472–7.

Liberatore, M.J. and Titus, G.J. (1983) *Management Science*, **29**, pp. 962–74.

Malcolm, D.G., Rosenboom, J.H., Clark, C.E. and Fazar, W. (1959) *Operations Research*, **7**(5), September–October, pp. 646–69.

Phadke, M.S. (1989) *Quality Engineering using Robust Design*, Prentice Hall, Englewood Cliffs, New Jersey.

Porter, M. (1985) *Competitive Advantage*, Free Press, NY.

Porter, M. (1990) *The Competitive Advantage of Nations*, Free Press, NY.

Pugh, S. (1981) *Proceedings of WDK5 International Conference on Engineering Design*, Rome, pp. 497–506.

Roush, G.B. (1991) *Journal of Business Strategy*, January/February.

Sarin, S. and Grief, I. (1985) *Computer*, **18**, pp. 33–45.

Sathi, A., Morton, T.E. and Roth, S.F. (1986) *AI Magazine*, Winter, pp. 34–52.

Scheel, C. (1991) in *Proceedings World Congress on Expert Systems* (ed. J. Liebowitz), Pergamon, New York, pp. 2388–95.

Taguchi, G. (1977, 1978) *Jiken Keikakuho*, (3rd Edn) **1** and **2**, Marzen, Tokyo, Japan. (in Japanese). English translation: G. Taguchi, *System of Experimental Design*, edited by Don Clausing.

Wiseman, C. (1988) *Strategic Information Systems*, Irwin, Homewood IL.

CHAPTER 23

Automating design for manufacturability through expert systems approaches

A.R. Venkatachalam, Joseph M. Mellichamp and David M. Miller

23.1 INTRODUCTION

In the light of growing global competition, organizations around the world today are constantly under pressure to produce high-quality products at an economical price. The need to manufacture products that perform efficiently in a cost effective manner is essential for the profitability and survival of organizations. Consequently, organizations have become increasingly aware of the importance of product design and are striving to improve manufacturing processes in order to introduce products to the market with the least development time.

Product design plays an important role in determining the cost and quality and thus the effective life of a product. In conventional manufacturing organizations, the activities involved – namely, market analysis, product design, production system design, manufacturing, and sales and distribution – take place in sequential order (Stoll, 1986). The serial nature prevents the integration of product and process design which often leads to poor manufacturing system design. Over the years, it has become the practice in manufacturing organizations for the design function to be isolated from the manufacturing operations. Product designs often get thrown 'over the wall' by designers, leaving the responsibility of building the products to manufacturing engineers.

Many companies have realized the disadvantages inherent in this sequential process. Product designs may result in higher cost, leading to reduced competitiveness. Forcing manufacturing to delay its operations until a design is released prolongs the product development time, resulting in missed market opportunities. The sequential nature of the process often leads to excessive complexity in designs by over-designing required features. Production restrictiveness is also a serious problem when features that are difficult to manufacture are designed with little or no manufacturing input.

The cost of the design process constitutes a small proportion of the total product cost, however, a large portion of the manufacturing cost is determined in the design phase of product development. Studies reveal that around 70% to 80% of the cost of a product is decided at the design phase (Andreasen *et al.*, 1983a; Corbett, 1986; Whitney, 1988; and Vogt, 1988). The remaining 20% to 30% of the cost is determined during actual production. It is apparent that in order to improve profitability and competitiveness, organizations need to address these issues in the design phase of a product rather than later in the production phase.

Design for manufacturability (DFM) is an approach to design that fosters simultaneous involvement of product design and process design. The primary objective of DFM is to produce a design at competitive cost by improving its manufacturability without affecting its functional and performance objectives. Implementation of the DFM approach can result in a number of benefits: simplification of product design, leading to reduction in product cost; integration of parts, resulting in a reduced number of parts, which not only reduces the product cost but also improves the reliability of the product; and improved productivity through standardization of components and lower inventory (Boothroyd and Dewhurst, 1988). Further, successful implementation of the DFM approach can result in breaking down the wall that exists between design and manufacturing, improving the morale of design and manufacturing personnel, and reducing warranty claims (CAD/CIM ALERT, 1987). Studies at NCR and IBM have shown that the application of design for assembly analysis (DFA) has resulted in significant savings through reduction in the number of assembly parts and manufacturing costs (Dwivedi and Klein, 1986; Sprague, 1989). In addition to the direct benefits, the DFM approach also provides several indirect benefits such as improved competitive edge through shorter development time, reduction in inventory through standardization of components, and improved quality and reliability. The DFM approach enables organizations to achieve the goal of 'getting it right the first time' in order to boost their competitive edge.

The purpose of this research is to show how expert systems methodology could be used to provide manufacturability expertise during the design phase of a product. An expert system which has the capability first to make process selection decisions based on a set of design and production parameters to achieve cost effective manufacture and second, to estimate manufacturing cost based on the identified processes was developed. In the following section a methodology for implementing the DFM approach using knowledge-based methodology is presented. The architecture of the process selection and cost estimation module of the expert system is described, and the capability of the expert system is demonstrated using an actual product design. Finally, the limitations of the research and direction for future research are discussed.

23.2 DFM – OVERVIEW

The DFM process aims at optimization of product and process concepts during the design phase of a product in order to ensure ease of manufacture. The optimization would then be followed by product simplification and conformance of product features to process compatibilities (Stoll, 1988). Product simplification is achieved through design of components for ease of assembly and handling to facilitate ease of manufacture, improved quality, and reduced manufacturing cost. Manufacturability is built into component designs through evaluation of design features for ease of manufacture with respect to process capabilities and limitations. The sequence of optimization and simplification of product design and its conformance with process design is repeated in a cycle, ensuring each time that the functional and performance objectives of the proposed product are not compromised or adversely affected by the factors introduced to maximize manufacturability.

Implementation of the DFM approach requires the application of a collection of tools and techniques. A number of tools are found in the literature and are briefly described below.

23.2.1 Axiomatic approach

The axiomatic approach aims at optimization of manufacturing systems through the application of good design principles or axioms (Suh *et al.*, 1978). Two steps are involved – specification of functional requirements of the end product and specification of constraints.

23.2.2 DFM guidelines

A frequently used approach to DFM has been the use of DFM guidelines. These guidelines are empirically derived from years of design and manufacturing experience. Following these guidelines usually leads to improved manufacturability (Andreasen *et al.*, 1983b; Dwivedi and Klein, 1986; Stoll, 1986; Billatos, 1988; Huthwaite, 1989). However, the application of these guidelines is neither straightforward nor easy.

23.2.3 Design for assembly

Design for assembly has been the focus of research by a number of researchers. The prominent work among these are Boothroyd and Dewhurst's method and Hitachi's assemblability evaluation method (Boothroyd and Dewhurst, 1983; Miyakawa and Ohashi, 1986). Both the methods aim at quantitative evaluation of designs with respect to ease of assembly. Another quantitative method used in industry is the GE/Hitachi method, which is basically the Hitachi method with the criteria for part count reduction from Boothroyd and Dewhurst's method added.

23.2.4 Taguchi method

The Taguchi method aims at the development of robust design (Taguchi and Yuin, 1979). The process involves identifying product and process concepts that are insensitive to variations and change and determining optimal values of design parameters that permit maximum robustness.

23.2.5 Process-driven design

Process-driven design methodology is based on the application of process specific expertise in the form of heuristics while designing a product to be manufactured using a specific process (Bralla, 1986). Examples include design for casting, design for forging, design for injection molding, and design for extrusion.

23.2.6 Facility-specific DFM

Facility-specific DFM enables design of products to be manufactured using specialized manufacturing facilities. Examples of such facilities include flexible assembly and manufacturing systems and design for weldments for production on flexible weld lines (Stoll, 1988).

23.2.7 Computer-aided DFM

Computer-aided DFM tools are software packages that aim at improving the quality of product/process design decisions early in the design stage. Most software packages are spreadsheet based and address only the assembly operation in product designs.

23.2.8 Traditional approaches

In addition to the methodologies and tools cited for DFM, traditional approaches including group technology, failure mode and effects analysis, and value analysis also incorporate DFM concepts. The group technology approach promotes manufacturability through standardization by identifying and exploiting similarity of parts based on geometric features and production processes (Ham, 1982). Failure mode and effects analysis helps to prevent failures and defects in product designs through a systematic approach in which causes and effects of failure are studied at the design phase (Dussault, 1983). Value analysis aims at eliminating unnecessary features and cost in a product design by analyzing each function and its cost (Demarle and Shillito, 1982).

Ensuring that the functional and performance objectives of the design are not violated while meeting cost and manufacturability objectives imposes a tremendous challenge for design engineers. Many reasons can be cited for the inability to implement the DFM approach effectively including: lack of interdisciplinary expertise of designers; inflexibility in organization structure which hinders interaction between design and manufacturing functions; lack of manufacturing cost information at the design phase; and absence of integrated engineering effort intended to maximize functional and manufacturability objectives.

The need for interdisciplinary expertise has been the main impediment to the wider use of the DFM approach. To design for economic manufacture requires knowledge of considerable breadth and complexity, which may take many years to accumulate. While design engineers are trained in the application of a variety of design techniques such as finite element analysis, solid modeling, etc., their level of expertise in performing process selection, cost estimation and manufacturability evaluation of designs is generally limited. To perform economic and technological assessment of product designs, designers must have expertise in many fields, including manufacturing engineering. Inherent in each manufacturing process are sets of good design practices which have been empirically derived over years of manufacturing practice. Though general guidelines are available for economic manufacturing, such guidelines by their very nature cannot be readily applied in evaluating product designs.

Researchers have also focused on the structural changes required in organizations for the effective implementation of the DFM approach. Dean and Susman proposed four types of structural changes in organizations in order to carry out the DFM approach (Dean and Susman, 1989):

1. Manufacturing sign-off – product designs have to be approved by manufacturing engineers before final release for initiating manufacture;
2. The integrator – a few individuals are entrusted the responsibility to incorporate producibility in product designs by coordinating the design and manufacturing functions;
3. Cross-functional teams – a team consisting of design and manufacturing engineers works toward improving manufacturability of designs; and
4. The product-process design department – a single department responsible for both product design and process design is created to address producibility issues during the design phase of a product.

All the suggested approaches require changes in the structure of organizations. This imposes a challenge to organizations through greater demand on people in terms of absorbing change and developing new skills, both technical and interpersonal.

The availability of product cost information at the design stage is a necessary prerequisite for making sound judgments concerning the choice of the most appropriate manufacturing process. The ability to perform cost estimation based on the design without detailed manufacturing planning is thus necessary. The availability of such manufacturing cost information would result in the selection of the most appropriate and cost effective process combinations to produce a given product design.

Very little work is reported in the literature in the area of cost estimation using the knowledge-based approach. Mathematical modeling and empirical formulas are the primary tools used for estimation of cost. Research work reported in the literature include Boothroyd and Dewhurst's mathematical model for estimation of machining cost and empirical model for injection molding cost (Dewhurst and Boothroyd, 1987), Apgar and Daschbach's parametric cost estimation model (Apgar and Daschbach, 1987), Boothroyd and Reynolds' costing model for machining rotational components on a CNC lathe (Boothroyd and. Reynolds, 1989), Knight and Poli's work on a cost analysis tool for forged components (Knight and Poli, 1985), and Poli *et al.*'s work on a relative cost model for injection molding and die casting processes (Poli *et al.*, 1991). London *et al.* have developed an expert system architecture that includes a cost estimation module and a tutorial module. The main research issue was customizability of the expert system to meet an organization's needs (London *et al.*, 1987).

In order to meet the objectives of the DFM approach, design and manufacturing planning activities have to be combined into a single engineering effort and applied throughout the life cycle of a product. Computer-aided design (CAD) systems offer powerful features such as the ability to develop complex solid models and perform engineering analyses, including stress analysis, interobject interference, collision detection, and inertial analysis. However, a prominent limitation faced by designers in CAD systems is the lack of 'intelligence'. Though designs could be developed, analyzed, and perfected from a functional viewpoint in CAD systems, manufacturability consideration may get little or no attention at all. As a result, product designs that are functionally sound may be developed at a high manufacturing cost. Thus, 'intelligence' should be incorporated in CAD systems, whereby product designs could not only be developed and analyzed but also evaluated for cost and manufacturability. The ability to detect design errors such as a nonstandard component or a component that is difficult to manufacture or cannot be manufactured at all would certainly enhance the productivity of the design function.

All the methodologies and tools discussed incorporate the DFM approach and aim for improved product designs that enhance the design team's ability to design for effective quality, cost, and delivery. However, implementing most of these methodologies and tools requires a high degree of involvement between the design and manufacturing personnel, interdisciplinary expertise, and alterations in the organization's structure. The key to successful implementation is the development of an integrated tool by which designers could develop product designs, select appropriate manufacturing processes, estimate first-order manufacturing costs and evaluate design features for manufacturability with respect to the identified processes. The application of expert systems methodology offers considerable promise in this direction. Since the expertise involved in these activities is in the form of heuristics empirically derived over years of design and manufacturing practice, it is logical to consider an expert systems methodology to implement the DFM approach.

23.3 EXPERT SYSTEMS APPROACH TO DFM

The selection of a combination of processes by which a proposed design is to be manufactured is an important step in the chronological sequence of events taking place from the concept stage of a product to its final launch in the market. Having identified the most appropriate primary and secondary processes, the designer is then able to evaluate the product design for manufacturability with respect to the identified processes.

23.3.1 Proposed DFM approach

The major factor that distinguishes the DFM approach from conventional manufacturing design is the presence of feedback at the various stages of the design activity. Decisions made during the design phase are evaluated with respect to cost, quality, and manufacturability objectives, and designs are modified so that cost-effective manufacturing can be realized with minimum design change during the life cycle of a product.

An overview of the DFM approach proposed in this research is shown in Fig. 23.1. This approach identifies types of manufacturing processes that would lead to cost-effective manufacturing of a proposed design and evaluates design features for manufacturability with respect to the processes identified earlier. The first step in this approach is design development. At this stage the designer is concerned with functional design of the product based on the product requirements. The next step involved is the selection of materials based on the product requirements and identification of the combinations of manufacturing processes for the proposed design.

The selection of manufacturing processes by which a product is to be manufactured depends on the choice of materials and limitations inherent in the processes. The selection process starts with a functionally sound design and a choice of material of construction. The primary process selection is then performed based on heuristic knowledge. Primary processes determine the basic shape of the product from which other finer features are produced.

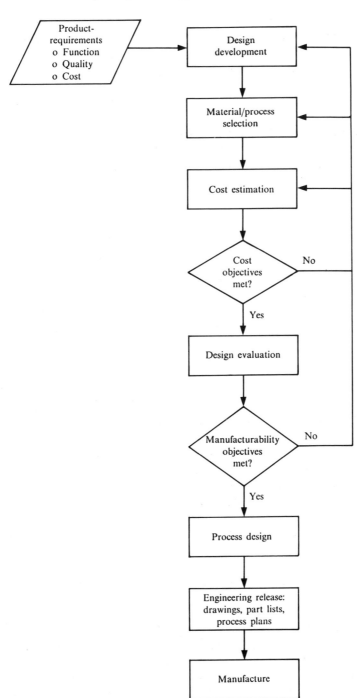

Fig. 23.1 Proposed DFM approach.

Examples of primary processes include casting and forging for metals and injection molding for plastics. The heuristic knowledge involved is based on a set of design and production parameters. Each primary process has its own limitations in terms of the size of product it can handle and the level of dimensional accuracy and surface finish that can be attained on the product. The minimum production volume also has a significant impact on the process selection from the viewpoint of economy of manufacture (Niebel and Draper, 1974; Dieter, 1983). Other factors considered significant in the selection of primary processes are nature of application, minimum wall thickness, and the weight of the component. The cost of primary processing is estimated using cost estimation heuristics specific to the selected process. The selection of secondary process is then performed based on the primary process selected earlier and design features required on the product. A list of design features considered significant in the manufacture of mechanical components include holes (through, blind, tapered, and square), threads (external and internal), slots, keys, and contoured (two- and three-dimensional) surfaces. The cost of secondary processing is then estimated for the identified processes.

Primary processes which are popular in the industry include casting and forging for metals, polymer processing such as injection molding for plastics, and particulate processing such as powder metallurgy for metals. Extrusion, forming, and spinning are also some of the frequently used primary processes. For ferrous and non-ferrous metals and alloys, casting and forging form the bulk of the primary processes.

In this research, the selection of primary processes is limited to casting (including sand, investment, and die casting) and forging (including conventional die, open die, and precision forging). The selection of secondary processes is limited to end milling and drilling operations performed on a CNC milling machine. The set of designs features considered includes through holes, blind holes, square holes, and slots. The presence of through and blind holes indicates the need for a drilling operation; square holes and slots indicate the need for a milling operation.

The process of first-order cost estimation is then performed based on the identified process combinations. If the cost objectives are not met, the cycle of functional design, material/process selection, and cost estimation is repeated.

The availability of manufacturing cost estimates is a necessary prerequisite for making judicious choices of processes to be adopted for the manufacture of a product. In this study, cost estimation expertise has been developed for the die casting process and is based on industry practice. In generating a cost estimate for die casting, a set of design parameters forms a basis for deriving process parameters, which in turn are translated into processing cost. In the die casting process, the weight of the component is recognized as the single most influential design parameter affecting the process cost. The weight of the casting is used to estimate the process parameters, including the number of cavities and rates per hour for shots, production, breaking gate, trimming, cleaning, and inspection, and scrap factor. Other factors considered in cost estimation include die life, die cost, die setup time, and component requirement – one time or recurring. The process parameter values are then translated into processing cost using the cost structure specific to the organization. The type of material used – aluminum or zinc – is used to estimate the material cost. The cost of primary processing is then computed by adding the material cost to the processing cost estimated earlier.

The cost involved in performing machining operations is based on the time required to set up the machine and the cycle time required to perform actual machining operations on each of the components in a given lot. The basic time elements involved in the estimation of setup time for a CNC milling machine include basic setup time, make part time, and tolerance checking time for 'cut and try' operations. The time elements involved in the estimation of cycle time include handling, cutting, tool replacement, and inspection. The estimation of setup and cycle time involves a number of parameters and is based on heuristic knowledge. The parameters considered for first-order cost estimation include the number of times the component has to be reoriented, the number of holes, dimensions of deepest and largest holes, and number of sizes of holes present in the design. Setup parameters considered include the type of cutting tool, type of work holding device, and required tolerance. Based on the setup and cycle times estimated, the cost of secondary processing is computed for a production volume of Q, using the formula $C_{sp} = C_k(t_s + Qt_c)$, where C_{sp} is the total cost of secondary processing, C_k is the cost of the machine and labor per hour, t_s is setup time, and t_c is cycle time.

The cost involved in performing assembly operations for mechanical assemblies is often significant, and hence has to be included in the manufacturing cost estimation. Past studies have shown that the proportion of assembly cost could be as high as 50% of the total manufacturing cost in the case of mechanical and electrical assemblies. In this research, the cost estimation expertise is limited to manual assembly operations and is based on the Boothroyd and Dewhurst methodology. The estimation of assembly time includes two time elements, namely, handling and insertion. Handling time estimation involves product geometry, including size, thickness, alpha symmetry, and beta symmetry of the component (Boothroyd and Dewhurst, 1983). Size and thickness refer to the length of the longest and smallest side respectively of the smallest rectangular prism enclosing a component. Alpha symmetry refers to the rotational symmetry of a component about an axis perpendicular to its axis of rotation. Beta symmetry refers to the rotational symmetry of a component about its axis of insertion. Insertion time estimation involves factors related to assembly, including securing of the component after insertion, the need to be held after assembly to maintain orientation and location, and the ability of the component or the assembly tool to reach the desired location.

When the cost objectives have been achieved (see Fig. 23.1), the focus shifts to evaluation of product design feaures for manufacturability with respect to the identified processes. By extracting design features, evaluating them with respect to established guidelines for manufacturability, and redesigning the features, the designer may be able to achieve reduced manufacturing cost and improved quality in a product. The evaluation of design features for maufacturability may sometimes lead to redesign of a product or may warrant selection of different materials or processes to provide for ease of manufacture and assembly. In such cases process selection, cost estimation, and design evaluation activities may be repeated in a cycle until the cost and quality objectives are met while satisfying the design guidelines specified for the selected processes.

Process design is the next logical step after a product design is developed, evaluated, and modified to satisfy cost, quality, and manufacturability objectives. Process design activities include specification, design, and layout of production equipment, process plans, inspection and testing programs, and tool and fixture design. In a computer-integrated environment, process design can be effectively carried out using part geometry information maintained in CAD databases.

The main thrust of the DFM approach is the designer's ability not only to perform functional analysis of a product design but also to improve its manufacturability by the selection of a combination of processes based on first-order cost estimates and evaluation and redesign of the design features for manufacturability. The implementation of the process selection, cost estimation, and manufacturability evaluation modules in computer-aided design systems would provide designers with an integrated environment to carry out functional design, DFM analysis, and process design.

23.3.2 Expert system architecture

The expert system developed is in two modules, one for process selection and cost estimation and the other for evaluation of design features for manufacturability in a CAD system. In this paper the discussion is limited to the process selection and cost estimation module.

The tool used in the development of these two modules is NEXPERT, a powerful expert system development tool built in the C language (Neuron Data, 1988). The power of this tool lies in its support for rule- and object-based knowledge representation schemes with various inference mechanisms. A schematic diagram of the expert system is given in Fig. 23.2. The following discussion will focus on how the knowledge representation schemes, the inference mechanism, and the user interface are incorporated in the expert system.

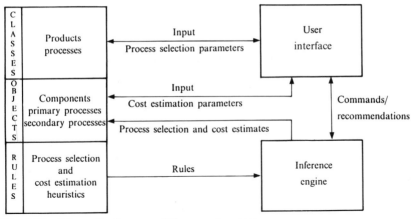

Fig. 23.2 Schematic diagram of the expert system.

(a) Knowledge representation

A hybrid object-oriented/rule knowledge representation scheme was used in the expert system. The product itself as well as candidate manufacturing processes are represented as classes in the knowledge base; the scope of the current application is limited to the consideration of two primary processes – casting and forging. Product components and the specialized casting and forging processes considered are represented as objects belonging to their respective classes, as shown in Fig. 23.2. Each object has a list of properties, each representing a data type. Secondary processes such as machining are represented as subobjects of primary processing objects; the setup object is considered as a subobject of the machining process object. This hierarchical relationship among the various objects enables property values to be inherited upwards or downwards, thus giving a powerful knowledge representation capability.

In addition to the objects, the expert system contains two categories of rules that provide process selection and cost estimation heuristics. The process selection rules are classified into two subgroups – primary and secondary process selection. The cost estimation rules are divided into three subgroups: die casting, machining process, and assembly operation. The guidelines involved in process selection and cost estimation are stored as production or IF-THEN rules in the knowledge base. The constraints used to translate product feature and production requirements into appropriate processes and cost estimates are represented as conditions or the IF side of rules. The conclusions or the THEN side of rules contain hypotheses which may represent process selection and actions such as creating objects (for example, die casting) under predefined classes (casting), assigning or modifying values of an object's properties, and evaluating new hypotheses. Table 23.1 gives examples of rules used in the expert system.

The knowledge base of the expert system has eighty-four rules for process selection and cost estimation. The rules were derived from knowledge

Table 23.1 Examples of rules

RULE PSP_8		
	IF	component.application is non_critical
	AND	component.prod_size is greater than or equal to 3000
	AND	component.weight is less than 66 pounds
	AND	component.surface_finish is greater than or equal to 32 microinches
	AND	component.dimen_tolerance is greater than or equal to 0.001 inches
	AND	component.wall_thickness is greater than or equal to 0.03 inches
	THEN	primary process is die casting and create a 'die_casting' object in 'casting' class
RULES PSS_1		
	IF	calculate_total_cost hypothesis is confirmed
	AND	value of component.through_holes is 'no'
	AND	value of component.square_holes is 'yes'
	AND	value of component.blind_holes is 'yes'
	THEN	evaluate_secondary_process hypothesis is confirmed and machining operation type is milling and drilling and evaluate machining_process.number_setup

obtained from the open literature as well as interviews with manufacturing experts in selected firms. The expert system also has four objects which represent the product, the diecasting process, the machining process, and machine setups. These objects have a total of sixty-six properties which can have symbolic or numeric values. The properties have a total of thirty-four metaslots associated them; each metaslot describes a set of actions to be performed if the value of its associated property is modified.

Table 23.2 gives examples of properties of the die casting object.

Table 23.2 Examples of properties of DIE_CASTING object

Name	Type	Description
cavities	numeric (I)	Number of cavities on the die used to make the component
break_gate_cost	numeric (F)	Cost piece involved in breaking gates
die_cost	numeric (F)	Cost of the die required to make the component
die_setup_cost	numeric (F)	Cost of the die setup process
clean_per_hr	numeric (I)	Die casting cleaning rate in pieces per hour

(b) Inference engine

The process selection and cost estimation expert system uses a backward chaining procedure for its inferencing process. Backward chaining is a hypothesis driven approach. The method focuses on working from a hypothesis and proceeding back to evidence.

The inference process is initiated with the testing of a user-suggested hypothesis which causes the creation of a component object, a die casting object, a machining process object, and a setup object. The properties of these objects are assigned values by the execution of rules and actions contained in metaslots. Cost estimates including primary processing cost, secondary processing cost, and assembly cost are derived as the values of the properties of the objects are computed. Finally, the cost estimates are inherited upwards to the component object and the process recommendations and cost estimates are displayed to the user.

(c) User interface

The expert system has an interactive menu-driven capability. Upon accessing the system the user encounters a set of introductory screens explaining its capabilities. Then the user is prompted for the values of the properties of the component object such as geometric volume, production volume, expected surface finish, dimensional tolerance, and minimum wall thickness, necessary to perform process selection. The recommendations of the system are displayed on separate screens with details of recommended process parameters. The cost estimates are displayed with breakdowns in terms of primary processing cost, secondary processing cost, and assembly cost. The interaction between the system and the user is entirely through menu-driven options; the user is prompted for data inputs whenever numerical values are required. Finally, the user is provided options to clear the working memory and restart the inference process, make a hard copy of the system recommendations, or quit the system.

23.4 APPLICATION OF THE EXPERT SYSTEM

In this section the role played by the process selection and cost estimation module of the expert system is demonstrated. The product design considered is a base hanger used in the construction of steel furniture. The material from which the base hanger is manufactured is aluminum and the volume to be produced is 5,000 units per year. The nature of the application is considered to be noncritical. An application is considered critical if a component is used in areas such as high pressure lines, nuclear plants, aircrafts, missiles, and instrumentation where the consequences of failure of the component could be serious. Noncritical application refers to areas other than those indicated for critical application. The finished volume of the proposed base hanger design is 2.95 cubic inches as obtained from the CAD database. Other relevant information obtained from the product drawing are required surface finish: 50 microinches; minimum section thickness: 0.125 inches; dimensional tolerance: 0.005 inches; number of holes: 1; depth of hole: 0.218 inches; and machining tolerance: 0.005 inches. All the values except the production volume are obtained from the product design drawing.

Figure 23.3 shows the first introductory message displayed by the system. The system then prompts the user for the material group, nature of application of the component, production volume, finished volume, surface finish, section thickness, and dimensional tolerance of the component. Based on the values of the design and production parameters, the system makes a recommendation (Fig. 23.4) that die casting is the appropriate process and the number of cavities required in the die is 4. The user is then prompted for information on cost standards used by the organization, cost of the metal, cost of the die required to make the base hanger, and the requirement nature – one time or recurring – of the product. For the purposes of cost estimation, a 600-ton die casting machine was considered. The primary process selection report displayed is shown in Fig. 23.5.

A M E X

Automated Manufacturability Evaluation Expert

Process Selection and Cost Estimation

Introduction

Automated Manufacturability Evaluation Expert (AMEX) is an expert system to assist design engineers in the selection of appropriate primary and secondary processes by which a product is to be manufactured. The expert system also provides a first-order cost estimate for manufacturing a proposed design. The cost estimate includes material cost, primary and secondary processing cost and assembly cost.

AMEX is a hybrid expert system developed for use in microcomputers. The heuristic knowledge used for the selection of processes and estimation of cost is represented in the form of objects and rules. A backward chaining inference engine is used to perform the inferencing process.

Fig. 23.3 Introduction.

The expert system then shifts its focus to secondary process selection and estimation of machining and assembly cost. The user is prompted for the presence or absence of through holes, blind holes, square holes, and slots to determine the machining processes required – milling or drilling or both. Then information on the number of times the component has to be reoriented, dimensions of the holes, type of cutting tool, type of work holding device, and machining tolerance are requested by the system. The machine considered is a CNC (BOSTOMATIC) milling machine. The manual assembly cost for the component is estimated by prompting the user

Automated Manufacturability Evaluation Expert

Fri 12/20/1991

Process Selection and Cost Estimation Expert

Based on the given design and production parameters, AMEX makes the following recommendation for the primary process:

Die Casting

Component material: Aluminum

Number of cavities : 4
required in the die

Press any key to continue

F1 - Clear WM   F2 + F3 - Restart   F4 - Quit to DOS   F5 - Print screen and Quit

Fig. 23.4   Primary process selection report.

Automated Manufacturability Evaluation Expert

Fri 12/20/1991

Primary Process Selection Report

Recommended primary process: Die Casting

Material type	: Aluminum
Nature of application	: Non-critical
Production volume	: 5000
Finished weight (lb.)	: 0.290
Material Cost ($)	: 0.21
Primary Processing cost ($)	: 0.42

Press any key to continue

F1 - Clear WM   F2 + F3 - Restart   F4 - Quit to DOS   F5 - Print screen and Quit

Fig. 23.5   Primary process cost estimation report.

for the 'size' and 'thickness' of the component. The user is also prompted for the alpha-symmetry and beta-symmetry of the product design and the level of difficulty involved in aligning, positioning, and inserting of the component during assembly. Figure 23.6 shows the final report prepared by the expert system for the base hanger. The total cost per unit is estimated to be 72 cents, including 21 cents for material cost, 42 cents for primary processing cost, 8 cents for machining cost, and 1 cent for assembly cost.

Automated Manufacturability Evaluation Expert

Fri 12/20/1991

Process Selection and Cost Estimation

Recommended primary process: Die Casting

Material cost ($ per piece)	: 0.21
Primary processing cost ($ per piece)	: 0.42
Secondary processing cost ($ per piece)	: 0.08
Assembly cost ($ per piece)	: 0.01
Total cost per piece ($)	: 0.72

Press any key to continue

F1 - Clear WM   F2 + F3 - Restart   F4 - Quit to DOS   F5 - Print screen and Quit

---

Fig. 23.6   Cost estimation summary report.

23.4.1 Comparison of estimated and actual costs

The expert system was validated using designs for two products used in the furniture industry. Manufacturing costs were estimated for the two product designs using the expert system and compared with the actual costs.

The estimated manufacturing cost for the base hanger using the expert system is 72 cents per unit. In comparison, the actual manufacturing cost for the base hanger, as reported by the manufacturing firm from which the example was taken, is 74 cents per piece, indicating that the manufacturing cost estimate is understated by a margin of 3.86%.

Manufacturing cost was also estimated for a hinge pin. The material cost (in this case, material of construction is zinc) and the cost of primary and secondary processing are estimated to be 48 cents per unit. The actual manufacturing cost for the product is 47 cents per unit. The manufacturing cost estimate represents a deviation of 2.3% from the actual cost.

The estimates of manufacturing cost provided by the process selection and cost estimation module of the expert system deviate from the actual manufacturing cost values by a small margin. Though this validation is based on a limited sample, the results suggest that the use of a knowledge-based approach to generate accurate early cost estimates for product designs is certainly feasible.

23.5 CONCLUSIONS

The major impediments to the successful implementation of the DFM approach are emphasis on the team approach, requirement of cross-functional and interpersonal skills for designers, and the need for structural changes in organizations. Most of the methods and tools for DFM cited in the literature insist on effective communication between design and manufacturing functions and the need for manufacturability expertise in designers. Due to divergent orientations of these functions, the realization of these requirements may require considerable training and take longer time for organizations, thus making these methods less effective.

In comparison, a knowledge-based approach to DFM may require less time to implement as the dependence on interaction between design and manufacturing functions is drastically reduced in this approach by the availability of manufacturability expertise to designers through knowledge bases. Though the present expert system has a number of limitations, conceptually the scope of such a system is unlimited. A variety of manufacturing processes may be included in the knowledge base for process selection and cost estimation. Also, the design evaluation expertise may be developed for complex features necessary for primary and secondary processes. A hybrid knowledge representation scheme in terms of rules and objects may be used for loading and unloading of knowledge bases and creation and deletion of objects as required, thus providing the capability to handle a wide variety of processes without encountering computer memory problems. The availability of manufacturability expertise in the form of rules and objects and the ability to separate domain-specific knowledge from problem-solving knowledge, thus providing designers with the flexibility to customize the heuristic knowledge required, would enhance the effectiveness of the knowledge-based approach. With the use of a robust expert system with the process selection, cost estimation, and design evaluation modules integrated into a single environment and the knowledge bases refined to take care of capacity limitations, the impediments to the DFM approach would be diminished considerably. Finally, and perhaps less apparent, is the use of the expert system as a tool for trailing designers in incorporating manufacturability in design decisions.

In this research an integrated and comprehensive framework has been proposed for implementing the DFM approach using expert systems methodology. An expert system has been developed in two modules to enable designers to carry out process selection and cost estimation on the basis of a set of design parameters and extraction and evaluation of design features for manufacturability with respect to the identified processes (Venkatachalam, 1990). The application of the two modules of the expert system to real-life examples demonstrates that the proposed approach using expert systems methodology is a feasible method for implementing the DFM concept.

Experimentation with the expert system demonstrates the feasibility of a knowledge-based approach to DFM. The system does, however, have a number of limitations. These limitations need to be resolved before the proposed approach can be judged as effective.

The primary issue has to do with the scope of the knowledge base. Specifically, the interdependence of material selection and process selection makes the process selection activity more complex. Though the geometry of a product is certainly one of the most important parameters that acts as a limiting restraint in the selection of processes, certain other aspects of material selection also play an important role in the selection of processes. Ordering information such as minimum order size, quantity breakpoints, sources of supply, and lead time required to procure materials has considerable impact on the selection of materials and hence on the processes to be selected. Ideally, the knowledge base should cover such interrelationships among materials and process selection.

The effectiveness of a knowledge-based approach for improving the manufacturability of product designs lies in the integration of process selection, cost estimation, and design evaluation activities into a single system. The ability to use a single database in a CAD system to perform process selection and cost estimation, in addition to evaluation for manufacturability, is one of the key factors for the success of this approach. Such integration would enable designers to achieve accuracy and consistency in cost and design features evaluation and reduced computing time for performing the analyses.

In the prototype expert system developed, the integration of the two modules would lead to automating some of the inputs required for process selection and cost estimation. Information on the design features of a solid model necessary to perform process selection – including volume, weight, surface area, and section thickness – could be obtained directly from the CAD database, thus eliminating user inputs for these design parameters. Also the presence of design features that have a significant impact on the selection of primary and secondary processes – such as two and three dimensional contoured surfaces, undercuts, webs, slots, holes pockets, keyways, threads, and corners – may be directly inferred by extracting such features from the CAD database.

The extraction of design features could be used for cost estimation as well. The values of design parameters such as volume and weight could be used to

derive process parameters specific to the process selected and subsequently translated into processing cost. The dimensions of the features to be machined could also be obtained from the mathematical blocks of the geometric entities associated with the features. The length of cut required to estimate cutting time for milling operations and diameter and depth of cut for drilling operations could be inferred from the properties associated with the features.

The estimation of assembly time and cost could be automated by the integration of the two modules. Information on the size and thickness of a rectangular envelope enclosing the solid model could be derived by interpreting the moment and products of inertia of the solid model.

23.6 REFERENCES

Andreasen, M.M., Kahler, S. and Lund, T. (1983a) *Design for Assembly*, IFS Publications Ltd., Bedford.

Andreasen, M.M., Kahler, S. and Lund. T. (1983b) *Design for Assembly*, IFS Publications Ltd., Bedford.

Apgar, H.E. and Daschbach, J.M. (1987) in *Proceedings of the 1987 International Conference on Engineering Design*, (Ed. W.E. Eder), ASME, New York, pp. 759–66.

Billatos, S.B. (1988) in *Proceedings of Manufacturing International '88 – Symposium on Manufacturing Systems – Design, Integration and Control*, (Eds. G. Chryssolouris, R. Von Turkovich, and P. Francis). ASME, New York, pp. 129–36.

Boothroyd, G. and Dewhurst, P. (1983) *Design for Assembly – A Designer's Handbook*, Department of Mechanical Engineering, University of Massachusetts, Amherst.

Boothroyd, G. and Dewhurst, P. (1988) *Manufacturing Engineering*, **100**(April), 42.

Boothroyd, G. and Reynolds, C. (1989) *Journal of Manufacturing Systems* **8**(3), 191.

Bralla, J.G. (ed.) (1986) *Handbook of Product Design for Manufacturing, A Practical Guide to Low-Cost Production*, McGray-Hill Book Company, New York, p. 1.5.

CAD/CIM ALERT (1987) in *Proceedings of the Conference on Design for Manufacturability: Getting it Right the First Time*, Chestnut Hill, pp. 5.1.–5.7.

Corbett, J. (1986) *Annals of CIRP*, **35**(1): 93.

Dean, Jr., J.W. and Susman, G.I. (1989) *Harvard Business Review* **67** (January–February), 28–36.

Demarle, D.J. and Shillito, M.L. (1982) in *Handbook of Industrial Engineering*, (Ed. G. Salvendy), John Wiley & Sons, Inc., New York, pp. 7.3.1–7.3.20.

Dewhurst, P. and Boothroyd, G. (1987) in *Proceedings of Second International Conference on Product Design for Manufacture and Assembly*, (Eds. G. Boothroyd, P. Dewhurst, and B. Huthwaite), Newport, pp. 1–15.

Dieter, G.E. (1983) *Engineering Design – A Materials and Processing Approach*, McGraw-Hill Book Company, New York, p. 166.

Dussault, H.B. (1983) *The Evolution and Practical Applications of Failure Modes and Effects Analyses*, Report Number RADC-TR-83-72, Rome Air Development Center, Griffiss Airforce Base, New York, p. 5.

Dwivedi, S.N. and Klein, B.R. (1986) *CIM Review* **3**(Sprig), 58.

Ham, I. (1982) in *Handbook of Industrial Engineering*, (Ed. G. Salvendy), John Wiley & Sons, Inc., New York, pp. 7.8.1–7.8.19.

Huthwaite, B. (1989) *Design for Competitiveness*, Institute for Competitive Design, Rochester, Michigan.

Knight, W.A. and Poli, C. (1985) *Machine Design* **57** (January 24), 94–9.
London, P., Hankins, B., Sapossnek, M. and Luby, S. (1987) in *Proceedings of the 1987 ASME International Computers in Engineering Conference and Exhibition*, (Eds.R. Raghavan and T.J. Cokonis), ASME, New York, pp. 125–9.
Miyakawa, S. and Ohashi, T. (1986) in *Proceedings of the First International Conference on Product Design for Assembly*, (Eds. G. Boothroyd, P. Dewhurst, and B. Huthwaite), Newport, pp. 1–13.
Neuron Data, (1988) *NEXPERT OBJECT Fundamentals, Version 1.1*, Neuron Data, Palo Alto, California.
Niebel, B.W. and Draper, A.B. (1974) *Product Design and Process Engineering*, McGraw-Hill Company, New York, pp. 296–409.
Poli, C., Fenoglio, F. and Shunmugasundaram, S. (1991) *Concurrent Engineering* **1** (March/April), 31–8.
Sprague, W.R. (1989) in *Proceedings of the 1989 International Industrial Engineering Conference and Societies' Manufacturing and Productivity Symposium*, Institute of Industrial Engineers, p. 672.
Stoll, H.W. (1986) *ASME Applied Mechanics Reviews* **39** (September): 1356–64.
Stoll, H.W. (1988) *Manufacturing Engineering* **100** (January): 67–73.
Stoll, H.W. (1988) in *Tool and Manufacturing Engineers Handbook – Volume V Manufacturing Management*, (Eds. R.F. Veillux and L.W. Petro) Society of Mechanical Engineers, Dearborn, p. 9.
Suh, N.P., Bell, A.C. and Gossard, D.C. (1978) *ASME Journal of Engineering for Industry* **100** (May): 127–30.
Taguchi, G. and Yuin, W. (1979) *Introduction to Off-line Quality Control*, Central Japan Quality Control Association, Nagaya, Japan.
Venkatachalam, A.R. (1990) *A Knowledge-Based Approach to Design for Manufacturability*, Ph.D. Dissertation, The University of Alabama.
Vogt Jr., C.F. (1988) *Design News* **44** (March 7): 18.
Whitney, D.E. (1988) *Harvard Business Review* **66** (July–August): 83.

CHAPTER 24

Modeling the design process with Petri nets

Andrew Kusiak and Hsu-Hao Yang

24.1 INTRODUCTION

Design of an object generally involves creating a formal model of the object. A design process can be seen as the process of creating a representation of the underlying object. This representation can then be analyzed to derive important characteristics of the design process.

A scheme for representation of a design process should capture major aspects of the process in a precise and concise manner. Furthermore, it is useful to have a direct relationship between the graphical and analytical representations. This is particularly advantageous in the analysis, simplification, and verification of a large system. Finally, the representation should be able to describe abstractions and refinements in the design process. In summary, the representation should integrate the basic principles and concepts of system design, such as analysis, modularity, and so on. A graphical representation appears to have the above-mentioned characteristics.

In this article, the application of Petri nets (PNs) to modeling of a design process is presented. Petri nets provide a useful representation of various issues in design, such as decomposition, multi-resource requirement, cycles, and so on.

24.2 THE CONCEPT OF PNs

Petri nets have originated from Carl Adam Petri's doctoral dissertation in 1962. Since then, considerable work has been done in the theory and applications in many areas (Peterson, 1981; Murata, 1989). PNs are a graphical and mathematical modeling tool applicable to many systems. They are useful for describing and studying concurrent systems, asynchronous, distributed, parallel, non-deterministic, and/or stochastic. Some reasons contributing to the popularity of PNs as a modeling tool for analysis and design of systems are: the ability to model systems graphically, representation of conflicts and hierarchies in modeling a system, and the ability to link structured properties (liveness, boundedness, etc.) to certain performance criteria. PNs have recently been applied to model flexible manufacturing systems (Gentina and Corbeel, 1987; Krogh and Sreenivas, 1987; Martinez *et al.*, 1987; Valavanis, 1990; Hatono *et al.*, 1991). They are also useful in modeling and analysis of discrete-event systems (Ichikawa and Hiraishi, 1987), distributed database systems (Ozsu, 1985), concurrent and parallel programs (Kasai and Miller; 1982, Genrich and Thiagarajan, 1984).

A Petri net is a four-tuple (P, T, F, W), where:

$P = \{p_1, p_2, \ldots, p_m\}$ is a finite set of places
$T = \{t_1, t_2, \ldots, t_n\}$ is a finite set of transitions
$F \subseteq (P \times T) \cup (T \times P)$ is a finite set of arcs (flow relation)
$W : F \to \{1, 2, 3, \ldots\}$ is a weight function
$P \cup T \neq \emptyset$, $P \cap T = \emptyset$, where the set of places and the set of transitions are disjoint.

The expression $(P \times T)$ defines directed arcs from places to transitions, whereas $(T \times P)$ defines directed arcs from transitions to places. Pictorially, places are represented by circles and transitions by bars or boxes. Some typical interpretations of transitions and their input places and output places are illustrated in Table 24.1.

Table 24.1 Illustration of the interpretations of transitions and places

Input place	Transition	Output place
Precondition	Event	Postcondition
Input data	Computation step	Output data
Resource needed	Task or job	Resource released

A marking, denoted by M, of a Petri net is a function from the set of places P to a non-negative number N; i.e., $M : P \to N$, $N \geq 0$. Marking means the assignment of tokens to the places of a Petri net. If the marking of a place p is a nonnegative integer k, then we say that p is marked with k tokens. An initial state of a Petri net is called the initial marking, M_0.

For example, consider a design process that consists of only one design task (see Fig. 24.1). To execute the design task, two resources, A and B, have to be used. As soon as the execution is over, the task is executed again provided that both resources are available. The interpretation of the symbols in Fig. 24.1 (a) is as follows:

p_1 Resource A
p_2 Resource B
p_3 Execute task
t_1 Begin execution
t_2 End execution

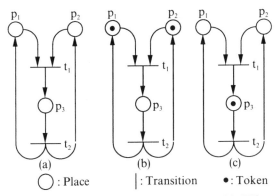

Fig. 24.1 A design task that requires two resources: (a) PN model, (b) initial marking, (c) marking after t_1 has been fired.

In the system, places p_1, p_2, and p_3 represent conditions, and transitions t_1 and t_2 represent the start and finish of tasks, respectively. For example, the place p_1 models the condition 'resource A'. The dots in places are called tokens. When each of a transition's input places has one or more tokens, the transition is enabled. Transition t_1 in Fig. 24.1(b) is enabled because both input places p_1 and p_2 have tokens. When a transition is enabled, it can be fired; therefore t_1 is ready to fire. After the transition t_1 has been fired, the output place, p_3, gets a token (see Fig. 24.1(c)). The initial marking, M_0, in this example is (1 1 0). After firing t_1, the marking changes to $M_1 = (0\ 0\ 1)$.

24.2.1 Properties of Petri nets

An advantage of using PNs comes from their support for analysis of many properties associated with the system being modeled. The most common properties associated with PNs that are useful in analysis are: *reachability, liveness, safeness, boundedness*, and *reversibility* (Peterson, 1981).

Reachability is a fundamental concept for studying dynamic properties of the system. A marking M_n is said to be reachable from a marking M_0 if there exists a firing sequence that transforms M_0 into M_n. The firing sequence is simply denoted by $\sigma = t_1 t_2 \dots t_n$. In Fig. 24.1, M_1 is reachable from M_0 by the firing sequences $\sigma_1 = t_1$, or $\sigma_2 = t_1 t_2 t_1$. Therefore, from the initial marking M_0, 'new' markings can be obtained by firing enabled transitions (e.g., M_1 is a new marking from M_0 by firing t_1 in Fig. 24.1). From each new marking, more markings can be reached. This process results in a tree representation of the markings. In this tree representation, nodes represent markings, and each arc represents a transition firing. Since the tree contains all possible reachable markings, it is called a 'reachability tree' (see Fig. 24.2). The set of all possible markings reachable from M_0 is denoted as $R(M_0)$. If the process is repeated a number of times, all reachable markings will eventually be produced. However, it is likely that the resulting reachability tree is infinite. To generate a finite tree, an algorithm has been developed (see Peterson, 1981).

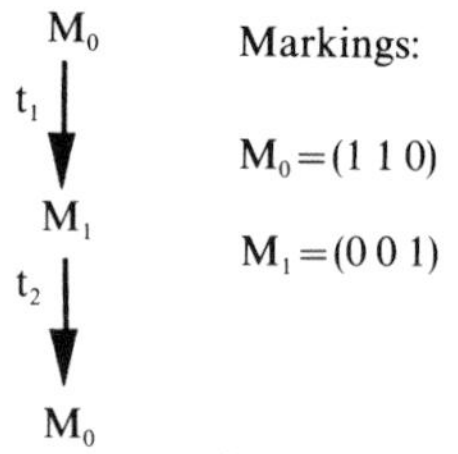

$$M_0$$
$$t_1 \downarrow$$
$$M_1$$
$$t_2 \downarrow$$
$$M_0$$

Markings:

$M_0 = (1\ 1\ 0)$

$M_1 = (0\ 0\ 1)$

Fig. 24.2 Reachability tree and the markings of the Petri net in Fig. 24.1 (b) and (c).

A Petri net is said to be *live*, if no matter what marking has been reached from the initial marking M_0, it is possible to fire any transition of the net. The liveness of a net implies the absence of the possibility of a deadlock. It also implies that any kind of event in a net is possible.

A Petri net is said to be *bounded*, if the number of tokens in each place does not exceed a finite number k for any marking reachable from the initial marking. A Petri net is said to be *safe*, if it is 1-bounded. In a system environment, the boundedness or safeness of a Petri net indicates the absence of overflow in the modeled system.

A Petri net is said to be *reversible*, if for every marking M, M_0 is reachable from M. In a reversible net, one can always return to the initial state. In many cases, it is not required to return to the initial state, but rather to a particular state only. This kind of state is said to be a *home state*. A marking is a home state if it is reachable from any other marking.

24.2.2 Incidence matrix and invariant theory

In this section, the incidence matrix and invariant theory are introduced (Memmi and Roucairol, 1980; Genrich and Lautenbach, 1981; Jensen, 1981; Lautenbach, 1987). Invariant theory is a powerful method which relies on linear algebraic operations for analyzing a Petri net. By considering the transformation of one marking into another marking as two separate linear operations (removal and replacement of tokens), it is possible to represent a PN with a matrix. To introduce the incidence matrix of a Petri net, the following notation is defined:

$\cdot p = \{p | (t, p) \in F\} = $ the set of input transitions of place p

$p\cdot = \{p | (p, t) \in F\} = $ the set of output transitions of place p

Then the incidence matrix $C = [c_{ij}]$ is an $m \times n$ matrix with entries c_{ij} given by $c_{ij} = c_{ij}^+ - c_{ij}^-$

where:

$$c_{ij}^+ = \begin{cases} w(i,j); & \text{if } t_j \in \cdot p_i \\ 0; & \text{otherwise} \end{cases}$$

$$c_{ij}^- = \begin{cases} w(i,j); & \text{if } t_j \in p_i \\ 0; & \text{otherwise} \end{cases}$$

$p_i = $ place i, $t_j = $ transition j, $w(i, j)$ is the weight of the arc from place i to transition j.

Using the above notation, the incidence matrix corresponding to the Petri net in Figure 1(a) is:

$$
c_{ij}^{+} = \begin{array}{c} \\ p_1 \\ p_2 \\ p_3 \end{array} \begin{array}{cc} t_1 & t_2 \\ \left[\begin{array}{cc} 0 & 1 \\ 0 & 1 \\ 1 & 0 \end{array}\right] \end{array}
$$

$$
c_{ij}^{-} = \begin{array}{c} \\ p_1 \\ p_2 \\ p_3 \end{array} \begin{array}{cc} t_1 & t_2 \\ \left[\begin{array}{cc} 1 & 0 \\ 1 & 0 \\ 0 & 1 \end{array}\right] \end{array}
$$

$$
C = c_{ij} = c_{ij}^{+} - c_{ij}^{-} = \left[\begin{array}{cc} 0 & 1 \\ 0 & 1 \\ 1 & 0 \end{array}\right] - \left[\begin{array}{cc} 1 & 0 \\ 1 & 0 \\ 0 & 1 \end{array}\right] = \left[\begin{array}{cc} -1 & 1 \\ -1 & 1 \\ 1 & -1 \end{array}\right]
$$

With this representation, the action of firing a transition in a PN can be viewed as multiplying the incidence matrix by a firing vector and adding the result to the marking vector. For example, $M_1 = (0\ 0\ 1)$ in Fig. 24.1(c) can be expressed as $M_1^{\mathrm{T}} = M_0^{\mathrm{T}} + C \cdot f$, where M_1^{T} denotes the transpose of M_1 and f indicates the firing vector which in this case equals to $(1\ 0)^{\mathrm{T}}$.

A non-zero vector x (or y) is said to be *p*-invariant (or *t*-invariant) if

$$
C^{\mathrm{T}} \cdot x = 0 \text{ or } C \cdot y = 0 \tag{24.1}
$$

Using the equation (1), $x = (1\ 1\ 1)^{\mathrm{T}}$ is a *p*-invariant vector and $y = (1\ 1)^{\mathrm{T}}$ is a *t*-invariant vector for the net in Fig. 24.1.

The *p*-invariant vector satisfies the following equation:

$$
M \cdot x = M_0 \cdot x \tag{24.2}
$$

Hence, it implies that the weighted sum of tokens in places of a *p*-invariant vector remains the same regardless of the firing sequence. The important application of this property is that it enables to verify the reachable markings. If there exists a *p*-invariant vector x such that $Mx \neq M_0 x$, then M is a nonreachable marking of M_0. For example, if one wants to verify whether $M = (1\ 1\ 1)$ is a reachable marking of M_0 for the net in Fig. 24.1, it is enough to calculate $M \cdot x = 3$ and $M_0 \cdot x = 2$. Since $3 \neq 2$, then M is a nonreachable marking of M_0. On the other hand, if equation (24.2) holds for M which is a reachable marking of M_0, then x is a *p*-invariant vector.

For a *t*-invariant vector, y, the following equation holds:

$$
M' = M + C \cdot y = M, \tag{24.3}
$$

where M and M' are reachable markings. Hence, the *t*-invariant vector is equal to the firing vector for a sequence, σ, that forms a cycle. The *i*th component is the number of times that the transition t_1 fires in the sequence σ to construct a cycle. For the Petri net in Fig. 24.1, the marking $M_1 = (0\ 0\ 1)$ is reachable from M_0, then $(0\ 0\ 1)^T + C \cdot (1\ 1)^T = (0\ 0\ 1)^T$, where $(1\ 1)^T$ is exactly a t-invariant vector and the sequence of transitions is $\sigma = t_1 t_2$. The t-invariant property is then useful in detecting cycles in a Petri net.

24.2.3 Colored Petri nets

A colored Petri net (CPN) is 5-tuple (P, T, F, W, C), where:

- P, T, F, and W are as previously defined;
- C is the color function defined from $P \cup T$ into non-empty sets; and
- $C(p)$ and $C(t)$ are the sets of colors associated with place $p \in P$ and transition $t \in T$.

A PN is a special case of a CPN where all the colors of tokens are identical, i.e. no distinction is made between the colors. The practical use of PNs to describe concurrent systems demands for nets that are easily manageable. In a CPN, information can be attached to each token as a token-color and each transition can occur in several ways represented by different occurrence-colors (Jensen, 1981). When a transition occurs, the relation between the occurrence-color and the token-colors involved is defined by functions or expressions attached to the arcs. The color attached to a token or to the occurrence of a transition may represent the content of a buffer area, different types of machines, processors, and so on. By using colors, it is possible to distinguish each member in the same group.

24.2.4 Time-based Petri nets

The concept of time was not explicitly considered in the original definition of a PN. However, it is useful to introduce time delays associated with either places or transitions for the modeling of dynamic systems and performance evaluation. Such a Petri net model is called a (deterministic) timed net if the delays are deterministic, or a stochastic net if the delays are probabilistic.

In the literature, there is no universal agreement on the definition of a timed Petri net (TPN), thus allowing each author to provide a definition that suits his/her application (Caspi and Halbwachs, 1985). One of the major distinctions is whether delays are associated with transitions or places. Another is whether a delay has a fixed duration (Ramchandani, 1974) or is in an interval (Merlin, 1976). A stochastic Petri net (SPN) is a net where each transition is associated with a distributed random variable that expresses the delay from the enabling to the firing of a transition. If the random variable has an exponential distribution (due to the memoryless property of the exponential distribution), it has been shown (Molloy, 1982) that the reachability graph of a live and bounded SPN is isomorphic to a Markov chain (MC). The state transition rate diagram of the MC corresponding to the

SPN is obtained by constructing the reachability graph, and by labeling arcs with the firing rate of the transition whose firing produces the marking changes.

24.2.5 Decomposition and refinement of Petri nets

It is desirable in a design process to use an abstract model for representing a large system. Therefore, techniques are developed to transform an abstract model into a more refined model in a hierarchical way. Decomposition and refinement techniques can be used together as a top-down strategy. The former divides a large net into smaller and understandable nets, whereas the latter adds details to the existing net by replacing a component of the net with a subnet. The refinement of PNs may be done by replacing a place or transition with a subnet as shown in Fig. 24.3.

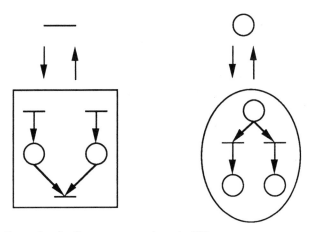

Fig. 24.3 Example of refinement operations in PNs.

The general procedures used in refinement of PNs are:

1. Construct a simple Petri net which describes the aggregate-level system; and
2. Proceed with a stepwise refinement by replacing places (or transitions) with subnets that describe more detailed logic.

Consider replacing a place with a subnet. Zhou *et al.* (1989) proposed to use following basic design modules (Fig. 24.4) for replacing a place with a subnet:

- A *sequential* Petri net represents a series of *n* successive design tasks;
- A *parallel* Petri net represents *n* design tasks that can be executed simultaneously; and
- A *conflict* Petri represents *n* choices of design tasks following a design task.

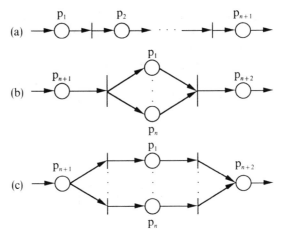

(a)

(b)

(c)

Fig. 24.4 Three types of PNs: (a) sequential, (b) parallel, and (c) conflict.

24.2.6 Modeling the design process with Petri nets

Consider a design process that includes a number of design tasks. A simple design process is represented in Fig. 24.5, where the circles denote design tasks involved in a design process and the arrows are input and output parameters, respectively. Each task may require or output multiple parameters. No specific meaning is assigned to any input or output parameter at this point. What they represent might be the precedence relations that impose a partial ordering among tasks. For example, Fig. 24.5 shows that the output from task 1 is the input to task 2, where x, y and z are parameters.

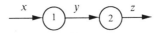

Fig. 24.5 A design process with two design tasks.

Hence, the arrow used here represents a precedence constraint, i.e., task 2 cannot begin until task 1 has been completed. It is assumed that there are two kinds of output parameters:

- Parallel output parameter – all of the design tasks following a parallel output can be executed simultaneously after their preceding tasks have been completed; and
- Conflict output parameter – on the other hand, only one of those tasks following a conflict output can be executed when their preceding tasks terminate.

To model a design process using PNs, the idea proposed by Krogh *et al.* (1987) and Valavanis (1990) has been adopted that a single place type is not adequate to express the possible complexity of a system. Therefore, two basic types of places are defined, task place (tk-place) and parameter place (pm-place), corresponding to the design task and parameter, respectively. To distinguish them, a tk-place is represented as a box and a pm-place as a circle. Figure 24.6 shows a simple design task with only one input and one output (I–O) parameter. A token in the tk-place implies that the task is being executed. A transition in a Petri net represents the initialization or termination of I–O parameters.

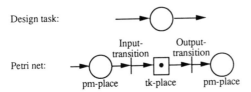

Fig. 24.6 Representation of a task place and parameter place in a PN.

Besides the tk-place and pm-place, there are other types of places defined in Valavanis (1990) which are also useful in representing a system: source place and sink place (Fig. 24.7). The source place denotes the initialization of a token, indicating the introduction of a new resource, information, and so on into the system. The sink place, on the other hand, indicates the termination of a token. This means that the resource or information leaves the system.

(a)   (b)

Fig. 24.7 Two types of places: (a) source place, and (b) sink place.

24.3 ILLUSTRATIVE EXAMPLE

In this section, a PN is applied to demonstrate the following aspects of a design process: decomposition, multi-resource requirements, and cycles. The example in Fig. 24.8 describes a design process that involves five different departments: A, B, C, D, and E. Each department performs some design tasks (see Fig. 24.8).

24.3.1 Decomposition of the design process

The design process may involve a number of subprocesses. Managing each subprocess might be easier due to a smaller problem space. In order to deal with complex problems, designers abstract and create models that represent only the essential features. Thus the abstraction of a system plays a key role in the development of a model. The applicability of PNs to model a system

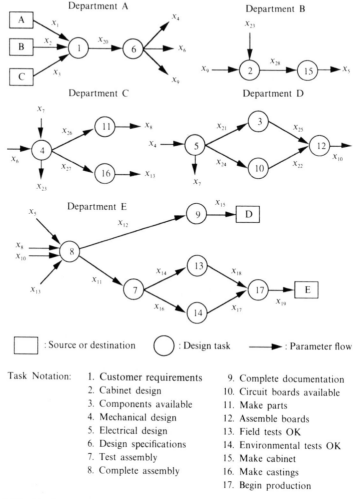

Fig. 24.8 Design example.

hierarchically derives from the fact that a net can be replaced with a single place or transition for modeling at a more abstract level, or places and transitions can be replaced with subnets to provide more detailed refinement.

For the design example in Fig. 24.8, its abstract PN model is shown in Fig. 24.9. To illustrate how to construct the net in Fig. 24.9, first, all five departments (excluding sources or destinations) in Fig. 24 8 are aggregated as a single design task and a tk-place, p_7, is associated with this aggregated task. Having represented the aggregated design task, the source or destination blocks are to be dealt with. This is done by classifying them as system initialization blocks (A, B and C) and system completion blocks (D and E),

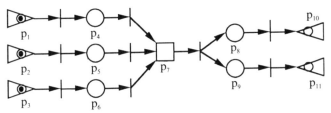

Fig. 24.9 The abstract PN model of the design process.

and by association with the corresponding places. For example, A is one of the system initialization blocks, it is then represented as a source place, p_1, and its output parameter (x_1) as a parameter place, p_4, Block D is associated with a destination place, p_{10}; a parameter place, p_8, represents its input parameter (x_{15}). Other elements can be obtained in a similar manner to complete the abstract PN. By doing so the net has a distinct view of the system initialization/completion against the system execution that adds to better graphical understanding. An alternative is to represent the system initialization/completion with an abstract place. It is not done in this case because of the small number of system initialization/completion parameters. Each of the three source places having a token denotes the initial state that input parameters are provided. If p_7 has a token, it denotes that the aggregated design task, which means the overall design process, is being executed.

When an abstract model is constructed, the next step is to replace a place with a more detailed subnet. In Fig. 24.9, the tk-place, p_7 is an aggregated task of five departments. Thus, it is to be substituted with a subnet that describes the interactions among departments. In Fig. 24.10, the subnet, circumscribed by dash lines, representing the five departments has replaced p_7 in Fig. 24.9. Department names as well as input/output parameters are provided in tk-places (boxes) and pm-places (circles). The refinement uses the notion that the subnet has an initial place (Department A), an inner part and a final place (Department E) to replace the abstract place. It is important that the new net after the refinement satisfies the relationship (e.g., the input/output parameters associated with each department) described in Fig. 24.8. To verify the consistency, Department A has three input (x_1, x_2, x_3) and three output (x_4, x_6, x_9) parameters in Fig. 24.8. Output parameters x_4, x_6, and x_9 become input parameters to Departments D, C, and B respectively. The modeling of this relationship is clearly satisfied in Fig. 24.10 (e.g., Department D is enabled only when x_4 is available). At this point, however, each department is represented by no more than a tk-place which is an abstract place as well. It means that the information within a department (e.g. x_{21}, x_{22}, x_{24}, and x_{25} within Department D) has not yet been expressed. One needs to refine each abstract place of a net in a similar way until further substitutions cannot be made.

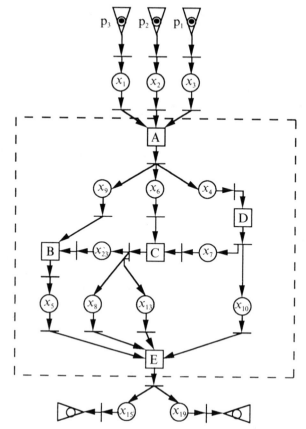

Fig. 24.10   Refinement of the abstract PN model in Fig. 24.9.

24.3.2 Cycles in the design process

A cycle in a graph is a path whose initial and terminal nodes are the same
(they coincide). The notion of a cycle is applicable in a PN that models a
design process because a place or a transition can be seen as a node in a
graph. In a Petri net, a cycle can be interpreted as the repetition of a specific
task after some tasks have been performed. One of the reasons for the
repetition of some tasks is to implement design changes. A design process
requires various resources, and one expects that whenever a change occurs,
the earlier it can be made, the less cost is involved. When a design change
occurs, one normally returns to some point and then repeats some design
tasks. The point may be the initial state or an intermediate state depending
on the extent of the change. The management of cycles is thus an important
issue in modeling of the design process.

   Consider the example in Fig. 24.11 that includes a number of design
cycles. Two outgoing arcs from test to mechanical system design and
electrical system design might be associated with the meaning that if test

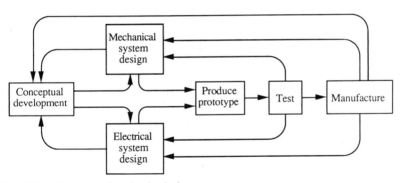

Fig. 24.11 Example of cycles in design.

fails, the information should be provided to these two design groups. This coincides with the principle of detecting and solving problems in the early stages of design. To model a system, one needs to capture the nature of the underlying problem rather than deal with a complex system. Therefore, only some of the possible cycles are considered in the model in Fig. 24.11. The PN model of Fig. 24.11 is represented in Fig. 24.12. In order to reduce the size of the PN, Fig. 24.12 does not distinguish a tk-place from a pm-place · defined before. Instead, a circle (place) is used to represent a design task. A token in p_1 denotes the initial status.

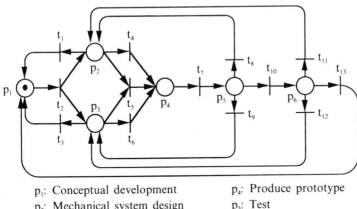

p_1: Conceptual development p_4: Produce prototype
p_2: Mechanical system design p_5: Test
p_3: Electrical system design p_6: Manufacture

Fig. 24.12 Representing cycles in design with PNs.

In Fig. 24.11, both mechanical system design and electrical system design have one arc leading to produce prototype. A question may arise whether the two arcs represent 'simultaneous' input? In other words, to proceed with produce prototype, should the outputs from both mechanical system design and electrical system design be provided at the same time? This vagueness can be easily resolved in a PN and the interpretation explicitly described. In

Fig. 24.12, it is easy to determine that there are two possible outcomes related to the underlying vagueness:

1. The two tasks mechanical system design and electrical system design should occur at the same time (to fire transition t_5); or
2. Either mechanical system design or electrical system design occurs (transition t_4 or t_6 fires).

Thus, the possible outcomes can be explicitly represented in accordance with the requirements. In this respect it is shown that one can take advantage of a PN model to resolve this vagueness. Moreover, without the loss of generality, the following assumptions are made: when t_2 fires, p_2 and p_3 get a token, each of them will not return to p_1 immediately by firing either t_1 or t_3. It is assumed that mechanical system design and electrical system design, when each of them receive the token from conceptual development, will be together to fire t_5 and proceed with produce prototype.

24.3.3 Multi-resource requirements of the design process

To accomplish a task, resources, e.g., human (designers, analysts), computer hardware, and CAD software are required. To model the resources, an additional type of place, called resource place (rs-place), was introduced by Krogh (1987). Pictorially, it is represented as a circle containing a box (see Fig. 24.13). A token in a rs-place indicates that the resource is available.

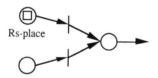

Fig. 24.13   Representing resource place in a PN.

In the previous example (Fig. 24.8), no resource requirement was specified. Consider department B, which involves two tasks, cabinet design (task 2) and cabinet made (task 15). Without the loss of generality, both tasks have been aggregated as a single task. In addition, in order to design and manufacture a product, three types of resources such as CAD software (for design), raw materials, and assembly line (for manufacturing) are needed (see Fig. 24.14).

A CPN keeps track of the status of resources, such as the availability of a specific resource between tasks, before-task, and after-task. In Fig. 24.15, RS is the set of colors expressing three different types of resources. The arc expression defined, rs, indicates that after a transition has been fired, a color $r \in RS$, will both remove the token from p_1 and add the token to place B. Below RS and rs, there are two expressions, (a, b, c) and $(2x+y)$. The triplet (a, b, c) specifies that place p_1 contains three different colors (resources), each

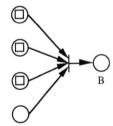

Fig. 24.14 Representation of multiple resources.

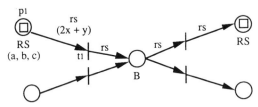

Fig. 24.15 Representing resources with a CPN.

of them with quantity 1. On the other hand, $(2x + y)$ means that in order to fire transition t_1, two different types of colors must be available, type x needs quantity 2 and type y needs quantity 1. Under the circumstances, t_1 cannot be fired because of the shortage of quantity 1 of each type of resource in place p_1.

The net in Fig. 24.12 can be extended to include two basic types of resources, namely, a mechanical resource and electrical resource (Fig. 24.16).

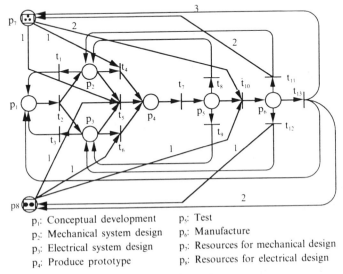

p_1: Conceptual development p_5: Test
p_2: Mechanical system design p_6: Manufacture
p_3: Electrical system design p_7: Resources for mechanical design
p_4: Produce prototype p_8: Resources for electrical design

Fig. 24.16 A PN representing design cycles and resources.

The mechanical (electrical) resource is a set that contains technical resources (computers and software) to human resources (designers and analysts) required to perform design and analysis tasks. The numbers along the arcs represent the exact units of each type of resource needed or returned. Having incorporated the resource requirements, the firing conditions of some transitions must take the resource aspect into account. For example, to proceed with manufacture (p_6), two units of mechanical resource and one unit of electrical resource must be available. This is equivalent to saying that at least two tokens must be in p_7 and one token in p_8 to fire t_{10}. The return of a resource means the reusability of that resource (assume it is not depleted) by some other tasks. For example, machines are occupied to perform manufacture task, after the task is done, they become free for use by other tasks.

A major difficulty with PNs is that industrial applications are likely to result in large systems consisting of many places and transitions (Peterson, 1981). Other shortcomings include the structural inflexibility, and the inability to identify individual tokens (He and Lee, 1991). The mentioned weakness can be overcome by CPNs. The major reason for the success of CPNs is that – without diminishing the possibility of formal analysis – they allow to create manageable models (Jensen, 1991). In CPNs, two basic types of a resource can be represented with a single place containing tokens associated with colors. In Fig. 24.17, p_6 is such the place with colored tokens. Another approach leading to the simplification of a large PN, that does not require tokens being colored, is to use some reduction rules (Hyung *et al.*, 1987; He and Lee, 1991). The reduction rules transform a net into a simpler one while preserving properties such as liveness and boundedness of the original net. Depending on the needs and the net structure an appropriate alternative (or both) should be selected. Deriving Fig. 24.17 from Fig. 24.16, the reduction rules were applied first and then the colors were associated with resource tokens.

The analysis of Figs. 24.16 and 24.17 shows that not only have the two resource places (p_7, p_8 in the former) been replaced with a single place (p_6 in

p_1: Develop concept
p_2: Design
p_3: Produce prototype
p_4: Test
p_5: Manufacture
p_6: Resources

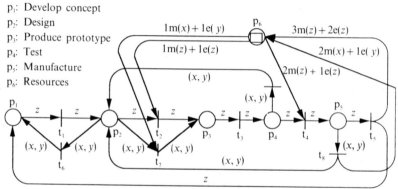

Fig. 24.17 Representing cycles and resource requirements in design with a CPN.

the latter), but also two design tasks (p_2, p_3 in the former) have been merged (p_2 in the latter). The number of places and transitions has been reduced. In the new net, however, functions or expressions are associated with the arcs to describe the flow relation. The arcs can be categorized into two classes:

1. Originating from the non-resource places. Expression (z) denotes that mechanical system design and electrical system design are 'binding' together. The other expression (x, y) means that either 'x' or 'y' is true, i.e. either mechanical system design or electrical system design, but not both at the same time, may happen. The set of colors, say design-color, can be defined to include three different types of designs; m (mechanical design), e (electrical design), and n (neutral, no distinction is made).
2. Originating from the resource place (p_6). Functions $m(v)$ and $e(v)$ (v is a variable) map from a design-color to the resource-color. The set of resource-color is defined to be $\{m, e\}$. These functions specify the type and quantity of resources needed to proceed with each task. For example, the function $m(v)$ means that one unit of mechanical resource is needed. Two arc expressions input the transition t_7, namely, $1m(x) + 1e(y)$ and (x, y). The former is from place p_6, while the latter is from place p_2. Since the expression (x, y) specifies that only either x or y can happen, one concludes that the type of resource needed for firing transition is either $m(x)$ or $e(y)$, which coincides with the original model.

24.4 CONCLUSIONS

In this chapter, PNs were applied to model a design process. A design process was hierarchically structured, involved various types of resources, and had cycles. It was demonstrated that PNs are useful in modeling the design process. Hierarchical decomposition was performed by replacing a place with a subnet that provides more detailed information. Multi-resource requirements were also modeled by CPNs. A cycle is equivalent to a firing sequence of transitions. Due to the graphical representation, a net is easy to understand by nonexperts.

24.5 REFERENCES

Caspi, P. and Halbwachs, N. (1985) *1985 International Workshop On Timed Petri Nets*, pp. 40–6.
Genrich, H.J. and Lautenbach, K. (1981) *Theoretical Computer Science*, **13**, 109–36.
Genrich, H.J. and Thiagarajan, P.S. (1984) *Theoretical Computer Science*, **30**, 241–318.
Gentina, J.C. and Corbeel, D. (1987) *Proceedings of IEEE International Conference on Robotics and Automation*, April, pp. 1166–73.
Hatono, I., Yamagata, K. and Tamura, H. (1991) *IEEE Transactions on Software Engineering*, **17**(2), 126–32.

He, X. and Lee, J.A.N. (1991) *Software: Practice and Experience*, **21**(8), 845–75.

Hyung, L.K., Favrel, J. and Baptise, P. (1987) *IEEE Transaction on Systems, Man, and Cybernetics*, **SMC-17**(2) 297–303.

Ichikawa, A. and Hiraishi, K. (1987) *Lecture Notes in Control and Information Sciences*, **103**, Direct Event Systems: Models and Applications, Springer-Verlag, New York, pp. 115–34.

Jensen, K. (1981) *Theoretical Computer Science*, **14**, 317–36.

Jensen, K. (1987) *Advances in Petri Nets* (eds W. Brauer, W. Reisig and G. Rosenberg) **254**, Springer-Verlag, New York, pp. 248–99.

Jensen, K. (1991) *Advances in Petri Nets* (ed. G. Rosenberg) **483**, Springer-Verlag, New York, pp. 342–416.

Kasai, T. and Miller, R.E. (1982) *Journal of Computer and System Sciences*, **25**(3), 285–331.

Krogh, B.H. and Sreenivas, R.S. (1987) *Proceedings of IEEE International Conference on Robotics and Automation*, April, 1005–011.

Lautenbach, K. (1987) *Advances in Petri Nets*, (eds. W. Brauer, W. Reisig and G. Rosenberg) **254**, Springer-Verlag, New York, pp. 142–67.

Martinez, J., Muro, P. and Silva, M. (1987) *Proceedings of IEEE International Conference on Robotics and Automation*, April, pp. 1180–5.

Memmi, G. and Roucairol, G. (1980) in *Net Theory and Applications* (ed. W. Brauer), Springer-Verlag, New York, pp. 213–23.

Merlin, P. (1976) *IEEE Transactions on Communications*, **COM-24**(6), 614–21.

Molloy, M.K. (1982) *IEEE Transactions on Computers*, **C-31**(9), 913–17.

Murata, T. (1989) *Proceedings of the IEEE*, **77**(4), 541–80.

Ozsu, M.T. (1985) *IEEE Transactions on Software Engineering*, **SE-11**(10), 1225–40.

Peterson, J.L. (1981) *Petri Net Theory and the Modeling of Systems*, Prentice Hall, Englewood Cliffs, NJ.

Ramchandani, C. (1974) Analysis of asynchronous concurrent systems by timed Petri nets, Ph.D. Dissertation, M.I.T., Cambridge, MA.

Valavanis, K.P. (1990) *IEEE Transactions on Systems, Man, and Cybernetics*, **20**(1), 94–110.

Zhou, M.C., Dicesare, F. and Desrochers, A. (1989) *IEEE International Conference on Robotics and Automation*, **1**, 534–9.

CHAPTER 25

Neuro-computing and concurrent engineering

Cihan H. Dagli, Pipatpong Poshyanonda and
Ali Bahrami

25.1 INTRODUCTION

The quest for building systems that can function automatically has attracted a lot of attention over the centuries and created continuous research activities. As users of these systems we have never been satisfied, and demand more from the artifacts that are designed and manufactured. The current trend is to build autonomous systems that can adapt to changes in their environment. While there is much to be done before we reach this point it is not possible to separate manufacturing systems from this trend. The desire to achieve fully automated manufacturing systems is here to stay.

Manufacturing systems of the twenty-first century will demand more flexibility in product design, process planning, scheduling and process control. This may well be achieved through integrated software and hardware, architectures that generate current decisions based on information collected from manufacturing systems, environment and execute these decisions by converting them into signals transferred through communication networks. Manufacturing technology has not yet reached this state. However, the urge for achieving this goal is transferred into the term 'intelligent systems' that we started to use more in the late 1980s.

Knowledge-based systems, our first efforts in this endeavor, were not sufficient to generate the 'intelligence' required – our quest still continues. The trick is to merge the old with the new, for example integrate mathematical programming with artificial intelligence, and keep adding new approaches and technologies. Artificial neural networks is another technology that needs to be integrated to improve our quest for 'intelligent systems' (Dagli, 1991). In this chapter, the impact of artificial neural networks as an emerging new technology is discussed in relation to concurrent engineering and the use of this technique is demonstrated in manufacturing feature recognition which is an essential part of effective system automation.

25.1.1 Intelligent manufacturing systems

Decision complexity is and will be an issue in manufacturing systems. This is due to the fact that excessive design and operation alternatives exist in the products to be produced and the choice among appropriate combinations is not an easy task. Recent changes in global economy and intense international communication capability has created the global market which necessitates flexibility in manufacturing systems to be able to compete effectively with companies emerging around the globe. There is a definite trend to move into customized product with short life-cycles and response to market changes pretty much instantaneously. During the 1990s, the flexibility is added to the list of competitive thrusts that has evolved since the 1960s, namely: cost, market, and quality. The manufacturing strategies for the 1990s can be summarized as: new production introduction, responsiveness, new organization forms, new accounting systems, international thrust, and extensive update in manufacturing systems. It is currently feasible to have an organization largely composed of specialists who direct their own performance through automated information and directions provided by peers, customers, and main offices located miles away. Manufacturing strategies for the 1990s will translate into tactical planning systems, decentralized systems, intelligent systems, computer integrated planning and simplification. The flexibility that we are searching for requires integration among basic functions of manufacturing, namely: product and part design, process planning, programming for machines, robots, automated guided vehicles production planning, manufacturing, receiving, storage and shipping. Manufacturing flexibility demands customized high quality and low cost products with inexpensive components. This translates into auto-nomous machine setup procedures, automation of design and process planning, and well integrated manufacturing information systems to get the first part right for the first time. Hence, flexibility requires intelligence, and this needs to be integrated with real time control to be able to adapt to changes both in market and manufacturing enviroment at the shop floor.

25.1.2 Concurrent engineering

The quest for designing intelligent manufacturing systems started in the 1980s through the use of knowledge-based systems. Research work done in the laboratories in the 1970s started to find its way into actual manufacturing operations during this time period. Today, these efforts are well documented in intelligent manufacturing books (Kerr, 1990 and Kusiak, 1990). However, the quest for intelligent systems still continues. This is due to the fact that most applications are concentrated in small units within the manufacturing systems and functions. It is not difficult to see a lot of application of this technology in scheduling (Dagli and Poshyanonda, 1990), process planning (Joneja, 1991 and Chang, 1990), group technology (Kusiak and Chow, 1987), quality control (Dagli and Stacey, 1988). Although these efforts were successful within their domains, they did not provide enough integration among functions due to difficulties associated with representation of manufacturing features. This fact restricted the ability to achieve ultimate automation and in turn intelligence. If the reasons of program changes are examined for manufacturing, it is not difficult to detect that lack of communication among manufacturing functions, namely: design and process planning. A survey conducted by Wright and Bourne (1988) indicated that roughly 70% of maufacturing programs changes are due to dimension error, numerical control programs, tooling, setup, function errors which can well be eliminated through effective product design, and intelligent process plans. Design process begins the manufacturing life cycle. The design alternative selected impacts product cost, ranging from 65% to 80% of the total cost of production (Swift, 1987; Khoshnevis and Park, 1988; Suri and Shimizu, 1989).

Hence, it is an obvious approach to include the restrictions imposed by manufacturing operations into the design process. Design needs to balance cost efficiency and manufacturability with quality and functionality and as such must satisfy many, sometimes conflicting, constraints. Concurrent engineering or design for manufacturability is the process of design that considers manufacturing, such that the products are designed in shorter periods of time with minimum cost of development, with an easy transition to production, and are assembled and tested with the minimum cost and time. Effective consideration of these attributes increases the quality and reliability of the product and gives utmost importance to the customer needs to be able to compete well in the global market place.

Concurrent engineering examines the manufacturing issues early in the design process to shorten the product development time and increase the capability of the company to respond to market changes effectively enabling manufacturing organizations to meet the competitive thrust of the 1990s 'flexibility' (Roberts, 1991). Concurrent engineering reduces the manufacturing cost, as approximately 80% of the cost of production is committed at the end of the design activity although the costs incurred hardly reach 10% (Anderson, 1990; Roberts, 1991). This is possible if products can be assembled quickly from fewer standard parts. Parts are designed for ease of fabrication and to use as much commonality as possible with other designs. This approach can create a larger product variety that can be manufactured by assembling common standard modules and parts into new products. This process heavily depends on manufacturing feature recognition. Hence, effective automation of the design process and concurrent engineering activities very much depend how manufacturing features are represented and retrieved. Early efforts based on knowledge-based systems did not provide effective solutions to this problem (Dagli and Lammers, 1989; Dagli, 1990; Dagli, Lammers and Vellanki, 1991).

In this chapter possible use of artificial neural networks for concurrent engineering is discussed. In Section 25.2 design automation in reference to design for manufacturability is examined. Section 25.3 summarizes the most used artificial neural network paradigm back propagation after the introduction of basic concepts. Use of artificial neural networks in manufacturing feature recognition is discussed extensively in Section 25.4. Section 25.5 summarizes other applications of this emerging technology in design, process planning, group technology and automated assembly and vision.

25.2 DESIGN FOR MANUFACTURING

25.2.1 Basic concepts

Design task can be considered as a mapping from the functional space to an attribute space. This process translates the customer requirements into a final product through continuous refinement. Generic steps can be classified as identify needs, set strategies, establish design concept, select feasible alternatives, select and specify parameters and evaluate and implement (Suri and Schimizu, 1989). Design, when completed, needs to specify for each part: shape, dimensions, materials, tolerances, static and kinetic interactions and assemblies. Although some of the engineering functions, such as: mass properties calculations, interference checking and geometry definitions for finite elements, drafting and numerical control are supported by Computer Aided Design (CAD) and Computer Aided Manufacturing (CAM), most of this process necessitates human interaction and team work (Shah and Rogers, 1988). Brown and Chandrasekaran (1988) identify three classes for design. Class 1 design is innovative, breaking barriers to develop a new product or process. Class 2 involves new techniques or requirements for known products or problems. Class 3 is routine, but it is also complex since there are many components and combinations. Current knowledge-based systems are capable of addressing only class 3 design effort to a certain degree.

Hence, effective understanding of the design problem and development of solution methods is a difficult problem. Research work is needed in the area for design automation and effective design models that will impact manufacturing. The complexity of this problem increases exponentially as soon as manufacturing considerations are added. However, as long as design of a product commits 80% of the total production costs the importance of this research problem will continue to attract manufacturing companies attention.

25.2.2 Manufacturing considerations in design

For a long time function of the product was the primary consideration in a design process. Although cost is considered, it did not appear as an important attribute at the beginning. There are many other considerations that should be taken into account in the design process, namely: delivery, quality, reliability, ease of assembly and test, ease of service and repair, shipping, human factors, appearance and style, safety, customer's needs, product customization, breadth of production line, time-to-market, expansion and upgrading, future designs, product pollution, processing pollution, ease of product recycling (Anderson, 1990). Consideration of these constraints and attributes during the design process is not an easy task and generally designers restrict attributes and alternatives available by making arbitrary decisions to eliminate some of the search space. In order to come up with good restrictions to solution space of the design problem, various strategies are developed that takes manufacturing restrictions into account. These are summarized by Anderson as standardization, assembly, fastening, test, motion of assembly, part shape, handling by automation, repair. Some of these strategies are (Anderson, 1990):

- Design product so that subsequent parts can be added to the base part;
- Make parts independently replaceable;
- Design the base part so that its manufacturing features provide quick and accurate positioning;
- Use the minimum number of total fasteners;
- Design screw assembly for downward motion;
- Specify proper tolerances for press fits;
- Minimize the number of types of cables;
- Fastened parts are located before fastener is applied;
- Standardize design features;
- Use standard parts;
- Combine parts and functions into a single part;
- Design parts with symmetry;
- Specify tolerances tight enough for automatic handling;
- Design and select parts that can be oriented by automation;
- Use modular design to allow replacement of modules; and
- Design in counters and times to aid preventive maintenance.

All of these strategies require efficient knowledge of manufacturing features, such as slots, steps, pockets etc. It is not difficult to conclude that effective concurrent engineering depends heavily on quick and accurate identification of these manufacturing features. This capability can increase the number of design alternatives generated and tested by the design engineer. Artificial neural networks can be used effectively for this purpose. This is demonstrated with an example in the following sections.

25.3 ARTIFICIAL NEURAL NETWORKS

25.3.1. Introduction

Artificial neural networks (ANN) are a new information processing technique, they simulate biological neurons using computers. Although biological details are eliminated by this computer model, artificial neural networks retain enough structure to work like a biological neural processing unit. They are mathematical models of theorized mind and brain activity.

ANN utilizes a parallel processing structure that has a large number of processors. These processors are much simpler than the typical central processing units (CPUs). Each processor is connected to many of its neighbors, thus, there are many more interconnections among processors than the number of processors itself. The damage to a few nodes or interconnections do not affect overall performance significantly.

ANN provides greater degrees of robustness or fault tolerance due to their massive parallelism in their design. Neural networks are used in the situations where only a few decisions are required from a massive amount of data, or a complex non-linear mapping needs to be learned.

ANN provides an effective approach for many applications in various fields such as engineering, science, finance. Some examples of neural networks applications are pattern recognition, image processing, design, statistical analysis, data compression, near-optimal solutions to combinatorial optimization problems, pattern matching, system modeling, function approximation, process planning, and loan approval.

25.3.2 Basic concepts

A neuron is the fundamental cellular unit of the nervous system. The neuron's nucleus is a simple processing unit which receives and combines signals from many other neurons through input paths called dendrites. If the combined signal is strong enough, it activates the firing of the neuron which produces an output signal. The path of the output signal is called the axon which splits up and connects to dendrites (input path) of other neurons through a function referred to as a synapse. The transmission across this junction is chemical in nature. The amount of signal transferred depends on the synaptic strength of the junction. In addition, synaptic strength is modified during the learning process of the brain, thus, it can be considered as the basic memory unit. ANN simulates neurons and their axons through processing elements and their interconnections.

(a) Processing elements

First, we need a node that approximates the functions of a biological neuron in order to do the simulation. Processing elements (PEs) are only the first step of this process (Chisvin and Duckworth, 1989). The PE handles many basic functions such as input signals evaluation, summation of signals and comparison to a threshold value to determine the output value (Fig. 25.1).

Inputs and outputs
There are many input signals come to a PE simultaneously, but there is only one output signal for each PE as is presented in a biological neuron. The PE, in response, gives either a positive or negative signal, depending on a threshold level for that PE. In addition, some networks provide an extra input called a bias term, or a forcing term for other influences such as a reinforcement from outside the network (Nelson and Illingworth, 1990).

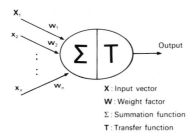

X : Input vector
W : Weight factor
Σ : Summation function
T : Transfer function

Fig. 25.1   Basic processing element unit.

Weighting factors
There is a relative weighting corresponding to each input presented to a PE. These weights are adaptive coefficients within the network which is used in order to determine the intensity of each input signal. Some inputs are more important than others in the way they combine to produce the output. Mathematically, input signals and weights corresponding to these inputs are vectors such as $x_k = (x_{1k}, \dots, x_{nk})$ and $w_j = (w_{j1}, \dots, w_{in})$. The value corresponding to each input that will affect the output is given as a vector product.

Summation function
After all inputs are received by a PE, they are summed and compared with the threshold value to determine the output. This summation function might be more complex than the simple summation. If the summed value is greater than the threshold value, PE will generate the output signal. If the summed value is less than the threshold value, PE will not generate any signal, or even an inhibitory signal. Both types of output signal are considered significant.

$$\text{output} = \sum_{i=1}^{n} w_i x_i - \theta$$

Transfer function

The transfer function receives the result from the summation function and generates the output for the PE. The transfer function can be a simple function, such as linear function, or even a complex function. Examples of transfer function are threshold logic, linear function, step function, sigmoid function, hyperbolic tangent function, and binary step function. The transfer function is usually continuous and non-linear. Linear functions are not very useful since the output value is simply proportional to the input (Fig. 25.2a). The step function outputs only binary values depending on the value received from the summation function. It gives output values only in the given range, which is between zero and one for the binary step function (Fig. 25.2b) The sigmoid function, which can be called as an S-shaped curve, approaches minimum and maximum values at the asymptotes (Fig. 25.2c). The derivatives of the sigmoid function are continuous which produces nonlinearities to the network. Therefore, the sigmoid function is often selected as a transfer function.

(b) Interconnections

An interconnection provides the data transfer from one PE to another, or even to a PE itself. A weight is assigned to each PE input connection. A connection is unidirectional. There are two primary connection types: excitory and inhibitory. Excitory connections enhance PE's input values, which are usually positive signals. On the other hand, inhibitory connections decrease PE's input values, and typically are negative signals. There are three primary interconnection schemes: intra-field, inter-field, and recurrent connections (Simpson, 1990). Intra-field connections connect PEs in the same layer, and inter-field connections connect them in different layers. Recurrent connections loop and connect back to the PE itself. If the information is allowed to flow among PEs only in one direction, these interconnections are called feedforward. Feedback allows the information to flow in any direction and/or recursively. The pattern of interconnections and learning rules (the procedure to update the weights corresponding to the interconnections) define the paradigm of the ANN.

(c) Learning

In artificial neural networks, learning is considered as the rate of change in the weights of interconnections, which is also called memory, of the network. There are two types of learning: supervised and unsupervised. Supervised learning requires a teacher which can be a training data set, or an observer who evaluates the performance of the network. On the other hand, unsupervised learning requires no external teacher. The network self-organizes using some internal rules.

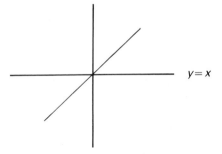

(a) Linear function

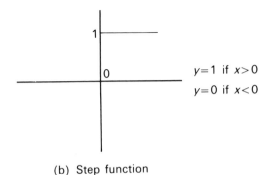

(b) Step function

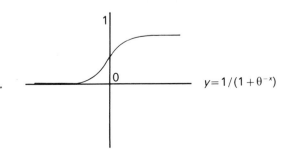

(c) Sigmoid function

Fig. 25.2 Examples of transfer functions.

25.3.3 Basic artificial neural network paradigms

In this section, some of the artificial neural network paradigms commonly used are introduced. One of the most commonly used paradigms is back propagation. This paradigm is discussed in detail in the following section. Some other neural network paradigms which are of interest are: Perceptron, ADALINE/MEDALINE, and Adaptive Resonance Theory.

Perceptron (Rosenblatt, 1957) is the earliest artificial neural network paradigm. The architecture of the perceptron is simple (see Fig. 25.3), it has processing elements connected to each other in the feedforward scheme and uses supervised learning approach. The outputs of the network are binary. Pattern classes are separated with linear boundaries. This major characteristic of perceptron is linear separation. However, perceptron is capable to generalize and handle noisy or variable inputs. Pattern recognition and image processing are the major applications area of this paradigm.

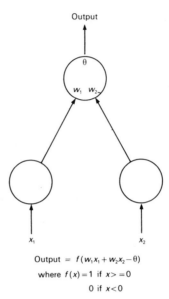

Output

$$\text{Output} = f(w_1 x_1 + w_2 x_2 - \theta)$$
$$\text{where } f(x) = 1 \text{ if } x > = 0$$
$$0 \text{ if } x < 0$$

Fig. 25.3 An elementary perceptron.

ADALINE stands for ADAptive LINear Elements (Widrow, 1959), and MEDALIN is Multiple ADALINEs in parallel. The topology of ADALINE is shown in Fig. 25.4. ADALINE can learn more quickly and accurately than Perceptron. It is also a feedforward network and uses supervised learning. The learning rule is based on a least mean square (LMS) algorithm. Even when the computed output is correct, heights are adjusted to improve reaction time (Nelson and Illingworth, 1990). Basic application areas are: control, speech recognition, weather forecasting.

Adaptive Resonance Theory (ART) is one of the feedback neural networks with unsupervised learning scheme (Carpenter and Grossberg, 1987). It learns on-line and operates in discrete time. ART uses the competitive learning, which is winner takes all. The basic features of the ART architecture are shown in Fig. 25.5. In an ART network, information in the PEs is transferred back and forth between layers. If the proper pattern is developed, the network becomes stable. There are two versions of ART: ART1 and ART2. ART1 operates with binary inputs only; on the other hand, ART2

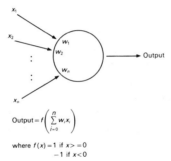

Fig. 25.4 Simple ADALINE processing unit.

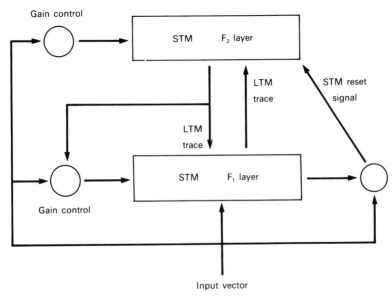

Fig. 25.5 ART system architecture.

can operate on both digital and analog patterns. ART networks have been widely used in many applications such as image processing, pattern classification, controls, diagnostics, and decision making under risk (Dagli and Lynch, 1991).

25.3.4 Back propagation

The elementary back propagation is a three-layer feedforward network. But in general, it is possible to have several hidden layers, connections that skip over layers, and lateral connections. Back propagation uses supervised learning and a multilayer gradient descent error correction encoding algorithm. It learns off-line and operates in discrete time. Output of the back propagation network is not restricted to binary (Simpson, 1990).

(a) Training

The back propagation algorithm performs the input-to-output mapping by minimizing a cost function to make connection weight adjustments according to the error between the computed and desired output. The cost function is minimized using the squared error which is the squared difference between the computed output value and the desired output value for each PE. An elementary back propagation topology is shown in Fig. 25.6.

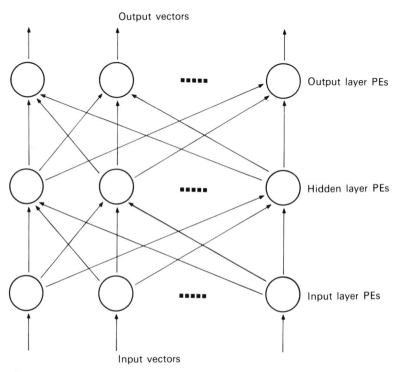

Output vectors

Output layer PEs

Hidden layer PEs

Input layer PEs

Input vectors

Fig. 25.6 Back propagation network architecture.

All the steps used in training of the three-layer back propagation algorithm neural network are summarized below. A network of four input neurons, three hidden layer neurons, and two input neurons are used as an example to demonstrate the algorithm. For simplicity, an input vector of [1 1 1 1], output vector of [1 1], and the threshold function is $f(x)=(1+e^{-x})^{-1}$, and learning rate $\alpha=0.3$, $\beta=0.4$ is assumed.

1. Initialize the network by assigning random values in the range [+1, −1] to all the input-to-hidden layer connection (w^h), hidden-to-output layer connections (w^o), and all threshold values in hidden and output layers (θ^h and θ^o). These initialized values, if they are nonrandom or structured, may cause the network to be trapped in a local minimum.

For simplicity, all weights are initialized to 0.5, all threshold values are 0.25.

$$w^h = \begin{bmatrix} 0.50 \; 0.50 \; 0.50 \\ 0.50 \; 0.50 \; 0.50 \\ 0.50 \; 0.50 \; 0.50 \\ 0.50 \; 0.50 \; 0.50 \end{bmatrix} \qquad w^0 = \begin{bmatrix} 0.50 \; 0.50 \\ 0.50 \; 0.50 \\ 0.50 \; 0.50 \end{bmatrix}$$

$$\theta^h = [0.25 \; 0.25 \; 0.25] \qquad \theta^0 = [0.25 \; 0.25]$$

2. For each input pattern pair do the following:
(a) Transfer all input values (x) to input layers, and calculate the new output values of the hidden layer using

$$h_j = f\left(\sum_{i=1}^{1} w_{ij}^h x_i + \theta_j^h \right) \qquad \text{for all } j = 1, 2, \ldots, m$$

where 1 is the number of PEs in input layer, m is the number of PEs in hidden layer, θ_j^h is the threshold of a PE in hidden layer, and $f()$ is the logistic threshold function.

$$h = [0.905 \; 0.905 \; 0.905]$$

(b) Calculate the output value from the hidden layer PEs' values calculated in (a)

$$o_k = f\left(\sum_{j=1}^{m} w_{jk}^0 h_j + \theta_k^0 \right) \qquad \text{for all } k = 1, 2, \ldots, n$$

where n is the number of PEs in output layer, θ_k^0 is the threshold of a PE in output layer.

$$o = [0.833 \; 0.833 \; 0.833]$$

(c) Compute error values for each of the output layer PEs.

$$e_k^0 = o_k(y_k - o_k)(1 - o_k) \qquad \text{for all } k = 1, 2, \ldots, n$$

where y_k is desired output of a PE in output layer.

$$e^0 = [0.0232 \; 0.0232]$$

(d) Compute error values for each of the hidden-layer PEs.

$$e_j^h = h_j(1 - h_j) \sum_{j=1}^{n} w_{jk}^h e_k^0 \qquad \text{for all } j = 1, 2, \ldots, m$$

$$e^h = [0.002 \; 0.002 \; 0.002]$$

(e) Adjust the weights of the hidden-to-output layer connections.

$$w_{jk}^0(t+1) = w_{jk}^0(t) + \alpha h_j e_k^0 \qquad \text{for all } j = 1, 2, \ldots, m \\ \text{and all } k = 1, 2, \ldots, n$$

α is a positive constant controlling the learning rate.

$$w^0(t+1)=\begin{bmatrix} 0.4937\,0.4937 \\ 0.4937\,0.4937 \\ 0.4937\,0.4937 \end{bmatrix}$$

(f) Adjust the threshold values of the output layer PEs.

$$\theta_k^0(t+1)=\theta_k^0(t)+\alpha e_k^0 \qquad \text{for all } k=1,2,\dots,n$$
$$\theta^0(t+1)=[0.243\,0.243]$$

(g) Adjust the weights of the input-to-hidden layer connections.

$$w_{ij}^h(t+1)=w_{ih}^h(t)+\beta x_i e_j^h \qquad \text{for all } i=1,2,\dots,1$$
$$\text{and all } j=1,2,\dots,m, \text{ and}$$

β is a positive constant controlling the learning rate.

$$w^h(t+1)=\begin{bmatrix} 0.4992\,0.4992\,0.4992 \\ 0.4992\,0.4992\,0.4992 \\ 0.4992\,0.4992\,0.4992 \\ 0.4992\,0.4992\,0.4992 \end{bmatrix}$$

(h) Adjust the threshold values of the hidden layer PEs.

$$\theta_j^h(t+1)=\theta_j^h(t)+\beta e_j^h \qquad \text{for all } j=1,2,\dots,m$$
$$\theta^h(t+1)=[0.2492\,0.0.2492\,0.2492]$$

(b) Recall

The recall process begins when the back propagation network receives an input at its input layer PEs and feeds it through the network and produces the corresponding output at the output layer. In order to demonstrate the recall process, all weights and threshold values are assumed to be the values of the trained network.

$$w^h=\begin{bmatrix} 0.22\,0.39\,0.85 \\ 0.77\,0.15\,0.42 \\ 0.09\,0.53\,0.66 \\ 0.31\,0.83\,0.05 \end{bmatrix} \qquad w^0=\begin{bmatrix} 0.23\,0.78 \\ 0.34\,0.93 \\ 0.75\,0.22 \end{bmatrix}$$

$$\theta^h=[0.32\,0.22\,0.57] \qquad \theta^0=[0.43\,0.62]$$

The feedforward operation produces the output of hidden layer PEs using the following equations.

$$h_j=f\left(\sum_{i=1}^{1} w_{ij}^h x_i+\theta_j^h\right) \qquad \text{for all } j=1,2,\dots,m$$

Thus, the output of hidden layer PEs are

$$h=[0.745\,0.843\,0.804]$$

Once all value of PEs in hidden layer are calculated, they are used to calculate the output value by the equation

$$o_k = f\left(\sum_{j=1}^{m} w_{jk}^0 h_j + \theta_k^2 \right) \quad \text{for all } k = 1, 2, \dots, n$$

$$o = [0.653\ 0.715]$$

25.4 MANUFACTURING FEATURE RECOGNITION

25.4.1 Manufacturing features

Manufacturing features are primitive or low level designs with their attributes, qualifiers and restrictions which affect functionality or manufacturability. Feaures can describe form (size and shape), precision (tolerances and finishing), or materials (type, grade, properties and treatment) (Shah and Rogers, 1988), and vary with product and manufacturing process. Cunningham and Dixon (1988) define manufacturing features based on basic manufacturing processes. For example, planes, ribs, bosses, fillers, webs, 3D corners, bends/twists are identified as forging features; slots, fingers, corners, holes, cut-outs, notches, windows define stamping features; circles, webs, walls, corner breaks, slots, hollowness, tongues constitute features for tools and dies; and finally, plates, shells, beams, frames, trusts, columns and arches define structural features (Fig. 25.7).

In the design process, there is a need to create, delete, modify, copy, move, detail, and add properties and structures to manufacturing features (Pratt, 1988). Most of these functions can easily be accomplished with existing CAD systems. However, these systems fall short in two important areas of design; feature recognition and synthesis, creativity and heuristic use. Neural networks and knowledge-based systems offer great opportunities to overcome this problem. Since the feature recognition process is still best done by humans, neural networks can be designed and trained to distinguish features into groups, as humans. Once these features are identified and categorized into groups, knowledge-based systems can be applied for heuristic use effectively. This capability of the neural network is demonstrated in a sample problem.

25.4.2 Sample feature recognition problem

In this example, two types of features are used to demonstrate how the neural network performs the feature recognition; bracket and circles. Our problem is how to recognize the differences between three similar brackets and three similar circles. Each bracket has different length of tongues (Fig. 25.8a). Each circle has a different number of holes (Fig. 25.9a). This problem is adapted from Smith and Dagli (1993).

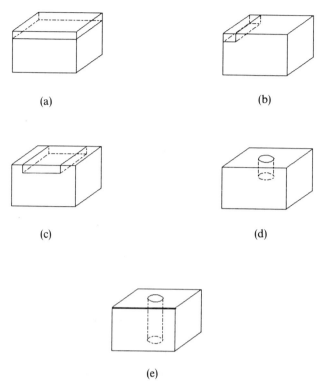

Fig. 25.7 Sample features; (a) plane, (b) step, (c) slot, (d) blind hole, (e) through hole.

25.4.3 Network architectures, inputs and outputs

In order to recognize these features, they must be an appropriate format for the neural network. The matrix representation (Dagli, 1990) is used to represent the features. Brackets are represented by binary numbers (Fig. 25.8b) while circles are represented by analog numbers (Fig. 25.9b). However, the network cannot understand this representation as a matrix, so these numbers are converted into a vector to be used as an input of the network (Table 25.1). Therefore, the desired accuracy of the representation is the major factor to be considered since it determines the number of input neurons of the network. The size of the network will be larger as the number of input neurons increases, and higher resolutions can provide better differentiation of details.

Outputs of the network should also be considered. Binary number is used as the output representation. Each binary number in the output string represents the corresponding feature. For example, in the problem of brackets recognition, the network is trained for the data set of three similar brackets. The output '100' means the input feature is recognized as the bracket #1; '010' means the bracket #2 is recognized; while '001' means the bracket #3 is recognized.

Table 25.1 Input vectors for brackets and circles

Feature	Input vector
Bracket #1	[1111111111111111111111111101110111101110111]
Bracket #2	[1111111111111111111111111111110111011111000000111]
Bracket #3	[111111111111111111111111111111101110001110000]
Circle #1	[.5.51.500.5.5.5.5.50.5.5.5.51.5.5.50.5.5.5.50.5.5.5.5.5000.511.500]
Circle #2	[00.511.5000.5.5.5.50.5.50.5.5.510.5.5.511.5.5110.511000.511.500]
Circle #3	[00.511.5000.5.5.5.5.50.5.50.5.50.5.5.51.500.5.5.5.500.5.5000.511.500]

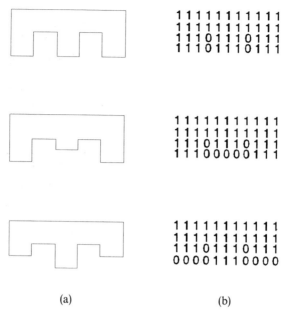

```
1 1 1 1 1 1 1 1 1 1
1 1 1 1 1 1 1 1 1 1
1 1 0 1 1 1 0 1 1 1
1 1 0 1 1 1 0 1 1 1
```

```
1 1 1 1 1 1 1 1 1 1
1 1 1 1 1 1 1 1 1 1
1 1 0 1 1 1 0 1 1 1
1 1 0 0 0 0 0 0 1 1
```

```
1 1 1 1 1 1 1 1 1 1
1 1 1 1 1 1 1 1 1 1
1 1 0 1 1 1 0 1 1 1
0 0 0 0 1 1 1 0 0 0 0
```

(a)   (b)

Fig. 25.8 Bracket features: (a) Three similar brackets; (b) Matrix representations of these three brackets.

25.4.4 Results

In our example, a back propagation paradigm is selected for both problems, brackets and circles, since this paradigm operates on both binary and analog inputs. The back propagation paradigm is also capable of distinguishing small differences among features. This is an important issue in the design process since the designed parts should be able to be oriented easily for full automation.

As reported in Smith and Dagli (1992), in the training process, similar three brackets with 50 input neurons and 3 output neurons use 100 iterations for 15 hidden layer neurons. Three similar circles with 49 input neurons and 3 output neurons; 12 hidden layer neurons require 1200 iterations; 25 hidden layer neurons require 700 iterations while 50 hidden layer neurons require 700 iterations. As the results, the number of iterations decreases when the number of hidden layer neurons is reduced. However, more computation time is required since there are more connections among the neurons (Smith and Dagli, 1993). All of these back propagation architectures were able to identify the manufacturing features effectively. As described in Section 25.3.4(b), only the weight matrix obtained after training is necessary to recognize these manufacturing features which are very similar to each other in each category.

As in the real manufacturing process, it is possible to have a partial image of the feature as an input to the network due to other factors in the operating environment such as noise, the feature is obstructed by another

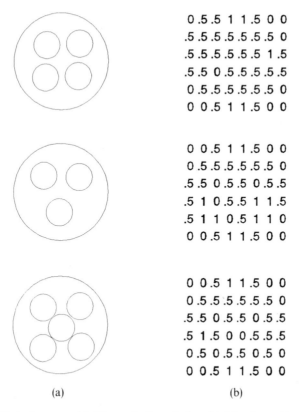

0 .5 .5 1 1 .5 0 0
.5 .5 .5 .5 .5 .5 .5 0
.5 .5 .5 .5 .5 .5 1 .5
.5 .5 0 .5 .5 .5 .5 .5
0 .5 .5 .5 .5 .5 .5 0
0 0.5 1 1 .5 0 0

0 0.5 1 1 .5 0 0
0 .5 .5 .5 .5 .5 .5 0
.5 .5 0 .5 .5 0 .5 .5
.5 1 0 .5 .5 1 1 .5
.5 1 1 0 .5 1 1 0
0 0.5 1 1 .5 0 0

0 0.5 1 1 .5 0 0
0 .5 .5 .5 .5 .5 .5 0
.5 .5 0 .5 .5 0 .5 .5
.5 1 .5 0 0 .5 .5 .5
0 .5 0 .5 .5 0 .5 0
0 0.5 1 1 .5 0 0

(a)   (b)

Fig. 25.9 Circle features: (a) Three similar circles; (b) Matrix representations of these circles.

object, etc. The neural network can still distinguish these corrupted patterns with a certain degree of obstruction. However, for other problems which necessitate on-line recognition of brackets and circles, the back propagation is not suitable, other paradigms such as counter propagation or ART2 can be used for this purpose since they learn on-line and they are able to distinguish small differences between input patterns. Another problem where back propagation can be used is the problem of feature matching; two features need to be recognized and placed next to each other with proper orientation and location. The network recognizes both features and gives a location and orientation for them during the matching process. The result of the matching is the combined features which occupy minimum placement area as shown in Fig. 25.10.

Artificial neural networks can be used in manufacturing feature recognition process effectively. The manufacturing features are used to train the network and their general features are reserved within the weight matrix. This matrix of weights is used in the automation process to recall appropriate manufacturing features. This recall process can be speeded up using neural network chips.

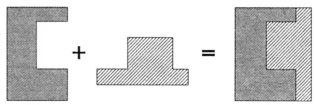

Fig. 25.10   An example of feature matching.

25.5 CONTRIBUTIONS OF ANN FOR
CONCURRENT ENGINEERING

In the previous section the use of artificial neural networks in manufacturing feature recognition is demonstrated with an example. Manufacturing systems provide fertile ground for artificial neural network applications. Some of these areas are: design, process planning, group technology, assembly automation, and vision. Some of the early results of these applications will appear in Dagli (1993). In this section, some basic approaches used in these applications are summarized.

The problem of conceptual design is studied using this new technology. Binary input-output fuzzy associative memory (BIOFAM), neural network paradigm is proposed by Bahrami and Dagli (1991) to map fuzzy functional requirements to set up physical structures to retrieve design solutions. Sample small problems gave satisfactory results.

Grouping similar parts and machines to exploit the similarity of items within the group for cellular manufacturing attracted a lot of attention during the 1980s. The processes could be done based on the binary machine-part matrix to form a basis for observing similarities by placing similar parts adjacently in the matrix. Normal algorithms are computationally inefficient in handling large matrices when several alternatives need to be considered. Dagli and Huggahalli (1991) used ART1 paradigm to solve this problem for large matrices and obtained satisfactory results.

Process planning is another area for artificial neural network applications. They not only greatly simplify the problem but also perfectly generate sophisticated process plans. Research work in this area is currently in progress.

Machine vision systems play an important role in the automated assembly and automated inspection of manufactured parts. In order to obtain an effective system, three-dimensional and accurate recognition of manufacturing features are essential. Lynch and Dagli (1991) proposed a neural network approach for this problem. The proposed vision system performs three-dimensional object recognition by using two stereoscopic cameras overlapping each other's views on the intensity maps to estimate the epipolar lines between images by back propagation, and an ART2 paradigm for classification. Another application of the vision system for printed circuit board (PCB) assembly is proposed by Vellanki and Dagli (1992) as a hybrid intelligent control system to control robot manipulators. The system uses ART1 paradigm for electronic part recognition.

All these efforts and the preliminary results obtained suggest that ANN are beginning to impact manufacturing. They will play an important part in our quest for designing manufacturing 'intelligence'.

25.6 SUMMARY

In this chapter, intelligent manufacturing, concurrent engineering, design for manufacturing, also their needs and relationships are discussed. In addition, the fundamental concepts of the artificial neural networks and some of the basic network paradigms, manufacturing applications of ANN, are given. The use of artificial neural networks in the manufacturing feature recognition problem is illustrated in detail. Finally, some other applications of the artificial neural networks for concurrent engineering are summarized and importance of manufacturing feature recognition in intelligent manufacturing system design is stressed.

25.7 REFERENCES

Anderson, D.M. (1990) *Design for Manufacturing: Optimizing Cost, Quality, and Time-to-Market*, CIM press, Lafayette, California.

Bahrami, A. and Dagli, C. H. (1991) in *Intelligent Engineering Systems Through Artificial Neural Networks* (eds C.H. Dagli, S.R.T. Kumara and Y.C. Shin), ASME Press, New York, pp. 745–50.

Brown, D.C. and Chandrasekaran, B. (1988) in *Expert Systems in Engineering*, (ed. D.T. Pham), Springer-Verlag, Exeter, UK.

Carpenter, G.A. and Grossberg, S. (1987) *Computer Vision, Graphics and Image Processing*, **37**, 54–115.

Chang, T. (1990) *Expert Process Planning for Manufacturing*, Addison Wesley, New York.

Chisvin, L. and Duckworth R.J. (1989) *IEEE Computer*, **22**(7), 51.

Cunningham, J.J. and Dixon, J.R. (1988), *Proceedings of the ASME Conference on Computers in Engineering*, 237.

Dagli, C.H. and Stacey, R. (1988) *International Journal of Production Research*, **26**(5), 987–96.

Dagli, C.H. and Lammers, S. (1989) *Proceedings of International Joint Conference on Neural Networks*, **2**, II-605.

Dagli, C.H., Poshyanonda, T. (1990) in *People and Product Management in Manufacturing*, (ed. J.A. Edosomwan), 221–39.

Dagli, C.H. (1990) *2nd International Conference on Computer Integrated Manufacturing*, 531–37.

Dagli, C.H. (1991) *Journal of Intelligent Manufacturing*, **2**, i.

Dagli, C.H. and Huggahalli, R. (1991) in *Knowledge-based Systems and Neural Networks: Techniques and Applications*, (eds R. Sharda, J.Y. Cheung, and N.J. Cochran), Elsevier, 213–28.

Dagli, C.H., Lammers, S. and Vellanki, M. (1991) *Journal of Neural Network Computing*, **2**(4), 4–10.

Dagli, C.H. and Lynch, M.B. (1991) *Proceeding of Workshop on Neural Networks Academic/Industrial/NASA/Defense*, 631–45.

Dagli, C.H. (eds) (1993) *Artificial Neural Networks for Intelligent Manufacturing*, Chapman & Hall, in preparation.

Joneja, A., and Chang, T.C. (1991) *Journal of Design and Manufacturing*, **1**, 5–15.

Kerr, R. (1990) *Knowledge-Based Manufacturing Management: Applications of Artificial Intelligence to the Effective Management of Manufacturing Companies*, Addison Wesley, Sydney.

Khoshnevis, B. and Park, J. (1988) Real time manufacturing process planning, *University of Southern California working paper*.

Kusiak, A. and Chow, W.S. (1987) *Journal of Manufacturing Systems*, **6**, 117–24.

Kusiak, A. (1990) *Intelligent Manufacturing Systems*, Prentice Hall, Englewood Cliffs, New Jersey.

Lynch, M.B. and Dagli, C.H. (1991) in *Intelligent Engineering Systems Through Artificial Neural Networks* (eds C.H. Dagli, S.R.T. Kumara and Y.C. Shin), ASME Press, New York, 495–500.

Nelson, M.M. and Illingworth W.T. (1990) *A Practical Guide to Neural Nets*, Addison Wesley, Massachusetts.

Pratt, M.J. (1988) *Proceedings of the ASME Conference on Computers in Engineering*, 263.

Roberts, R.S. (1991) *Concurrent Engineering*, **1**(1), 29–36.

Rosenblatt, F. (1957) *Cornell Aeronautical Laboratory Report*, 85–460–1.

Shah, J.J. and Rogers, M.T. (1988) *Proceedings of the ASME Conference on Computers in Engineering*, 255.

Simpson, P.K. (1990) *Artificial Neural Systems*, Pergamon Press, Elmsford, New York.

Smith, A. and Dagli, C.H. (1993) in *Intelligent Systems in Design and Manufacturing*, (eds C.H. Dagli and A. Kusiak), ASME Press (In Print).

Suri, R. and Shimizu, M. (1989) *University of Wisconsin –Madison technical report*, 89–3.

Swift, K.G. (1987) *Knowledge Based Design for Manufacture*, Englewood Cliffs, NJ, Prentice Hall.

Widrow, B. (1959) *WESCON Convention Record: Part 4*, 74–85.

Wright, P.K. and Bourne, D.A. (1988) *Manufacturing Intelligence*, Addison Wesley, Reading, Massachusetts.

Vellanki, M. and Dagli, C.H. (1992) in *New Directions for Operations Research in Manufacturing*, (eds A. Jones, T. Gulledge and G. Fandel), Springer-Verlag, 346–368.

Index

487